I0809047

LAST HOPE

The Blood Chit Story

R.E. Baldwin & Thomas Wm. McGarry

Schiffer Military History
Atglen, PA

來華助戰
洋人(美)國
軍民一體
救護
航空委員會
字第 3038 號
RICHARD GRIGGS

LAST HOPE
The Blood Chit Story

Book Design by Robert Biondi.
Photos by R.E. Baldwin unless otherwise noted.

Text copyright © 1997 by R.E. Baldwin and Thomas Wm. McGarry.
All other material, including tables, photos, and illustrations, copyright by R.E. Baldwin.
Library of Congress Catalog Number: 96-70415.

All rights reserved. No part of this work may be reproduced or used in any forms or by any means – graphic, electronic or mechanical, including photocopying or information storage and retrieval systems – without written permission from the copyright holder.

Printed in China.
ISBN: 0-7643-0222-1

We are interested in hearing from authors with book ideas on related topics.

Published by Schiffer Publishing Ltd.
77 Lower Valley Road
Atglen, PA 19310
Phone: (610) 593-1777
FAX: (610) 593-2002
Please write for a free catalog.
This book may be purchased from the publisher.
Please include $2.95 postage.
Try your bookstore first.

Preface

More than three-quarters of a century ago, in the skies over what are now Afghanistan and Pakistan, Royal Air Force flyers began carrying documents written in languages of the area promising to reward native peoples for the safe return of airmen who were forced down among them. The descendants of those documents, known today as Blood Chits, are routinely carried by both British and American air crews on operations worldwide. They are a part of a sophisticated Survival, Evasion, Resistance, and Escape (SERE) system instituted by these governments to effect the safe return of their airmen. This book is the fully illustrated history of Blood Chits from their infancy at the dawn of the air age through their maturity at the close of World War II.

The military is not a tradition in my family, nor could I in good conscience have followed that path, yet I came to understand that this story needs to be recorded and shared, and that it must be done now while those who know it are still with us. This book is the result of my realization that the history of Blood Chits might be lost for all time, replaced with colorful but ultimately false rumors and speculation, unless I undertook the task of recording it.

When I first became interested in Blood Chits nearly a decade ago, I found little more than rumors. As the true history of these lifesaving artifacts was uncovered, a more powerful story than the popular mythology surrounding them was revealed, a story that begged to be told.

Although it contains the most comprehensive collection of color photographs of Blood Chits ever assembled for publication, and provides readers an unparalleled opportunity to discover the beauty and variety of these lifesaving artifacts, this book is more than an illustrated guide to a colorful collectible. It is an introduction to the very real story of the often heroic efforts by governments to provide their airmen with the tools to evade capture should they be downed in enemy controlled territory.

This book has been meticulously researched, and contains information provided by Royal Air Force veterans who were involved in colonial policing operations in the early part of this century, by veterans of the American Volunteer Group who served under Claire Chennault in China, and by British, American, Canadian, Australian, and other veterans of World War II. Material gathered from veterans of military intelligence agencies that produced Blood Chits and distributed them to combat crews is also included.

Written material from declassified government documents, most never before published, provides readers with specific details regarding the functioning of evasion and escape organizations as well as highly detailed information on when and how Blood Chits were designed and produced, and how they were distributed to and used by operational air crews. The inclusion of instructions to air crews gives readers an essential understanding of the conditions under which Blood Chits and other evasion aids were used. I have attempted to balance the material in the book so that the history of Blood Chits and language-related evasion aids will be enhanced and illuminated by other historical information, but not overwhelmed by it.

One difficult aspect of writing this book was combining the text with photos and captions. At the suggestion of the publisher, I have attempted to include a substantial amount of information in the photo captions. This tends to disrupt the flow of the text somewhat, but it allows readers to see the artifact while reading some of the information about it, and it keeps the captions from merely being repetitions of the text. Readers are encouraged to read the text, and to refer to each photo and caption when called for in the text.

Some of the terms used in this work do not conform to conventional English usage, but they do conform to other historical literature or to military usage. An example is the use of word "escaper", rather than "escapee", because the former appears in numerous World War II era government documents as well as in contemporary histories. Another example is the occasional use of the term "Middle East" in reference to both the Near East and the Middle East, because the term was used in this manner by the British military establishment. In instances where meanings differ between military and civilian usage, it can be assumed that the military usage is intended. The terms "evasion and escape" and "escape and evasion" both appear in government documents dating from World War II. I have adopted the former for use in this book.

The Pinyin system for the anglicization of Chinese, preferred by the current regime in China, is not used in this work. For consistency, the Wade system is used because it was used during the historical period covered by this work, it is used in the written material from which this work was compiled, and it is used by the veterans who served in China that contributed to this work.

Masculine nouns and pronouns such as he, him, airman, and crewman are used in this work. The increasingly active role of women in modern military forces is recognized and appreciated, but the masculine gender is used in this work for purposes of brevity and simplicity. There is no depreciatory connotation.

Sources will be quoted verbatim in this work, including those from the World War II period containing offensive terms such as "Jap." Likewise, illustrations that may stereotype some nationalities will be used without alteration. The difficult decision not to censor the material was made in the belief that the unsavory parts of our past should not be covered up, that readers deserve access to references, and that readers are intelligent enough to separate the prejudices of the past from relevant historical material. No offense is intended by the inclusion of this material.

All government documents used in the preparation of this book are unclassified. The original classifications (restricted, confidential, secret, etc.) of documents are given, when known.

When original artifacts were not available to be photographed, several methods were used to produce illustrations. These include the use of color photocopies, and computer enhancement of photocopies and black-and-white photos. Endnotes will identify illustrations produced in this manner. The primary consideration throughout was to provide the most authentic, best possible illustrations.

Collectors will notice that I have included some common items, such as phrase books, in this work. The book is not only a guide for collectors but a detailed history, and while some items may be common to collectors, they are generally unknown to the public at large. Moreover, these items have a rightful place in the history being recorded here.

I have introduced my own numbering system for Blood Chits and related items in order to implement a standardized system of identification for these artifacts. Reference to these numbers when discussing Blood Chits will facilitate communication between museum curators, researchers, and collectors. The numbers, preceded by the designation *REB*, appear in captions for illustrations, and in a table at the end of the book that gives the relative scarcity of the artifacts. Artifacts that are already easily identified have not been given numbers under this system.

I do not pretend to be an expert on the history of Blood Chits and related evasion aids. I consider myself a student of that history, and I welcome this opportunity to share some of what I have learned of it with others. Also, although I have accumulated a great deal of information regarding the Blood Chits illustrated in this book, I am considerably less knowledgeable about the theater-made souvenir patches, also shown.

It is my intent in this work to accurately report history. I leave the analysis of it to others. I have made every attempt to verify the accuracy of the material included in this work, and any errors in it are the results of my own shortcomings.

History is not preserved by accumulating objects, or by perpetuating the popular rumors of the day, but by efforts to unearth and record facts, and to make those facts available to others. It is my sincere hope that this book will prove useful in preserving the history of a small yet important aspect of our military aviation heritage.

R. E. Baldwin
September 1996

Contents

Acknowledgments

This book came into being through the efforts of many dedicated people, and I consider myself privileged to be the person who brings the results of their work to print.

John A. Scott and Lloyd R. Shoemaker, to whose memory this work is dedicated, are responsible for a large portion of the research material used in preparing this book. Both generously provided me with copies of declassified government documents dealing with Blood Chits that they obtained from the National Archives and other sources during research for their individual projects. Both imparted to me their own enthusiasm for this little known part of history, and did everything within their power to ensure that this work would be accurate and definitive.

From December 1991 until November 1995, Thomas Wm. McGarry joined me in the effort to research and write this book. His work and his friendship are both deeply appreciated. His research was vital in uncovering key elements in the history of Blood Chits, his many contributions are a large and essential part this work, and he rightly shares the credit for its authorship. Thomas wishes to thank his wife Laura for her encouragement and support during his work on this project.

Terry Aitkin and Scott Ferguson of the United States Air Force Museum provided valuable assistance to this project, arranging for access to the Museum's holdings for research by John A. Scott and myself. Nigel Steele of the Imperial War Museum and Richard Simpson of the Royal Air Force Museum searched the holdings of their respective museums for relevant artifacts and information. These people have duties far more important than assisting an unknown researcher on an obscure quest, and their efforts on my behalf are especially appreciated.

Thanks are in order for the veterans who sent their personal mementos through the mail to be photographed, and especially for those like Ben L. Donahue, John L. Cella, and William F. Beck who supported this project with both donated artifacts and information. Ben was also a constant source of encouragement, and his support and friendship are valued beyond words. Ken Jernstedt and J.R. "Dick" Rossi are also singled out for special thanks, for providing vital information regarding the AVG and their Rescue Patches. Ken proofread the chapter on early China to ensure that it is accurate, and allowed me and my wife the privilege of attending the 1995 AVG reunion, a very special experience for us. His assistance and his friendship are especially valued.

Thanks also to collectors Bob Gill, Randy Graham, and Walter Willett III, and to writer David Vincent, who loaned artifacts from their own collections to be photographed, and to collector Jeff Guidry who provided copies of several rare early British chits. I also want to express my thanks for the support, encouragement, and friendship of Steve Griffith and William S. Miller.

Especially appreciated is the work of Agnes Chan, who spent many hours translating and explaining the Chinese Blood Chit texts. Her expertise in both the Chinese language and modern Chinese history provided insights into small but important details in the texts that would otherwise have been overlooked. The assistance of Vince Lau, who provided translations of both Chinese and Burmese material, is also greatly appreciated. I am also grateful for the kind assistance of Amir Kaifi, who translated the Farsi texts, of Salem Attiga for his translation of Arabic texts, of Tsion Arrega for her translation of the Amharic language of Ethiopia, of Paul Tran for his translation of Vietnamese text, and of Peter B. Adeney who translated Old Malay material from the Netherlands East Indies. Thanks also to Homayun Karim, Haidook Hovanesian, Shirlaine Baldwin, Matt Baldwin, and Aimée Baldwin for their assistance in finding some of these translators, and for seeing that the translations found their way to me.

Aimée Baldwin, through computer enhancement of marginal and substandard material, produced some of the quality color illustrations absolutely essential to this book. Her work allows me to give readers access to more material than I thought possible, and is especially valued. The work of Kim Stone, who salvaged some marginal material photographically, is also greatly appreciated.

Geoff Cantrell, a dear friend for more than 35 years, proofread the final manuscript. His insights and suggestions were vital to this project, and his assistance is deeply appreciated.

My special thanks to Robert C. Mikesh, author and retired curator of the National Air and Space Museum, who, over lunch one day in 1995, gave me some desperately needed encouragement to complete this project despite major difficulties and disappointments.

I also wish to thank my father, Ed Baldwin, an engineer in Kelly Johnson's *Skunk Works* for more than 42 years, for passing on to me his love of aviation, and my mother, Jean Baldwin, a former teacher, for giving me her love of history. Last, but far from least, deepest thanks to my wife, Shirlaine, and my children Matt and Aimée, for their steadfast support and their long-suffering patience with what must seem an obsession to bring this obscure part of history to light.

All those who assisted this effort have my everlasting gratitude for their valuable contributions. This book was made possible by the kindness and generosity of the following people:

Terry Aitkin, U.S. Air Force Museum (USA); Anonymous by request (USA); C.H. Anderson, RAF (UK); Tsion Arega (USA); R.G. Auckland, Psywar Society (UK); J.R. Ashworth, RCAF (Canada); Salem Attiga (USA); Aimée J. Baldwin (USA); Matt Baldwin (USA); Shirlaine Baldwin (USA); Russell A. "Rusty" Bales, USAAF (USA); K.S. Batchelor CBE DFC, RAF (UK); G. Beauchamp (Canada); William F. Beck, USAAF (USA); Horace O. Bedsole, USAAF (USA); Jimmy Bell, RAF (UK); Josef Beyer (Germany); Marvin C. Bigelow, USAAF (USA); Gen. Charles Bond, AVG, USAF (USA); Chaz Boyer (UK); H. Paul Brehm, USN (USA); Jack Butterfield, RAF (UK); Geoffrey S. Cantrell (USA); Warren Carroll (Canada); John L. Cella, USAAF (USA); Agnes Chan (USA); Eddie S. Cheek (UK); E.J. Wesley Coe, RAF (UK); R.A. Collins DFC, RAF (UK); Paul Cornish, Imperial War Museum (UK); Malcolm E. Corum, USAAF (USA); Ben L. Donahue, USAAF (USA); Joe Donato, USAAF (USA); William Dreeland (USA); Air-Vice Marshall A.G. Dudgeon CBE DFC, RAF (UK); Jim Eng (USA); G.L. Ferguson, USMC (USA); Scott Ferguson, U.S. Air Force Museum (USA); Santiago A. Flores (Mexico); M.R.D. Foot (UK); Eldon Ray Ford, USAAF (USA); William Fulcher, RAF (USA); Raymond Garvey (USA); Maj. Donald J. Gibb, USAAF, USAF (USA); Bob Gill (USA); Bob Gilroy, RAF (UK); Ian Glen DFC, RCAF (Canada); Randy Graham (USA); Jeffrey D. Guidry (USA); Gary Hammill (USA); Mark A. Hardisty, USN (USA); Stephen J. Hasegawa (USA); David L. "Tex" Hill, AVG, USAF (USA); Charles F. Hodges (USA); Haidook Hovanesian (USA); Christopher Hunt, Imperial War Museum (UK); Robert L. Hunter, M.D. (USA); Jack Ilfrey, USAAF (USA); Mrs. Walter D. Innis (USA); Ken Jernstedt, AVG (USA); Amir Kaifi; Homayun Karim (USA); Jeff Knecht (USA); Vince Lau (USA); Ed Leiser, Curator, San Diego Aerospace Museum (USA); Jan Madarasz (Germany); Raymond Magnani (France); Bruce Magnuson (USA); Robert Maslen, RCAF (Canada); Hugh A. MacArthur (USA); Robert S. McCarter, USAAF (USA); Thomas Wm. McGarry, US Army (USA); Merv M. McGrane, RAAF (Australia); Robert C. Mikesh, USAF (USA); William Miller, AGAS, (West Indies); Mrs. Carmen Mullin (USA); John J. Munro, USMC (USA); David B. Nash, Imperial War Museum (UK); Donald R. Neate (UK); J.F. Notman, RAF (UK); Terence P. O'Halloran (UK); Jerry Pahl, Kalamazoo Aviation History Museum (USA); Roger Perreault, USMC (USA); Kirsten Oftedahl, Pima Air Museum (USA); A.C. Rawlinson, RAAF, RAF (Australia); Duane Reed, Special Collections Library, USAF Academy (USA); Bill Robie (USA); William A. Rooney, USAAF (USA); J.R. "Dick" Rossi, AVG (USA); John F. Russell, RAAF (Australia); John A. Scott, US Army, Oregon Military Museum (USA); Mark Seaman, Imperial War Museum (UK); Erik Shilling, AVG (USA); Lloyd R. Shoemaker, MIS-X (USA); Richard Simpson, RAF Museum (UK); Lee Siwecki (USA); R.T. Smith, AVG (USA); Nigel Steele, Imperial War Museum (UK); Ralph L. Stevenson, Jr. (USA); John and Kim Stone; Everett E. Swanson, USAAF (USA); Col. Gabriel Taschereau DFC, RCAF (Canada); Owen Thetford (UK); Don Townend, RAF (UK); Paul Tran (USA); J.R. Twine, RAAF (Australia); David Underwood (UK); D.G. Upward, RAF (UK); David Vincent (Australia); L.D. Volkmer, USAAF (USA); Ray Wagner, Archivist, San Diego Aerospace Museum (USA); D.W. Walker, RAF (UK); George H. Wheeler (UK); Walter Willett III (USA); Angela M. Wooton, Imperial War Museum (UK).

R. E. Baldwin
September, 1996

Dedication

This book is Dedicated to the Memory of John A. Scott and Lloyd R. Shoemaker

John A. Scott (1934-1995)

The life of John A. Scott was one of passion. John loved skydiving, white-water rafting, military vehicles, air crew evasion and survival aids, and history.

John joined the U.S. Army in 1955, and served with the 11th Airborne Division and one of the very early Special Forces Groups. After completing his active duty, he kept a reserve commission and continued to pursue skydiving. He traveled and competed extensively as a member of the National Parachute Team, and became a judge for national skydiving competitions. The D License is the most senior license available to sport parachutists, and John's skill and devotion to the sport earned him the 93rd such License issued by the U.S. Parachute Association. With typical enthusiasm, John celebrated his 40th birthday by making 40 jumps. John's skydiving career ended abruptly when he was seriously injured by a falling log while selflessly assisting a novice jumper who had landed in a lumber mill.

John became a pre-eminent collector of air crew survival and evasion aids, and of cloth maps in particular. John's love of collecting was infused with a thirst for knowledge. He was driven to find out everything about each artifact, and to discover all of its variations. In building a collection or display, he insisted that everything be correct and complete. John's desire for knowledge was balanced by his devotion to teaching others who shared his interests, and he routinely loaned items from his personal collection so that others could learn about them firsthand.

In his role as curator of the Oregon Military Museum, in the company of the stray cats he adopted, and surrounded by the history and artifacts he knew and loved, John was at home. Museums and collectors from around the world sought out John to identify and explain the items in their collections. If each of us has a place in the universe, John's was clearly at the museum.

At the memorial service for John, his daughter Heather received his Oregon Exceptional Service Medal from the Oregon National Guard. Regrettably, there are no medals for sharing an infectious enthusiasm for learning everything there is to know about something, for the gift of gentle humor, or for strong, warm, and genuine friendship. John would have earned all of these too, though he would have been too modest to wear them.

This book is part of the legacy of John A. Scott. It is hoped that the quality of the material assembled here does honor to the standards he always insisted upon.

Lloyd R. Shoemaker (1921-1995)

Lloyd R. Shoemaker was one of the quiet heroes who helped the United States and its Allies win World War II. Shoemaker and the other members of the American evasion and escape organization MIS-X, and its British counterpart MI9, were responsible for saving thousands of Allied service members' lives. Many Allied airmen who were forced down survived World War II because of the little-noticed but often ingenious and heroic efforts of Lloyd and men like him.

The work done by Shoemaker and MIS-X was the beginning of the current Survival, Evasion, Resistance and Escape (SERE) training that American, NATO, and other air crew members now receive. The survival and evasion aids developed and distributed by MIS-X have direct descendants in the survival vests, blood chits, and miniature transceivers carried by contemporary pilots. Unfortunately, because of the shortsightedness of the military at the end of World War II, much of what Shoemaker and his associates knew had to be re-learned at the cost of human life on the frozen ground of Korea and in the humid jungles and fetid paddies of Vietnam.

Like his work, Lloyd was not widely known by the public, but to those who deal in SERE matters from offices in Washington, D.C., and places like Arlington and Fort Belvoir, a trip or call to consult him was not infrequent. Lloyd was a national resource, a quiet hero who let his accomplishments – in having helped others survive to enjoy their freedom and futures – speak volumes.

Lloyd was an extremely caring person who, although never married, took it upon himself to care for three young men in need and raise them into exemplary adults. He was always ready, even at the end of his life during a severe and difficult illness, to lend a helping hand whenever and wherever it was needed.

Without the assistance, encouragement, and friendship of Lloyd R. Shoemaker, this book would not have been written.

Introduction

By Lloyd R. Shoemaker

Veteran of the Military Intelligence Service, Evasion & Escape Section (MIS-X)

Governments around the globe have acknowledged the possibility that their men-at-arms, when engaged in conflicts in foreign territories, could become isolated from their comrades or captured by the enemy. The instillation of a philosophy of "escape and evasion" into soldiers, and later pilots and air crews, eventually reaped great benefits.

In ancient times, only the wealthiest soldiers – princes and knights and the like – were bought out of captivity by their relatives or subjects. The poorer soldiers, the ancient equivalent of infantrymen, were less fortunate. They were often slaughtered out of hand because they were a burden to their captors. Blood Chits are a modern consequence of this tradition.

I first saw a Blood Chit when I climbed on a plane, probably a C-47, to go into Burma. Someone had a flight jacket with a chit on it, a chit with the royal coat of arms if my memory serves me right, but it was dark and I just had a fleeting glance. We were preoccupied with our mission and it was a long time ago.

Our supplies were air-dropped to us in the jungle. With one drop there was a fistful of paper chits. I was told to keep one; I folded it up and put it in my pocket, but I can't remember what happened to it. We carried everything we needed for the mission plus our personal gear, and if you could get rid of something to lighten your load, you did.

As a member of the American Army's Military Intelligence Service (MIS) during World War II, I handled thousands of Blood Chits, often spending days in a small, cramped back office of the Chinese Embassy in Washington, D.C., offered endless cups of steaming tea in dirty cups and stamping chits with the "chop", or the official seal, of the Nationalist Chinese Ambassador to the United States.

Like many of the people who handled them, I paid little attention to chits except for noticing the variety of designs as they passed through our warehouse. They had then, and have now, a unique beauty.

A Blood Chit is really a government's written promise to pay, to reward, someone for aiding a warrior of that nation evade capture by his enemies and return to his own forces.

Thousands of dollars paid for liberations have gone virtually unnoticed. The single most expensive payment, $100,000, was made to the son of a Korean fisherman who in June of 1950 aided six members of a B-29 crew in avoiding capture by North Korean forces. The dollar amount was based on the established payment at the time, plus interest for more than 40 years. Despite the best efforts of actuaries, lawyers and bureaucrats, it is still difficult to place a dollar value on human life.

At MIS-X a part of our mission was to provide indoctrination to Army Air Force, and later U.S. Navy and U.S. Marine Corps Intelligence Officers. These men would in turn instruct air crews in the various Theaters of Operation on evasion techniques to avoid capture when forced down in enemy or unknown territory.

We also were to assist in the preparation and distribution of escape kits and emergency kits containing maps, money and other necessities to be furnished to air crews on missions, and to incorporate new ideas and improvements in such equipment, which included Blood Chits.

Sometimes the youthful optimism and bravado of the first American servicemen countered our efforts, and I expect that the British and our other allies had similar problems. Overconfident pilots and air crew frequently left their E&E kits and materials behind in their barracks, and many who were shot down early in the war were consequently taken prisoner. As stories of POW camp life spread and as the cost of combat came home to the airmen, these lapses were minimized.

The reports we received of evasions and escapes were vital: they showed us when we were successful and when we needed to make changes.

We maintained close liaison with the British MI9. We started with little, mostly the knowledge and experiences of the British. By the end of the war, our efforts in the field in China, in the Pacific, in South East Asia, in Europe and the Balkans, in Italy and North Africa had taught us much.

The volume of activity was significant: by mid-1944, for example, we had 55,000 evasion kits at Pearl Harbor for pilots going to the Pacific. There were also MIS-X "briefers", about 40 of these, on U.S. Navy aircraft carriers to instruct their pilots and crews.

We worked with many other agencies, military units and political and diplomatic services. Very little was easy. Experience, the full mobilization of America's industrial and human potential, and finally the growing success of the Allies in the war, improved our skills and equipment. The development and expansion of services like the Air-Sea Rescue Service, the network of coast watchers in the Pacific, and a line of outposts along the route flown by the "Hump" pilots, as well as partisan and resistance groups in Europe, aided our efforts and the work of other groups to help downed air crew survive, successfully evade capture and return to friendly control.

I know little about the work of MIS-X in the China Theater where it was known as the Air Ground Aid Section (AGAS). I do know that we worked more with the American Navy in the Pacific than with the Army because of the strong personalities and egos of some of the higher commanders.

Perhaps someday a musty box of records will be found in a deserted area of the National Archives or an Air Force base, and the complete story will be told.

This book is a part of that story, a part whose history has not been told before. Blood Chits have always been in the shadow of secrecy, though never secret in themselves. The authors have ferreted out much of the colorful and hidden history of these rare and beautiful artifacts of war, and the small but nonetheless important part they played in helping people survive what was doubtless the most perilous experience of their lives. Read and enjoy this, the Blood Chit story.

Lloyd R. Shoemaker
27 June 1994

LAST HOPE
The Blood Chit Story

CHAPTER 1

The Evader's Last Hope An Overview

Being forced down over enemy or unknown territory is probably one of a combat pilot's worst fears. Separated from his[1] aircraft and wingmen, no longer in control of a graceful and deadly machine, often alone and perhaps injured on the ground of a foreign, unfamiliar territory, his nightmare has become reality.

Although pilots and air crews train for this contingency, their survival and safe return is far from certain. A downed pilot's survival and his return to friendly forces may eventually depend on the ability to communicate with strangers, to convince them to overcome their fears, and to seek their active assistance. At best, the local inhabitants may be willing to help, but there may be no common language other than gestures. At worst, they may be angry, afraid, and hostile, possibly having suffered from brutal occupation by a foreign force, or even from air attacks by the now-grounded pilot himself.

If all else fails, the pilot's last hope for evading capture and returning to friendly forces lies in a simple device known as a Blood Chit – a few square inches of fabric promising a reward for the necessities of survival, assistance in evading capture, and safe return to friendly hands.

For most of this century these simple devices have been used by military aviators of Great Britain, the United States, Canada, Australia, New Zealand, The Netherlands, and other nations to communicate their need for assistance in languages they don't speak to cultures they don't understand.

Colloquially, a chit is an IOU. As David B. Nash of the Imperial War Museum has noted, "the word *chit* has Hindustani origins."[2] The English word *chit* is derived from an earlier Anglo-Indian word, *chitty*, which was derived from the Hindustani word *citthi*, meaning "pass", which itself was derived from the Sanskrit word *citra*, meaning "mark". Through its evolutionary Anglicization it has come to mean "a note of a sum owed."[3]

A 1945 report by the Army Air Forces Training Command makes a logical summation by stating, "It has been pointed out that the term 'blood' could be interpreted as 'life' – hence the term 'Blood Chit' meaning a sort of IOU for one's life."[4] A more concise definition appears in the 1956 edition of *The United States Air Force Dictionary*: "BLOOD CHIT. A piece of cloth carried by an air crew member bearing an inscription in several languages that identifies him as a member of the U.S. military establishment and promises a reward for his return to U.S. or friendly control."[5]

This definition is somewhat dated as it refers to "a piece of cloth," which in 1956 was the material of choice for American Blood Chits. Blood Chits have been and continue to be printed on materials other than cloth. They are also issued to military forces of nations other than the United States. For the purposes of this work, the *Air Force Dictionary* definition with some modification is offered as the definition of the classic Blood Chit:

> **BLOOD CHIT**: a device issued by a government to an air crew member bearing a message in one or more languages that identifies him as a member of a particular military establishment, and promises a reward for his return to friendly control.[6]

The words "device" and "message" are deliberately chosen. They do not limit the type of material used, and allow for possible future developments of chits such as microchip based devices encoded with spoken messages for use in situations when non-literate peoples are encountered.

It has been argued that the promise of a reward in return for assistance is a primary property of a true Blood Chit. This defining characteristic excludes some items that have come to be referred to as Blood Chits through usage, as well as items actually referred to as Blood Chits in official government documents. Rescue Patches issued by the Nationalist Chinese government to American airmen during World War II, such as that sewn to the lining of the A-2 jacket shown in Fig. 1.1, directing the Chinese people to aid the flyer without promising a reward, have come to be referred to as Blood Chits. The documents issued to British and American airmen in the European Theater of Operations during World War II requesting that the bearer's particulars be communicated to the relevant military mission in Moscow, documents that do not offer a reward, are referred to in official military communications of the time as "Russian Blood Chits". The above definition of a classic Blood Chit notwithstanding, documents issued by a government that identify the bearer as a member of its military forces and request assistance to effect his safe return will, in general, be referred to as "chits" in this work regardless of whether a reward is offered.

Fig. 1.1: Shown here is perhaps the most famous A-2 jacket in existence — that of Gen. Claire L. Chennault, who commanded American air forces in China throughout World War II. It is thought to have been issued to him when he was commissioned Brigadier General in 1942. Chinese Rescue Patches, like the one sewn to the inside lining of the jacket in accordance with orders issued to Army Air Force personnel, have come to epitomize the Blood Chit in the mind of the American public. *Photo Courtesy of National Air and Space Museum, Smithsonian Institution (SI Neg. No 81-317). Thanks also to Robert B.T. and Martha Chennault Rogers., Max T. and Anne Chennault, and Robert and Ruth Chennault.*

The payment of a Blood Chit reward is a promise kept. If a promise is not fulfilled, then all other chits may become worthless in the eyes of local populations and their effectiveness negated. The reward is an expression of appreciation and gratitude, and an attempt to compensate one person for courageous acts in helping another, often at great risk to self and family.

Although this book primarily concerns Blood Chits, other language-related evasion aids carried by air crew members will be described and illustrated, including Pointie Talkies, Phrase Cards, and Phrase Books.

One publication dating from World War II describes the Pointie Talkie and its use: "The 'Pointie Talkie' . . . is exactly what its name implies. It is a collection of pertinent questions and answers in . . . languages spoken and sometimes read . . . with their English equivalents. If the user wishes to get an idea across to a native . . . he points to the native language for the question to which he wishes a reply. The native in turn points to the appropriate phrase for the answer."[7]

Phrase Cards are printed documents, generally folded accordion-style, containing a collection of phrases in English and their equivalents in native languages. They are not specifically designed for questions and answers like the Pointie Talkie, but communication is facilitated in the same manner, by pointing.

Evading capture by an enemy has been an essential element of combat since the first soldier found himself cut off from his comrades behind enemy lines. Some of the same characteristics that make for a good pilot – initiative, courage, skill, concentration, and psychological toughness – also are key elements required for survival and evasion. These factors were recognized by World War II survival experts who stated, "No amount of survival equipment will help some men, while others, through sheer ingenuity and determination, and under similar circumstances, will fare much better, and return regardless of whether they possessed such equipment or not. It's largely a mental outlook – the will to survive – that is the deciding factor."[8]

Historically, downed aviators seeking to evade capture by enemy forces have required assistance from local populations. Blood Chits and other language related evasion aids are an attempt to bridge seemingly insurmountable linguistic and cultural chasms between the evader and the local people in order to affect his safe return.

To those with a sense of history, it will not be surprising to learn that Blood Chits were developed and matured by the English in colonial India and Africa. In a long series of colonial wars, British troops were for years engaged in combat with peoples who were resisting both the British themselves and the figurehead rulers the Crown wished to install. In the 1920s and 1930s, while the American military establishment was endlessly debating the role of aviation in warfare, the Royal Air Force was routinely using air power as a means of colonial control in the far-flung reaches of the Empire. The employment of Blood Chits by RAF flyers in these areas was standard practice.

The history of Blood Chits stretches directly from the skies over India's North West Frontier Province in 1917 to the cockpits of American and NATO aircraft patrolling the sweltering "no-fly zone" over southern Iraq and the and chilled landscape of Bosnia in the mid-1990s.

Since their inception, Blood Chits have changed relatively little compared to the world around them. During the period that aircraft evolved from fragile biplanes made of fabric stretched across wooden ribs and stringers to fourth generation military jets made of space-age metals and composites, the essence of these lifesaving documents has remained constant. They have changed very little, essentially from ink on paper to ink on Tyvek, a nearly indestructible fabric created by DuPont. Regardless of the material, a Blood Chit with its promise of reward may still be considered the downed pilot's last hope.

Blood Chits are essentially utilitarian, yet these colorful items have acquired a patina of mystery and romance similar to that sometimes associated with flight itself. This work seeks not to remove that romance from the lore of Blood Chits, but to enhance the public's appreciation of these items, and of their true value in the vital work of rescuing downed air crews, by telling their very real story.

It is hoped that this book will encourage those of you who are veterans to rediscover your Blood Chits, to show them to your your children and grandchildren, and to tell them how you received and used these mysterious objects that once offered you a last hope.

CHAPTER 2

Innovation
The Age of Balloons and Biplanes

The peril to soldiers cut off behind enemy lines is ancient and universal. The need for nonverbal methods of communication in languages one does not speak or understand is likewise ancient, and is not limited to periods of war or to the development of aviation. The military aviator's Blood Chit, developed in the early part of the 20th century, was neither the first nor the only response to these needs.

With even the earliest realistic efforts at flight in balloons it was clear that aviators and aircraft would not, or could not, respect national borders or linguistic boundaries. Pioneering aeronauts speaking only their native language faced difficulties in communicating with local inhabitants after making a successful descent.

Jean-Pierre Blanchard, a Frenchman who was considered to be Europe's pre-eminent balloonist, visited the United States for the express purpose of promoting ballooning in America. On 9 January 1793, Blanchard lifted off from the yard of Philadelphia's Walnut Street Prison on his 45th flight – the first piloted hydrogen balloon flight in America.[1]

The flight was witnessed by some 4,800 spectators who had purchased tickets to the event, including President George Washington and the four men who would succeed him in office: John Adams, Thomas Jefferson, James Madison, and James Monroe. Also present were Secretary of the Treasury Alexander Hamilton and John Jay, first Chief Justice of the U.S. Supreme Court.

Blanchard, who by differing accounts spoke no English or only broken English, was given a "passport" that morning by President Washington. This document was given to guarantee his safety and help him identify himself upon landing. According to a display card in the Museum of American History in Washington, D.C., the passport, which contained some elements of today's Blood Chit, introduced Blanchard as a French citizen, and informed the reader that he was attempting to fly his balloon from Philadelphia that day. It further recommended that all U.S. citizens and others neither molest nor hinder him in his endeavor, that they receive him with good will, and that they provide him with any necessary aid.[2]

With an American flag unfurled from his gondola and accompanied by a small black dog, Blanchard completed a 45-minute flight covering 15 miles, landing near Woodbury, New Jersey.

According to one account, upon landing he presented his *passport* to a farmer, who unfortunately could not read. The farmer did understand Blanchard's friendly intentions when Blanchard offered him a bottle of wine.

In another description of his landing, a crowd gathered soon after his descent. It may be that this crowd had observed his flight and landing, and arrived after Blanchard's encounter with the farmer. In the American History Museum's exhibit, a statement attributed to Blanchard describes the reading of his passport to the crowd, wherein he noted that the passport served him well because he could not answer all the friendly questions that were put to him, and that the crowd was respectfully silent when the passport was read loudly and clearly by one of them.[3]

Blanchard's was the first recorded instance of an aeronaut using a written document to communicate with local inhabitants in a language he did not speak. More than 100 years before the first successful powered flight and introduction of fixed-wing aircraft as a weapon, Blanchard's situation neatly summarized the dilemma that both airmen and evasion experts would later attempt to resolve.

Despite an explanatory message in the common language from a popular and distinguished leader, the pilot's ability to express his peaceful intentions and immediate needs were thwarted by a farmer's inability to read. The scenario would be repeated many times in the 20th Century when members of different cultures met, despite attempts to overcome the language barrier with Blood Chits and other language material.

Blanchard's quick wit and his gesture of friendship in offering the farmer a bottle of wine, human acts that cannot be taught, illustrate the need for a downed airman to think creatively and act positively in a way that needs no translation. Later, military pilots facing potentially hostile situations would need to understand this, and evasion and escape (E&E) training would seek to give them the skills to make the appropriate decisions and take appropriate actions.

Barely one and one-half years after Blanchard's flight from Philadelphia, on 2 June 1794, hot air balloons were used by the French army to observe movements by Austrian forces at Maubeige near the Belgian border. The Austrians responded by putting their cannons in ditches and scored two near misses, one damaging a balloon, with the first anti-aircraft fire. Military aviation had arrived on the battlefield, complete with countermeasures.

Not long thereafter, in 1808, the world learned that pilots could survive despite the destruction of their aircraft when an aeronaut, using a primitive parachute, successfully abandoned a crippled hot air balloon in the first recorded bailout. Despite this advance, and the continuing use of balloons by the military, little progress seems to have been made in the field of pilot safety and survival during the nineteenth century.

An important step toward the development of the first Blood Chit may have occurred on 13 January 1842, in the mountains of Afghanistan. Captain Souter of the British Army's 44th Regiment of Foot found himself with the remnants of his command on a hilltop near of the village of Gundamuk. The thin red line of British soldiers was surrounded by an overwhelmingly superior force of angry and brilliantly led Ghilzee tribesmen revolting against an infidel occupation and the British-installed puppet government.

Less than two weeks before, on 6 January, Souter and 4,500 other regulars and mercenaries, their families, and more than 12,200 Indian camp-followers, had begun a 90-mile withdrawal from Kabul to Jallalabad. The retreat was a disaster.

With his force shattered, Souter waited while several of his fellow officers negotiated with the Afghans. Captain Souter had taken the regiment's colors from a dying lieutenant the night before, and to save the colors from capture he had wrapped them around his body, beneath his coat. As the officers were negotiating, they were cut down by Afghan rifle fire. The British positions were then assaulted.

The regiment was overcome, and its men slaughtered. Convinced that he was a very important man and worth a large ransom since the flag was wrapped around him, Souter was spared the fate of his men. The flag, a valued symbol of national identity, had played an important part in saving Captain Souter's life. He was among only 39 soldiers taken hostage.[4]

A valuable lesson learned from this incident was that a soldier might be valued for ransom not because he came from a rich family, but simply because of something he had on his person, and that a national symbol such as a flag, seal, or coat of arms might confirm his worth.

There are persistent rumors that some form of chit was first used by British ground forces more than 100 years ago. One informant reports that while traveling in Spain he read in a European produced English-language magazine, regrettably discarded, that chits were used by British officers in the 1860s during operations against Pathan tribesmen in the Khyber Pass region.[5] Attempts to identify the publication, and to otherwise confirm this speculation, have been unsuccessful. Given the timeless practice of holding hostages for ransom in time of war, and with the obvious long-standing difficulties of communicating in a foreign language, it is possible that some precursor of the aviator's Blood Chit saw use during the 19th Century.

The first device developed specifically to aid downed aviators came not from the military, but from a civilian aviation organization. Not surprisingly it was, in the tradition of George Washington and Jean-Pierre Blanchard, a language aid.

The growth of aviation led to the formation of the Paris-based Federation Aeronautique Internationale (FAI) on 14 October 1905. A month later its American representative, the Aero Club of America, was formed in New York City. The FAI established guidelines for aviators, and beginning in 1909 pilots who met these guidelines were issued certificates by the various national branches of the organization. These certificates allowed pilots to compete in FAI sanctioned events and to attempt record-breaking flights. Although not pilot licenses, these certificates were accepted by many governments as proof of the flying ability of those seeking admission to military flying corps. Included on the certificate was a request in six European languages that the pilot be given assistance . Bill Robie, author of *For the Greatest Achievement: A History of the Aero Club of America and the National Aeronautic Society*, details the reasoning that led to the inclusion of this material:

> Originally, international aeronautic competition took the form of balloon "racing". These races were actually distance rather than speed events, and were most often held in Europe, the winning nation of one year's competition would sponsor the following year's race. Often, the distances were substantial, sometimes exceeding 1,000 miles.
>
> The circumstances of competitors from many nations (and speaking many different languages) flying over the borders of possibly several different countries, were combined with the fact that these early balloonists might land in a remote area where the populace had never before seen a person flying. It is for this reason that the . . . certificates issued by the various national branches of the FAI included the request for assis-

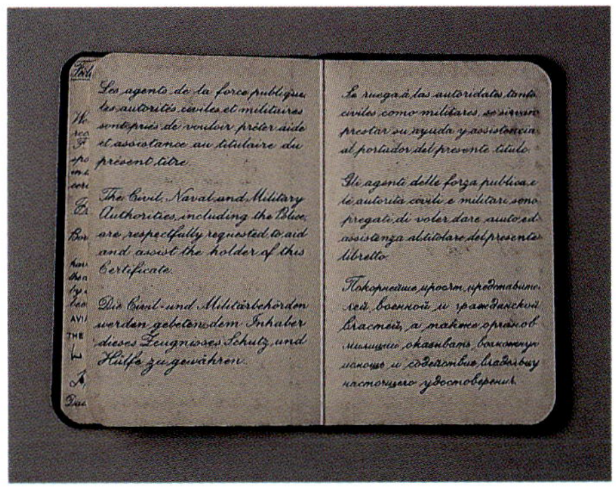

Fig. 2.1: British Empire FAI Aviator's Certificate no. 894 in both English and French, issued to Francis Thomas Courtney in 1914. The FAI, recognizing the far reaching effects aviation would have on international and linguistic boundaries, included on the last two pages a request for aid in six European languages: English, French, German, Italian, Spanish and Russian. Although not Blood Chits, these certificates were the first mass produced devices to request assistance for pilots in languages they did not speak. *Courtesy of the San Diego Aerospace Museum*

> tance . . . It is my belief that all FAI certificates (regardless of issuing nation) included such statements – probably from the time that the first certificates were issued.[6]

These certificates were the earliest mass-produced devices specifically designed to help aviators communicate their particulars should they come down in a region where the language was not their own or one they could speak, and they are recognized as the civilian predecessor of the military aviator's Blood Chit.

Two examples of FAI certificates are illustrated. That in Fig. 2.1 was issued by the British Empire to Francis Thomas Courtney in 1914, and that in Fig. 2.2 was issued by the Aero-Club of America to Lt. W. A. Gates.

The first use of airplanes in warfare may have occurred on 1 February 1911, when René Simon and Roland Garros, two French citizens who had emigrated to the United States, flew their Bleriot aircraft for the Mexican government. Hired by the Mexican military to observe rebel movements along the Rio Grande near Juarez, the two pilots did not face any antiaircraft fire and dropped oranges and cigarettes to the rebels while observing them.

On 1 November 1911 an Italian aviator, Lt. Giulio Gavotti, flew the first aerial combat mission during the First Balkans War. Flying one of the Austrian *Taube* (Dove) monoplanes the Italians had in Libya, Gavotti dropped four 4.4 lb. Swedish grenades on Turkish positions and returned without apparent incident.

Shortly after Gavotti's flight, Carlo Montu, the leader of the Italian corps of civilian volunteer pilots, became the first aviator wounded in action when he was struck by a Turkish rifle bullet fired against his aircraft.[7]

The earliest reference to the possible fate of military aviators if they were to be captured by an enemy dates from 1912. Flying for the Italians against the Turks were German, French, and British pilots, and possibly an American or two. While some of these pilots would not drop bombs for their own ethical or political reasons, "these distinctions meant nothing to the Turks, who had sent word by wireless that they would summarily execute any enemy airmen . . . who fell into their hands."[8]

The lives of air crew members were clearly at risk not only in the air during missions, but on the ground if forced down behind enemy lines. While an airman's survival and escape from a dam-

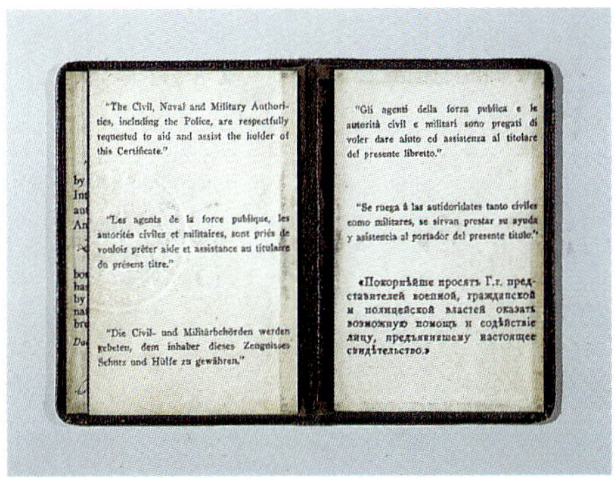

Fig. 2.2: FAI Aviator's Certificate no. 2534, issued to military aviator W. A. Gates of Louisville, Kentucky, on 20 August 1918 by the Aero-Club of America. The photo on the FAI card shows Lt. Gates in uniform. Although of a slightly different style than the British Empire FAI card, the standard request for aid in six European languages appears on the last two pages. This request appeared on all FAI Certificates, regardless of the issuing nation. *International Blood Chit Museum*

aged aircraft was a genuine concern to some military planners during this period, the fate of these men once on the ground and faced with capture by an enemy was not as carefully considered. Despite the increasing use of aircraft in warfare, there were virtually no developments in the field of aircrew evasion in the years between the creation of the FAI Certificate with its language aid and World War I.

Before its entry into World War I, the American army responded to a series of border crossings by Mexican revolutionaries led by Pancho Villa by sending troops commanded by General John J. "Black Jack" Pershing into Mexico. The 1st Aero Squadron began operations based in Columbus, New Mexico, with eight Curtiss JN-3 aircraft. The Squadron's pilots included Benjamin Foulois, a former enlisted man who earned a battlefield commission leading a rifle company in the Philippines during the Spanish-American War, and who then became one of the Army's earliest pilots and an outspoken supporter of air power. In March 1916, Foulois landed his JN-3 in unfriendly territory to complete a reconnaissance on foot, per standing instructions. He was seized by Villa's rurales, becoming the first American airman to be captured, and was held until the American consul negotiated his release.

There is no record of special survival or evasion aids being used during the American air operations against Villa. Pilots in this campaign assembled their own emergency equipment such as maps, compasses, water, rations, and small arms, and stowed this gear in their aircraft.[9] There is no record that any language aids were included among this equipment.

For Americans, World War I was the first major involvement in international warfare since U.S. troops had gone abroad in the Spanish-American War. English-only speaking American aviators and soldiers heading for Europe during World War I had access to a number of French-English/English-French language books. It is difficult to estimate how common these books were, but they turn up frequently enough on the antique and collector's market to assume that they were not uncommon. These books are among the few language aids that were available to aviators during the First World War.

Several examples of these books are shown in Fig. 2.3. These books are primarily intended to teach the French language, and are organized for that purpose. None is organized for easy use in communicating in the field.

Some aviators carried their nation's flag with them. On display at the United States Air Force Museum[10] is an American flag which was first carried in aerial combat in 1917 by H. MacMenigall, and subsequently carried in the air wars of World War II, the Korean conflict, and the Vietnam War. Although probably begun for reasons of patriotic pride, this practice was also practical because flags could be used to immediately identify the nationality or affiliation of the bearer before a single word was spoken. Two small silk flags dating from World War I, one British and one French, are illustrated in Fig. 2.4.

Fig. 2.3: The recognition of a need for language material by Americans headed overseas during World War I resulted in the production of a variety of instructional booklets, several of which are illustrated here. The top row contains, from left to right, *War French* by Cornélis DeWitt Willcox, Col. U.S. Army, printed by MacMillan Co., New York, in 1918, and *First Lessons in Spoken French for Men in Military Service* by Ernest H. Wilkins, Algernon Coleman and Howard R. Huse, printed by the University of Chicago Press in 1917. Displayed in the second row are the *Cortina French-English Soldier's Handbook*, written by Jean A. Picard of the French Army, published in 1917 by The Cortina Academy of Languages of New York and printed by The International Press, and *Speak French - A Book for the Soldiers*, subtitled *Easy Lessons in French*, printed in 1917 by The Goldsmith Publishing Co. of Cleveland, Ohio. Several of these books give useful information other than language lessons or phrases. The *Soldier's Handbook*, for example, contains information on Grade and Insignia of the French Army, the French Rifle, and Soldiers' Slang expressions. *International Blood Chit Museum*

Although some progress was made in the fields of aircrew selection and safety equipment, official interest in evasion and escape remained almost nonexistent throughout most of World War I. Late in the war, the British used men who had evaded or escaped from the Germans to deliver lectures to aviators, telling of their experiences and offering advice. There was, however, no official interest in evasion aids, and aviators were left on their own in this regard. First World War aviator Arch Whitehouse recalls:

Fig. 2.4: Small flags such as these were originally carried by aviators for patriotic reasons, but the flags would have also been useful for quick, non-verbal identification of the bearer's nationality or affiliation. The two flags shown date from World War I and are made of a sheer silk. They could easily be folded and kept in a convenient pocket. The flag on the left is French, and that on the right is English. *International Blood Chit Museum*

Fig. 2.5: There being no official governmental interest in evasion and escape for most of World War I, some aviators made their own kits. Among the items available for use by American airmen were pocket sized flags, compasses, small scale maps showing major cities, rivers and railway routes, FAI Certificates containing requests for assistance in six European languages, and language material such as *The Soldiers' French Phrase Book* produced for free distribution to American soldiers by Felt & Tarrant Mfg. Co. of Chicago. The items are displayed on a period French military map. These items were light and compact, and could be easily carried in a single pocket, although E&E items were often secreted in clothing and shoes. The Phrase Book, detailed in Fig. 2.6, is designed for ease of use in the field. FAI Certificates could not have been used in those cases when it was forbidden to carry identification material on a mission. *International Blood Chit Museum*

> we did our best to prepare for such an eventuality . . . We obtained small scale maps of Germany which we stitched inside some section of our tunics. We purchased small compasses in nearby towns and hid them away by hollowing out a heel of one boot . . . we borrowed pieces of hacksaw blade from the mechanics and secreted them in the seams of our jackets. Knife blades of various sorts were saved and added to these escape kits. Small pieces of mirror . . . for "signaling" . . . were included . . .[11]

Shown in Fig. 2.5 is an assortment of items available to American airmen during World War I that could have been used in assembling evasion kits. Included is a booklet designed for quick and easy use in the field, *The Soldiers' French Phrase Book*, detailed in Fig. 2.6, and an FAI certificate.

Regarding the use of FAI certificates during the First World War, Bill Robie writes that the Aero Club of America maintained a branch in Paris under the aegis of the Foreign Service Committee. After the United States joined the war, "a separate membership classification for members of the armed forces" was formed "and it appears to have been common practice for Aero Club pilots – civilian and military – to use their FAI certificates for identification purposes" although there is "no documentation of an American pilot being aided by French or Belgian citizens, based on the statements accompanying the certificate."[12] Other sources indicate that, in many instances, it was forbidden to carry identification documents into combat. This would have precluded the use of FAI Certificates with their plea for assistance.

While inhabitants of countries occupied by the Germans were generally friendly to Allied airmen, statistics showed that most airmen downed in German held territory were likely to be captured and interned in Prisoner of War (POW) camps. In spite of this, official interest in E&E remained nearly nonexistent, and airmen continued to assemble their own E&E kits.

It would take different kind of war, waged half a world away, to produce the catalyst that sparked the development of the first Blood Chits for military aviators.

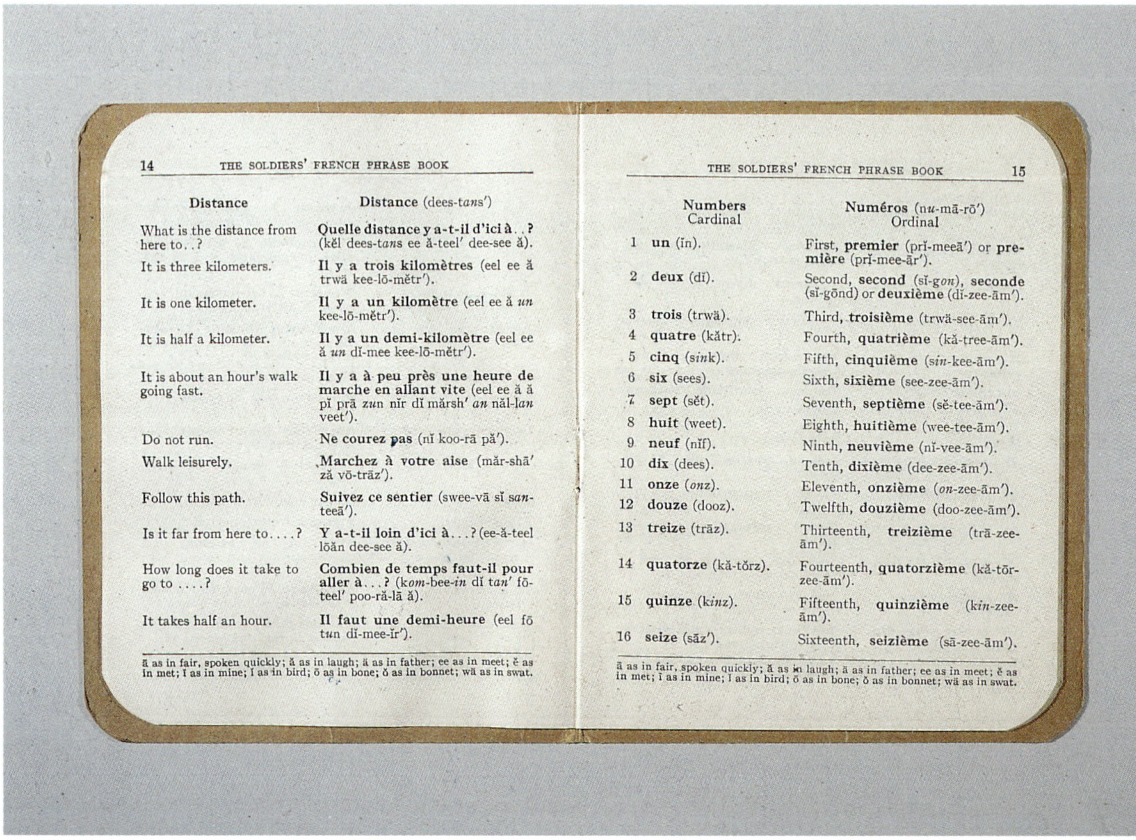

14 THE SOLDIERS' FRENCH PHRASE BOOK

Distance	Distance (dees-*tans*')
What is the distance from here to. . ?	**Quelle distance y a-t-il d'ici à. . ?** (kěl dees-*tans* ee ă-teel' dee-see ă).
It is three kilometers.	**Il y a trois kilomètres** (eel ee ă trwä kee-lō-mětr').
It is one kilometer.	**Il y a un kilomètre** (eel ee ă *un* kee-lō-mětr').
It is half a kilometer.	**Il y a un demi-kilomètre** (eel ee ă *un* dĭ-mee kee-lō-mětr').
It is about an hour's walk going fast.	**Il y a à peu près une heure de marche en allant vite** (eel ee ă ă pĭ prā *zun* nĭr dĭ mărsh' *an* năl-*lan* veet').
Do not run.	**Ne courez pas** (nĭ koo-rā pă').
Walk leisurely.	**Marchez à votre aise** (măr-shā' ză vō-trāz').
Follow this path.	**Suivez ce sentier** (swee-vā sĭ *san*-teeā').
Is it far from here to. . . . ?	**Y a-t-il loin d'ici à. . . ?** (ee-ă-teel lōăn dee-see ă).
How long does it take to go to . . . . ?	**Combien de temps faut-il pour aller à. . . ?** (k*om*-bee-*in* dĭ t*an*' fō-teel' poo-ră-lā ă).
It takes half an hour.	**Il faut une demi-heure** (eel fō t*un* dĭ-mee-ĭr').

ā as in fair, spoken quickly; ă as in laugh; ä as in father; ee as in meet; ĕ as in met; ī as in mine; ĭ as in bird; ō as in bone; ŏ as in bonnet; wä as in swat.

THE SOLDIERS' FRENCH PHRASE BOOK 15

	Numbers Cardinal	Numéros (n*u*-mā-rō') Ordinal
1	**un** (ĭn).	First, **premier** (prĭ-meeā') or **première** (prĭ-mee-ār').
2	**deux** (dĭ).	Second, **second** (sĭ-g*on*), **seconde** (sĭ-gōnd) or **deuxième** (dĭ-zee-ām').
3	**trois** (trwä).	Third, **troisième** (trwä-see-ām').
4	**quatre** (kătr).	Fourth, **quatrième** (kă-tree-ām').
5	**cinq** (s*in*k).	Fifth, **cinquième** (s*in*-kee-ām').
6	**six** (sees).	Sixth, **sixième** (see-zee-ām').
7	**sept** (sĕt).	Seventh, **septième** (sĕ-tee-ām').
8	**huit** (weet).	Eighth, **huitième** (wee-tee-ām').
9	**neuf** (nĭf).	Ninth, **neuvième** (nĭ-vee-ām').
10	**dix** (dees).	Tenth, **dixième** (dee-zee-ām').
11	**onze** (*onz*).	Eleventh, **onzième** (*on*-zee-ām').
12	**douze** (dooz).	Twelfth, **douzième** (doo-zee-ām').
13	**treize** (trāz).	Thirteenth, **treizième** (trā-zee-ām').
14	**quatorze** (kă-tŏrz).	Fourteenth, **quatorzième** (kă-tōr-zee-ām').
15	**quinze** (k*inz*).	Fifteenth, **quinzième** (k*in*-zee-ām').
16	**seize** (sāz').	Sixteenth, **seizième** (sā-zee-ām').

ā as in fair, spoken quickly; ă as in laugh; ä as in father; ee as in meet; ĕ as in met; ī as in mine; ĭ as in bird; ō as in bone; ŏ as in bonnet; wä as in swat.

Fig. 2.6: The layout of *The Soldiers' French Phrase Book* is ideal for use by two people pointing to phrases in order to communicate, and sets the pattern for future Phrase Books and Phrase Cards which would be produced for air crew members in later years. It is organized with English phrases on the left side of a page and the equivalent phrases in both French and phonetic spelling on the right side. At the bottom of each page is a concise pronunciation guide so that the user does not need to refer to a separate page for pronunciation hints. The booklet is not intended specifically for evasion purposes, and the phrases in it cover a wide variety of possible situations. *International Blood Chit Museum*

CHAPTER 3

Blood Chit Genesis Policing the Empire

American popular mythology to the contrary, Blood Chits did not originate in China with American aviators flying under Claire Chennault with the 14th International Squadron in 1937-38 or the American Volunteer Group (AVG) in 1941-42.

Speculation about the origins of Blood Chits has produced volumes of erroneous information in the United States. The American propensity to perpetuate the myth that Blood Chits originated with American airmen in China, despite numerous references to their use by Royal Flying Corps crews beginning as early as 1917 and by Royal Air Force crews throughout the 1920s and 1930s, is particularly difficult to explain. It appears to be an instance of repeating what was once stated in ignorance, although with perfect sincerity, in lieu of doing basic research.

Repeated and embellished, this "Official American Blood Chit Myth" has found its way into untold histories and even contemporary government documents. A statement recently printed by the Defense Mapping Agency Aerospace Center (DMAAC) on the back of a replica of one of the current American Blood Chits perpetuates the myth: "The concept of the Blood Chit originated in China during the early days of World War II. It was created by members of General Chennault's American Volunteer Group – the Flying Tigers." What is most unfortunate about this statement is that, by virtue of it being printed by the U.S. government agency that produces contemporary American Blood Chits, it implies official recognition of a story which is laughable to scores of living British airmen who carried Blood Chits during the two decades preceding 1941.

The often cited American document from World War II that might rightly be credited with originating this myth reports:

> The "Blood Chit" sprang up in various areas from Europe to the South Pacific, but the name originated in China . . . In the early days of the war, airmen who had been forced down found it possible to pay helpful natives with an IOU scribbled on a piece of paper, cloth or anything else that happened to be handy or convenient. Subsequently, after wide semiunofficial use of the IOU or "Blood Chit", various governments recognized the practice officially and established payment as a general policy. At the same time, more formal "Chits" began to spring up.[1]

The above account is highly fictionalized, if not entirely incorrect. In fairness, it must be noted that it was written during the war when information regarding evasion aids was highly classified, and by an American who could not be expected to know details of the British use of chits during the preceding two decades. Today there is less excuse for ignorance.

Royal Air Force squadrons serving in India during the latter years of World War I were the first military units to use Blood Chits in a systematic way.

Before the introduction of aircraft, the British used their army and locally established forces to control their colonial territories and populations. This sometimes meant mounting large, costly, difficult, and not always successful punitive expeditions into remote and hostile regions. The weaknesses of the British ground forces were well-known by the opposition. These liabilities, particularly the forces' long supply and communications lines, were exploited by the hostile forces as fully as their resources allowed.

Aircraft offered an immediate and less costly solution to the problem of controlling those who defied the Empire but who refused to join battle or who were difficult to reach. In the Sudan, for example, an uprising in 1916 occurred 400 miles from the nearest railhead, making the deployment of ground forces costly and difficult.

Hugh Trenchard, a veteran of the Boer War who learned to fly in 1912 and former Officer Commanding-In-Chief of the Royal Flying Corps, became chief of the air staff of the newly re-named Royal Air Force in April of 1918. Trenchard became the advocate of what the Royal Air Force and the British government euphemistically referred to as "Air Control": the use of aircraft, particularly bombers, to police the British Empire in the Middle East and India. This policy of controlling populations without occupying territory relied upon air power to deal quickly and effectively with any uprising, assuming that native peoples would choose to obey their colonial rulers rather than face a weapon against which they had little or no defense.

British military aircraft were in service in India and in the Middle East at least two years before the Royal Flying Corps (RFC) and the Royal Navy Air Service (RNAS) were merged into the Royal Air Force. 31 Squadron of the RFC arrived in India in December

1915 with 12 aircraft, 28 officers, and 84 other ranks. The continuing internal warfare in India and in neighboring countries kept 31 Squadron on the sub-continent despite the end of the war in Europe. Following the 11 November 1918 Armistice, at least five other RAF squadrons were assigned to air control duties, including Squadrons No. 5, 20, 27, 28, and 60.

Although the precise genesis of Blood Chits, originally called Ransom Notes, is unknown, a reference to their earliest use appears in No. 31 Squadron's published history, *First In The Indian Skies* by Norman L.R. Franks:

> In the new year of 1917, 31 Squadron found itself operating more and more over Waziristan, that strip of land between the North West Frontier and Afghanistan. It is peopled by the Wazirs and the Mahsuds, both very warlike tribes, and danger surrounded any flying men who might fall into their hands. This was when Ransom Notes began to be carried by the pilots and observers, promising considerable monetary rewards for the return of airmen to the nearest British outpost . . .[2]

That the RAF first referred to Blood Chits as "Ransom Notes", is appropriate, for in essence they are. While the reward offered in a Blood Chit is technically not a ransom, it is certainly a direct result of the ancient practice of holding prisoners of war for ransom. Ransoms are detailed in the earliest written set of laws, the Code of Hammurabi the Assyrian, dating to about 2000 years before the Christian era. Narrative accounts of this practice are numerous, including that of Cervantes, author of *Don Quixote*, who remarked that during the years of his imprisonment following capture by the Moors he dared not let them know that he was from a poor family lest he be killed outright.

Richard Simpson of the Royal Air Force Museum clearly establishes the link between chits and the traditional ransom of prisoners in this concise statement regarding the rationale for issuing the first Ransom Notes:

> The British armed forces were acting in a series of colonial policing operations against native tribes. Inter tribe warfare had been the norm for centuries but so had been the holding of certain prisoners for ransom. Women and children had little use in these societies except as slaves, but dignitaries or soldiers might be held against payment in gold. The so called civilised codes of war did not apply so the object of the written chit was to impress on the native leader the man's increased value if returned unharmed.[3]

In RAF slang, the Ransom Note became known as a "Goolie Chit". "Goolie", or in some references "Goli" or "Gooli", is the Anglicized Hindustani word *goli*, meaning "bullet", "ball", or "pill". It was the RAF barracks, or slang, term for a testicle.[4] The significance of this term is explained by one member of 31 Squadron who describes one use prior to the introduction of Chits for another piece of equipment the crews carried: "I should mention that on the Frontier we carried revolvers to destroy ourselves in event of a forced landing in tribal territory, as it was said the tribal women, after extracting gold from false teeth, would remove one's genitals while still alive."[5]

Retired Air Vice-Marshall A.G."Tony" Dudgeon, CBE, DFC, author of *Luck of the Devil* and *The War That Never Was*, served with 11 Squadron RAF in "The Raj" during the mid and late 1930s. He elaborates on the situation faced by airmen:

> Although we flew above the range of rifle fire almost all the time, we were not out of danger. The consequences of an engine failure over tribal territory were not attractive.
>
> We know that in the twenties they handed over a downed crew to the women of the tribe. They, with due care, removed what are known in colloquial French as the *bijoux de famille* – the family jewels. One portion was sewn into the man's mouth and the two other were for sewing into his eye sockets. Naturally, as the men would enter Paradise as they left the earth, the women took considerable pains to ensure they did not die before the operations were complete. This made certain that in spite of the best ministrations of any houris (heavenly female attendants blessed with a renewable virginity) that they were booked for a pretty rotten eternity – unable to eat, unable to speak, unable to see – and due for nothing else of much importance.[6]

Although chits were originally referred to variously as "Ransom Notes", "Protection Certificates" and "Safety Chits", it is the term "Goolie Chit" that in the eyes of the airmen best describes their most valuable use, and it is the term by which they are known among U.K. and Commonwealth aviators even today.

Eric Nicholls, an armourer and sometimes gunner on Hawker Audaxes with 28 Squadron on the Northwest Frontier, uses the term and explains its importance when writing of a Winter mission in the late 1930s to drop warm clothes to an Army unit isolated by snows high in the Hindu Kush. In picking up a message from a loop cord suspended between two poles on the ground, his pilot had flown so low that he had caught the loop in the aircraft's undercarriage rather than "on hook". Nicholls writes:

> Thankfully, the cable did not foul anything apart from the undercarriage so that we were able to return safely to Miramshah and did not require use of the Goolie Chit which was kept in a pocket doped to the fuselage behind the rear cockpit.
>
> This document was a ransom note written in Urdu and Pushto languages and was used by airmen who were unfortunate enough to force land in Waziristan. It promised a generous cash reward as long as the aircrew was returned to a British camp intact. (The tribal women were prone to separate a fellow from his manhood, for the probable reason that it would prevent him from producing more "infidels".)[7]

Of the origin of another slang term which was eventually adopted officially by the militaries of both Britain and United States – the term *Blood Chit* – Air Vice-Marshall Dudgeon writes, "Because of its most valued potential, it was always known as a goolie chit. 'Blood Chit' was merely a watered down or cleaned up term used solely when sensitive ladies were listening, which was almost never."[8]

Considerable detail is available regarding how chits were issued and carried operationally. According to Dudgeon, the government determined how many chits were issued to each squadron and how they were to be used. Each squadron had, at full authorized strength, 12 aircraft and 18 pilots and 18 crews. There was a chit for each aircraft and a chit for each crew, so a squadron could have been issued as many as 50 chits, though Dudgeon relates that in reality only one chit per aircraft was issued:

> The chits were on the left outside of the aircraft, just behind and level with the pilot's shoulder - folded in four and put under a square of linen like the wings were covered with, and doped over with silver dope just like a patch over a tear in the wing would be. There was some lettering on the patch, but I cannot remember what it was.
>
> We wanted them in our parachutes, so if you bailed out, or if you crashed and it caught fire, or if one of your crew-members was wounded and you became separated, you still had your goolie chit . . . But the government said "no" because someone could pinch a parachute – but not an aircraft. This was stupid because it was simpler to enter a hangar and pull the patch off an aircraft than to get a 'chute out of a locked cabinet in a locked office. Bureaucrats don't change.
>
> They were allegedly paid "on sight" so that the Indian Government, who rented out RAF squadrons to look after their security on the NW Frontier, and would have to pay out on goolie chits used, were very parsimonious with their issue . . . The amount to be paid out was graded downwards according to how unhealthy or incomplete the user was.[9]

Records regarding the use of Goolie Chits in colonial India are sparse. On 14 May 1918, one of 31 Squadron's aircraft was lost when it crash landed in the Bazar Valley during a reconnaissance mission. The pilot and observer, Lieutenants G. C. Barker and L. T. Hoare, were captured by Afridi tribesmen and eventually ransomed back to the RAF for 30,000 rupees.[10] Though it apparently became standard practice for members of 31 Squadron to carry chits beginning in early 1917, there is no record of whether Lts. Barker and Hoare used their Ransom Note.

31 Squadron operated from Dardoni/Miranshah in Waziristan for a year, beginning in 1923, and one documented use of Goolie Chits comes from this era: "When flying over Waziristan we always carried what we called 'gooli chits', small documents written in Urdu and Pushto, which promised a ransom should we be returned unharmed after a forced landing. They certainly worked in the one incident I remember of two officers, Foster and Ridley, though they had a rough time."[11]

Examples of the chits from India during this period are rare. Presented here, for the first time together, are illustrations of six varieties.

An example of what is thought to be the very first Goolie Chit, written in two languages, is shown in Fig. 3.1.[12] At the top is the seal of the Chief Commissioner, N. W. Frontier Province. The chit is written in Farsi on the left side and Pashto on the right side. The text of this and of other early chits does not mention the government of India. One translation of this chit is quite flowery, referring to the Chief Commissioner of the North West Frontier Province, Sir Hamilton Grant who signed the chit as A.H. Grant, as "Sahib

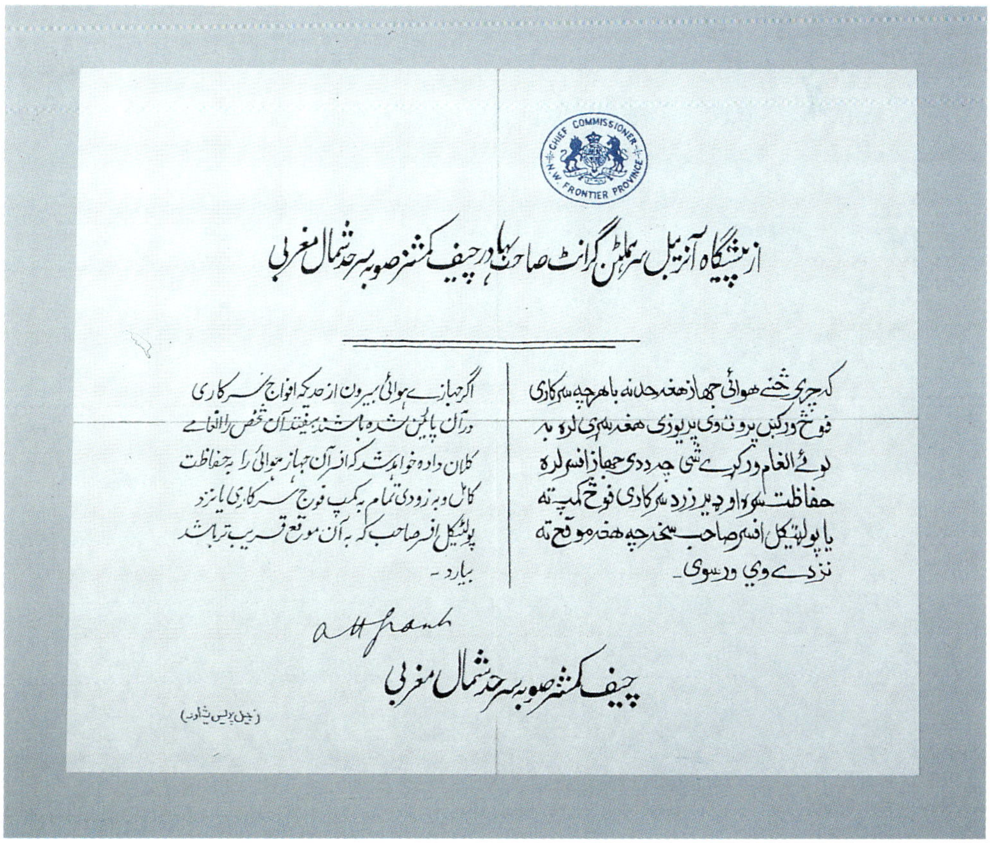

Fig.3.1: Goolie Chits like this example were issued to airmen flying Colonial policing missions in the North West Frontier Province of India. Thought to be the first military aviator's Blood Chit, it may have been issued as early as 1917, and is known to have been issued in quantity during 1919. One translation of this chit reads, "If the crew whose aircraft landed in your region are taken safely to the nearest government post or military base, you will be given a large reward. /s/ A. H. Grant, Chief Commissioner of the North West Frontier Province." The writing in the lower left corner reads, "Chief of Police, Peshawar." Peshawar is located in present day Pakistan, just southwest of Risalpur, north of Kohat, and west of the Khyber Pass. (REB 1110) *Courtesy of Jeffrey D. Guidry*

Bahidur." It also specifies that the person bringing the pilot of a downed aircraft to a British Army camp or political officer will be well rewarded.[13]

Another translation gives a less flowery interpretation of the text:

> If the crew whose aircraft landed in your region are taken safely to the nearest government post or military base, you will be given a large reward. /s/ A. H. Grant, Chief Commissioner of the North West Frontier Province.[14]

Differences in translations are often due to the inability to translate dissimilar languages literally. Translations of dissimilar languages must often be done by meaning, not word-for-word.

One early chit of this type is named to a Captain Foster and marked with a serial number, 2, and is said to have been issued in 1919.[15]

The two chits shown in Figs. 3.2 and 3.3 are handwritten notes, and were apparently intended for a special mission into Afghanistan. Although Afghanistan was an independent nation and officially at peace with its British neighbors by the time these chits were used, there was still a risk to British flyers who might be forced down. These chits, beneath the seal of the Chief Commissioner, mention the British government but not the Indian government. The text of the Farsi chit (Fig. 3.2) and that of the Pashto chit (Fig. 3.3) are nearly identical:

> This letter is written by the Chief Commissioner of the North West Frontier Province. The bearers of this letter are British officers. The aircraft which they are flying belongs to the British Royal Air Force. This aircraft is flying between Kabul and Peshawar. They go to Kabul to evacuate the family members of the Embassies in Kabul and bring them to Peshawar. Anyone who helps and protects these officers will be given a substantial reward by the British Government. Signed by Chief Commissioner of the North West Frontier Province. Date: 26 December 1928.[16]

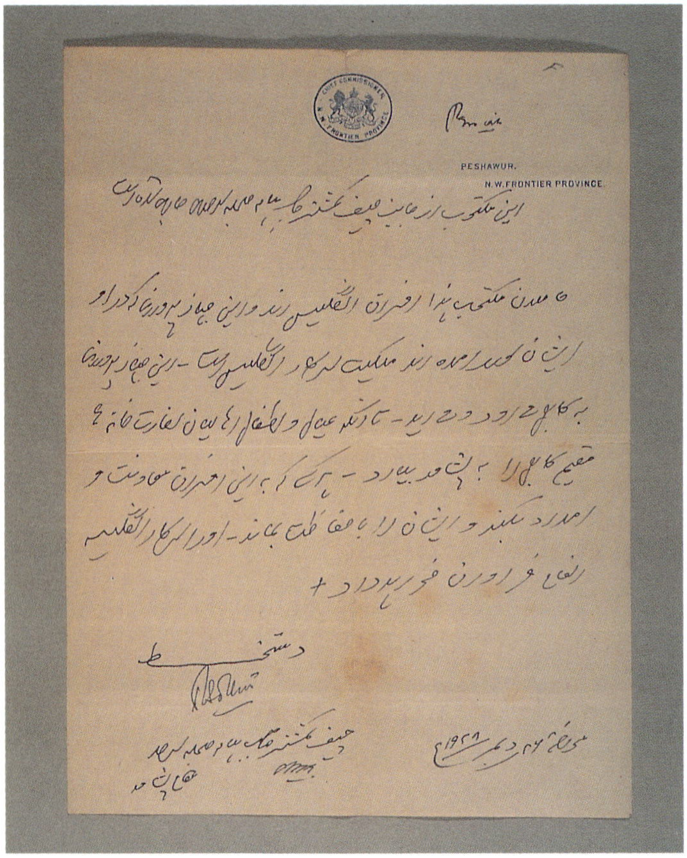

Fig. 3.2: Shown here is one of two handwritten chits issued in December 1928 for a mission to evacuate families from the British Embassy in Kabul, Afghanistan, to Peshawar. This chit is written in Farsi, and reads, "This letter is written by the Chief Commissioner of the North West Frontier Province. The bearers of this letter are British officers. The aircraft which they are flying belongs to the British Royal Air Force. This aircraft is flying between Kabul and Peshawar. They go to Kabul to evacuate the family members of the Embassies in Kabul and bring them to Peshawar. Anyone who helps and protects these officers will be given a substantial reward by the British Government. Signed by Chief Commissioner of the North West Frontier Province. Date: 26 December 1928." (REB 1112) © *British Crown copyright. Imperial War Museum photo, printed by permission of the Trustees of the Imperial War Museum.*

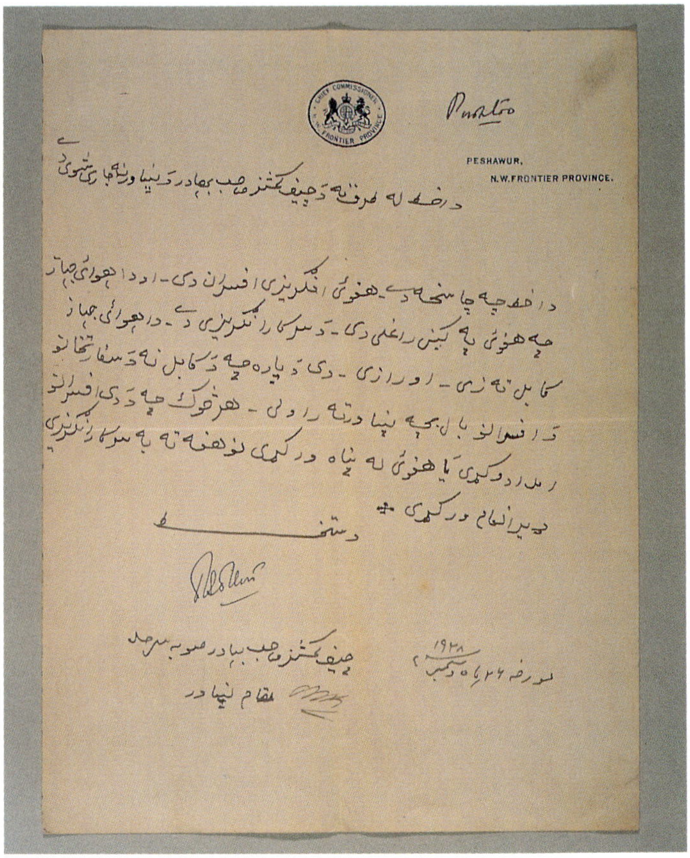

Fig. 3.3: The companion to the handwritten chit in Farsi (Fig. 3.2), this letter written in the Pashto language was also used on the mission to evacuate families from the British Embassy in Kabul, Afghanistan, in December 1928. The translation of its text is essentially identical to that given for its companion chit written in Farsi. (REB 1113) © *British Crown copyright. Imperial War Museum photo, printed by permission of the Trustees of the Imperial War Museum.*

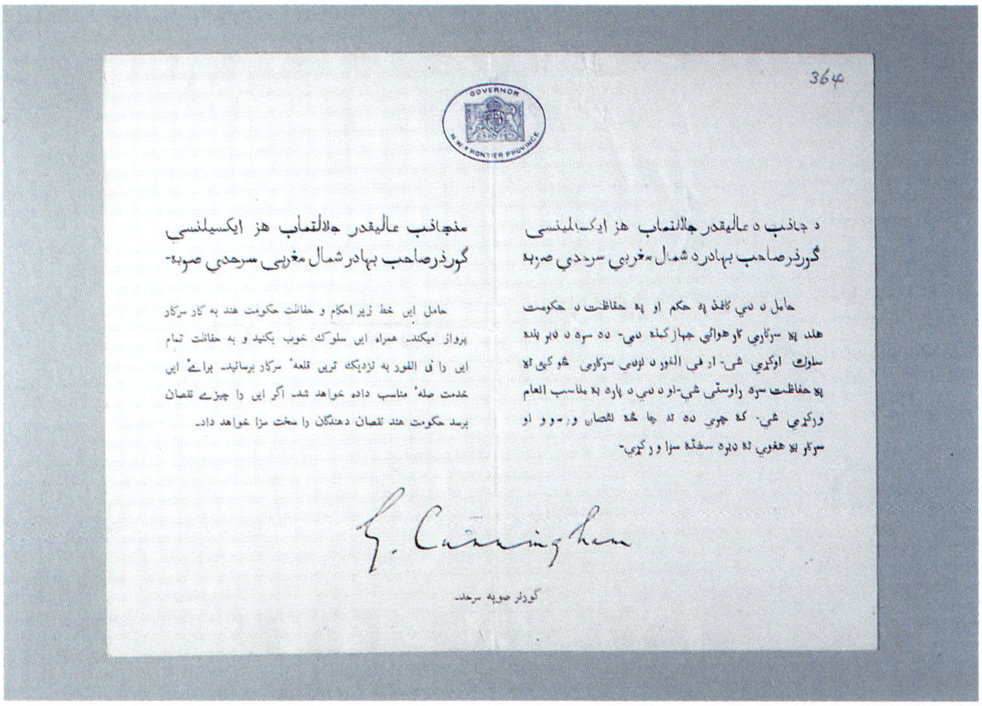

Fig. 3.4: The Goolie Chit illustrated here bears the seal of the Governor of the N. W. Frontier Province, rather than that of the Chief Commissioner. This example was issued by the Government of India rather than by the local government of the N.W. Frontier Province. This chit was found in the stores of RAF 20 Squadron during World War II. It was purchased from the Squadron's Administrative Staff in March 1943 by Flying Officer James R. Ashworth, RCAF, attached to 20 Squadron at "Hove-"George", Arakan, south of Chittagong, Imphal, India. It is written in Farsi and Pashto, and the Farsi text reads, "From His Excellency, Mr. Bahadur, The Governor of the North West Frontier Province: The bearer of this letter is flying under the orders and protection of the Government of India. The aircraft belongs to the Royal Air Force. He must be treated properly and taken to the nearest government post at once. If this is done, a suitable reward will be given. If he is harmed, the ones responsible will be severely punished." (REB 1116) *Courtesy of J.R. Ashworth.*

The two chits were donated to the Imperial War Museum, and the note from the donor that accompanied them states that he was an officer in the Indian Army before he became a Flight Lieutenant and RAF pilot in the North West Frontier Province with No. 5 Squadron. He noted that the squadron flew over territory controlled by Pathan tribesmen, and that the chits in Farsi and Pashto were issued because downed airmen could be subject to castration if captured. He also remarked that the chits were called Goli chits, from the Hindustani word for ball, that they offered a reward to tribesmen who returned airmen intact, and that the chits were highly valued by pilots for obvious reasons.[17]

At the top of the Goolie Chit shown in Fig. 3.4 is the seal of the Governor of the N. W. Frontier Province, rather than that of the Chief Commissioner, and it is signed by G. Cunningham. This is an important distinction, for it reveals this to be of a later vintage than the chits bearing the Chief Commissioner's seal. The translation of the Farsi text reads:

> From His Excellency, Mr. Bahadur, The Governor of the North West Frontier Province: The bearer of this letter is flying under the orders and protection of the Government of India. The aircraft belongs to the Royal Air Force. He must be treated properly and taken to the nearest government post at once. If this is done, a suitable reward will be given. If he is harmed, the ones responsible will be severely punished.[18]

The terms "Sahib Bahadur" and "Mr. Bahadur" are used in the text of several chits. This appears to be a formal title applied to, in these cases, A.H. Grant, Chief Commissioner, and G. Cunningham, Governor, of the N.W. Frontier Province.

Fig. 3.5 shows a Goolie Chit with the Royal coat of arms, often called simply the "Royal Arms", and the words "North West Frontier Province" at the top, rather than the seal of either the Commissioner or Governor. The date of this chit is unknown but the inclusion of the Royal Arms and the mention of the Indian Government would appear to date it later than the preceding illustrated chits. It was issued to a member of 31 Squadron, and appears in their squadron history, *First in the Indian Skies*. The text, written in Pashto, reads:

> Notice: An aircraft unsuccessfully landed in your territory. The crew of this aircraft belongs to the Royal Air Force. They are flying under the orders and protection of the Indian Government. They must be treated properly and immediately turned in safely to the nearest government post. If this is done, a proper reward will be awarded. Should they be harmed, the ones responsible will be severely punished.[19]

Fig. 3.5: This Goolie Chit in the Pashto language issued by the Indian government appears to be the only example from this era to display the Royal Coat of Arms. The King's monogram, "GR", is incorporated into the Royal Arms. The text reads, "Notice: An aircraft unsuccessfully landed in your territory. The crew of this aircraft belongs to the Royal Air Force. They are flying under the orders and protection of the Indian Government. They must be treated properly and immediately turned in safely to the nearest government post. If this is done, a proper reward will be awarded. Should they be harmed, the ones responsible will be severely punished." (REB 1118) *From* First in the Indian Skies *by Norman L..R. Franks, p. 14. Reprinted by permission of Life Publications, Ltd./31 Squadron RAF Association.*

Shown in Fig. 3.6 is a two-sided Goolie Chit written in English and Farsi.[20] The English text is a good translation of the Farsi, and the writing in the lower left corner reads "Albert, Quita Police". Since it shows neither the seals of the NW Frontier government nor the Royal Arms, it is thought be of a later, though indeterminate date. It may be the only chit dating from the Indian colonial period containing an English translation of the message.

One unusual instance of a non-military crew being issued a Goolie Chit by the British government in India is recorded. Peter McMillan, who in 1994 piloted a replica Vickers *Vimy* over the route of the first flight from England to Australia, and who authored *The Greatest Flight* detailing his adventure, noted that a Goolie Chit was provided to the first *Vimy* crew in 1919. In his talk to the Northern California Chapter of the American Aviation Historical Society, McMillan related that Capt. Ross Smith and his brother Keith Smith, pilots of the original *Vimy*, were given a Goolie Chit by the British Consul for use in late November 1919 on the leg from Bundar Abbas, Iran, to Karachi, now in Pakistan but then a part of India. This leg, considered the most dangerous part of the entire journey to Australia, was over more than 700 miles of inhospitable territory populated by hostile tribesmen, and the danger to the infidel airmen was real and immediate.

India was not the only colonial outpost where the need for Goolie Chits arose. RAF squadrons performed similar duties in the areas of the former Ottoman Empire which were under British mandate by terms of the Treaty of Versailles. At the end of World War I, Britain had assumed responsibility for the political integrity of Mesopotamia (Iraq), Palestine, and the Trans-jordan, and had to patrol these areas in addition to its own colonial territory. This duty fell primarily to a garrison based at Hinaidi, near Baghdad. The force included four RAF squadrons, one British and one Indian division, and a cavalry brigade of Iraqis. Even after Iraq gained its independence in 1932, the RAF continued its presence to ensure the protection of British citizens, property, and political interests, and to support the Iraqi Army against internal rebels or invasion.

The early history of flight in Africa and the Middle East emphasizes the need for a system to effect the safe return of downed air crews. In the year 1926, for example, there are instances of airmen being taken hostage, ransomed, and even murdered. The crew of an RAF DH.9A forced down in the desert by engine failure on 14 June 1926 was held hostage in Iraq after being captured by a group of Kurdish nationalists led by Sheik Mahmud. One month later, in July, Jean Mermoz, a French pilot, was captured and held hostage by Moors after his aircraft crashed in the desert between Casablanca and Dakar. Mermoz was released only after an Arab chieftain agreed to accept a ransom for his safe return. Later that year, in November, North African tribesmen killed the three-man crew of a French aircraft forced down while flying an airmail route for the French company CGEA.

As the RAF's operational experience in desert and remote areas grew, it began to understand more clearly the dangers to which its air crews were exposed and the value of preparing them for survival in the event they were forced or shot down. Given the relatively high cost of training and outfitting airmen, and given the paucity of funding for the RAF during the 1920s and 1930s, efforts to assist downed air crews survive were sound fiscal efforts.

RAF air crews assigned to duties in Mesopotamia flew a variety of aircraft, including the Vickers Vincent, of particular interest because it seems to mark the first time that an aircraft manufacturer took into account the possibility of its crew being forced down. The Vincent's design incorporated a first aid kit, a supply of survival rations and water, a sleeping bag, a flare pistol and flares, and an emergency radio/telephone that could be set up and used on the ground.[21]

The aircraft of 55 Squadron based at Hinaidi were likewise stocked with emergency rations and supplies,as described by Wing Commander S. J. "Beau" Carr, DFC, who writes, "there were charguls, water bottles made of canvas . . . emergency rations in the locker: tins of Maconachies meat and veg., sardines, bullybeef, hard tack biscuits, condensed milk and Horlicks malted milk tablets."[22]

Goolie Chits may not have been included in the first emergency equipment carried by airmen flying over the desert because

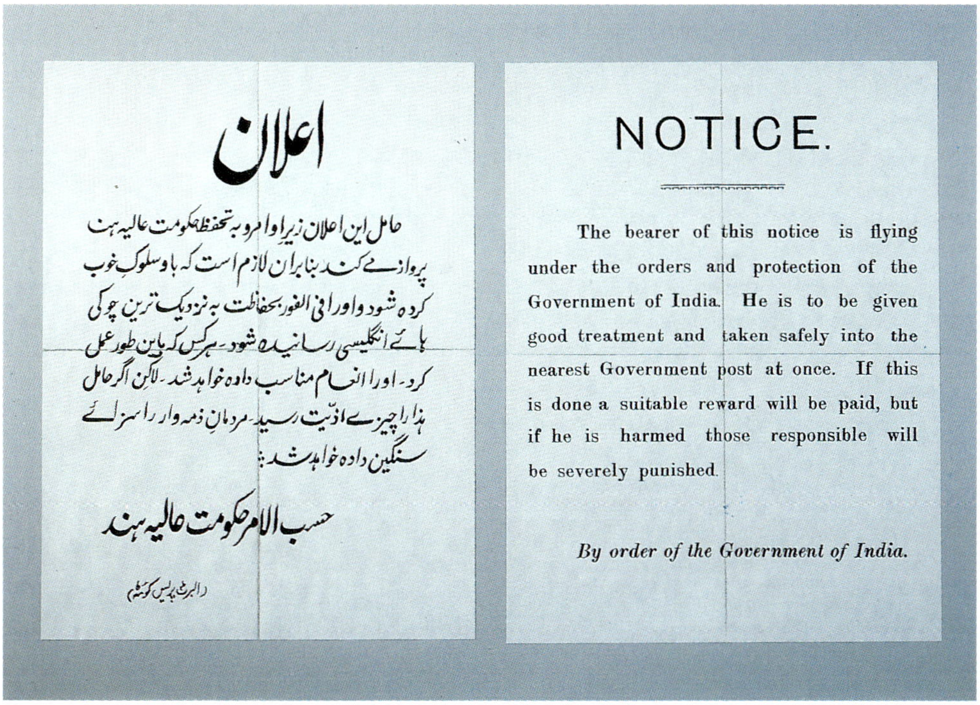
اعلان

حامل این اعلان زیرا وامر و بہ تحفظ حکومت عالیہ ہند
پرواز مے کند بنابران لازم است کہ با وسلوک خوب
کردہ شود و او را فی الفور بحفاظت بہ نزدیک ترین چوکی
ہائے انگلیسی رسانیدہ شود۔ ہر کس کہ باین طور عمل
کرد۔ او را انعام مناسب دادہ خواہد شد۔ لاکن اگر حامل
ہذا را چیزے اذیت رسید۔ مردمان ذمہ وار را سزائے
سنگین دادہ خواہد شد:

حسب الامر حکومت عالیہ ہند

NOTICE.

The bearer of this notice is flying under the orders and protection of the Government of India. He is to be given good treatment and taken safely into the nearest Government post at once. If this is done a suitable reward will be paid, but if he is harmed those responsible will be severely punished.

By order of the Government of India.

Fig. 3.6: A two-sided Goolie Chit in English and Farsi, issued by the Indian government and probably dating from the 1930s, is shown here. This is the only example from the colonial period known to have a message in English in addition to one of the native languages. It has been confirmed that the message in Farsi is identical to that given in English. (REB 1120) *Courtesy of Jeffrey D. Guidry.*

Hugh Trenchard, commander of the RAF, initially would not allow RAF crews to carry Goolie Chits promising rewards to Arabs and Kurds.[23] This early policy was logical in light of the Arabs' historical disposition against written chits.

In his classic autobiographical works, *Revolt in the Desert* and *Seven Pillars of Wisdom*, T. E. Lawrence writes that in 1917 he issued the first promissory notes in Arabia. These were written in pencil on Army telegraph forms at Bair, Guweira, and Jefer, in support of Sherif Nasir of Medina. Until that time, no one had issued written chits because of the traditional Arab disapproval of such items. Lawrence also noted that immediately following his notification to General Clayton that the notes were outstanding, £16,000 in gold was dispatched for payment.[24]

The British realization of the urgency in making good Lawrence's promissory notes doubtless did much to establish the credibility of written chits among the Arabs, and their use during the Arab Revolt laid the groundwork for the later use of Goolie Chits by airmen in the area. Trenchard eventually reversed his original position regarding their use, and by 1923 it was standard practice for the RAF to issue Goolie Chits to its crewmen in the Middle East.

The chits recognized that air crews faced the strong probability of being captured by a hostile tribe, tortured and then killed if their aircraft were forced down through engine failure or combat damage. Don R. Neate, author of 84 Squadron's history, writes that the torture was "usually performed by the tribe's women . . . castrating . . . and . . . finally disemboweling them," and that the threat of torture and death brought "a stronger relation between air crew and ground crew, ensuring a higher degree of trust" that their maintenance and repair tasks on the aircraft were done at the highest level despite the paucity of RAF funding and a continuing problem with spare parts and supplies. Neate also notes the most notorious drawback of attempting to communicate in writing: "One gentleman told me that 'the problem was finding someone who could read'."[25]

Wing Commander S. J. "Beau" Carr, DFC, a veteran of 31 years of RAF service from 1929 to 1960, describes, in his book *You Are Not Sparrows*, his reaction to his friends' taunts regarding his possible fate after learning of his assignment to 55 Squadron at Hinaidi in Iraq:

> Surely it couldn't be true, could it? They were joking, but I wasn't sure. I was even less sure when I did get to the squadron and found, among the few items issued, a document described irreverently as a "Goli chit." It was in Turkish, Arabic and Persian, and promised to my captors the magnificent sum of twenty gold Turkish Mejidies if I was returned unharmed to the nearest British military unit. So that's all I was worth to His Majesty's Government. A miserable twenty quid. It was no comfort to learn that none of the tribesmen could read, but all that came later.[26]

To Carr and the other pilots, the threat of castration and death at the hands of the tribes they bombed and strafed was very real. Former RAF Group Captain K.S. Batchelor, CBE, DFC, a vice president of the Bomber Command Association, carried a Goolie Chit while serving with No. 84 Squadron at Shaibah, in pre-World War II Iraq, and recalls that "several RAF officers captured by the Arabs were emasculated. Handed over to the women who cut off their testicles . . ."[27] He elaborates:

> Before I arrived there was a war with the Marsh Arabs in the Hammer Lakes in which several RAF aircrews were either killed or captured. The alleged treatment of a captive was to

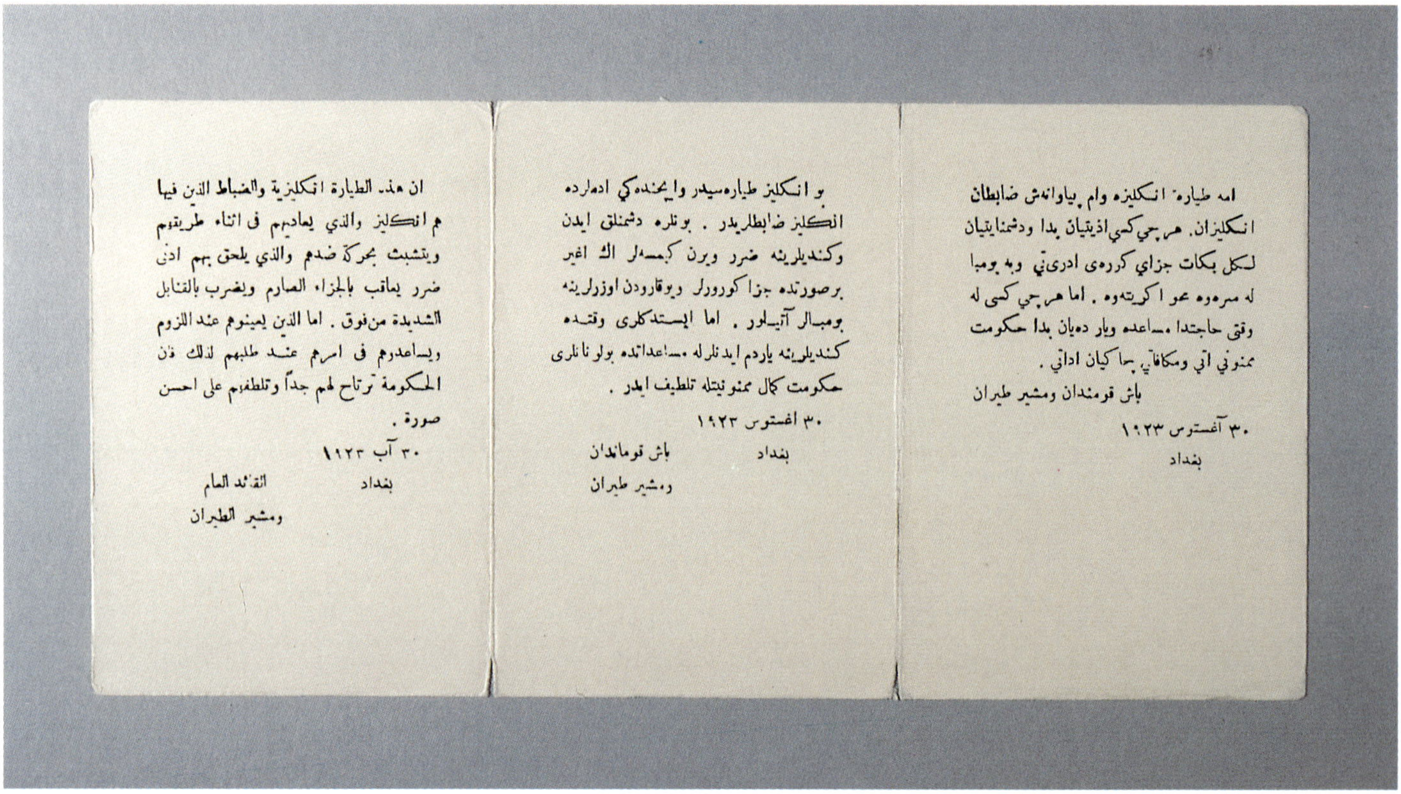

Fig. 3.7: Three panel Goolie Chits were carried by members of 84 Squadron in Mesopotamia (Iraq) during the 1920s. These were written in three languages: Arabic, Turkoman script, and old Persian. The Arabic text reads, "This is an English airplane flown by English officers. Anyone who tries to stop them or to harm them in any way will be severely punished and bombed from the air. However, the British government will reward those who help them and provide for their needs if possible. Air Force General Commander, 30 August 1923, Baghdad." (REB 1122) *Courtesy of Donald R. Neate.*

hand him over to the women who banged his testicles until they swelled up and then cut them off.

Certainly I knew of three who were emasculated. My chief instructor at the AFTS [Advanced Flying Training School] at Abu Sueir in Egypt . . . had been so treated. He had to have regular insulin injections. When he got excited he had a high pitched voice. He was killed in a German bombing raid during the Battle of Britain when serving as Group Commander at a Fighter Command base in Detling, England.[28]

Because of the very real threat to captured pilots, the Goolie Chits carried by Batchelor and his fellow airmen in Mesopotamia and elsewhere were deadly serious documents. Regarding their particulars, Batchelor writes:

> As far as I can recall, we carried a folded card printed in several Arabic dialects. Until I had it translated, I thought it read, in effect, "This bod belongs to His Majesty's Government and if he is returned intact you will be rewarded with a King's ransom." Alas! I believe it read, "This bod belongs to His Majesty's Government. Should any harm befall him you will be bombed from the air." A sure guarantee to receive a rude gesture![29]

The chit shown in Fig. 3.7 dates from 1923, and was held by a member of the RAF serving with 84 Squadron in Mesopotamia.[30] It fits well the recollections of both Batchelor and Carr: a three-panel two-fold card written in three languages, Arabic on the left, Turkoman script, and old Persian on the right. The Arabic text reads:

> This is an English airplane flown by English officers. Anyone who tries to stop them or to harm them in any way will be severely punished and bombed from the air. However, the British government will reward those who help them and provide for their needs if possible. Air Force General Commander, 30 August 1923, Baghdad.[31]

At least one other language aid, shown in Fig. 3.8, was issued to RAF crews on operations in the Middle East.[32] This item, entitled Short Phrases in 'Iraq Arabic for Use in the Event of a Forced Landing, is thought to be the first Phrase Card prepared specifically for airmen. The card is designed for easy use in the field, and contains phrases in both phonetic and written Arabic.

In addition to Mesopotamia and the Northwest Frontier Province in India, many areas of the British Empire were threatened by local leaders, tribes and groups resisting colonial rule. These eventually caused the RAF to mount aerial policing duties in Aden and

Sudan. Lesser known were RAF air control operations in Ireland, South Africa, and in 1931, Burma. There are no known references of the use of chits in these areas.

Even less known than the RAF's colonial air operations are their involvement with the anti-communist "White Russian" forces in post-World War I Russia, or the work of the Kosciuszko Squadron, made up of American volunteers, that fought from 1919 until May of 1921 to defeat the Bolsheviks in Poland. It is not known whether chits or other language aids were available to any of those air crews.

The participation of foreign airmen in the Spanish Civil War from 1936 to 1939 is well documented, but no evidence has been found to indicate that language aids were used by flyers on either side during that conflict.

While the Royal Air Force was flying ground-attack missions against dissident tribesmen in the Middle East and on the Indian subcontinent in the 1930s, in another part of the world foreign pilots, air crews, and mechanics were engaged in an air war against a strong, modern, and aggressive military power.

When the Empire of Japan invaded China in 1932, the Chinese Air force was no match for Japanese air power. The Chinese government often relied on foreign pilots to bear the brunt of aerial combat during the 1930s while it struggled to rescue its own air force from years of corruption and neglect.

When forced down, these foreign airmen would require the assistance of the Chinese people in order to return to their units, yet they could not speak Chinese. The solution to this familiar dilemma would come in the late 1930s, and the solution would likewise be familiar.

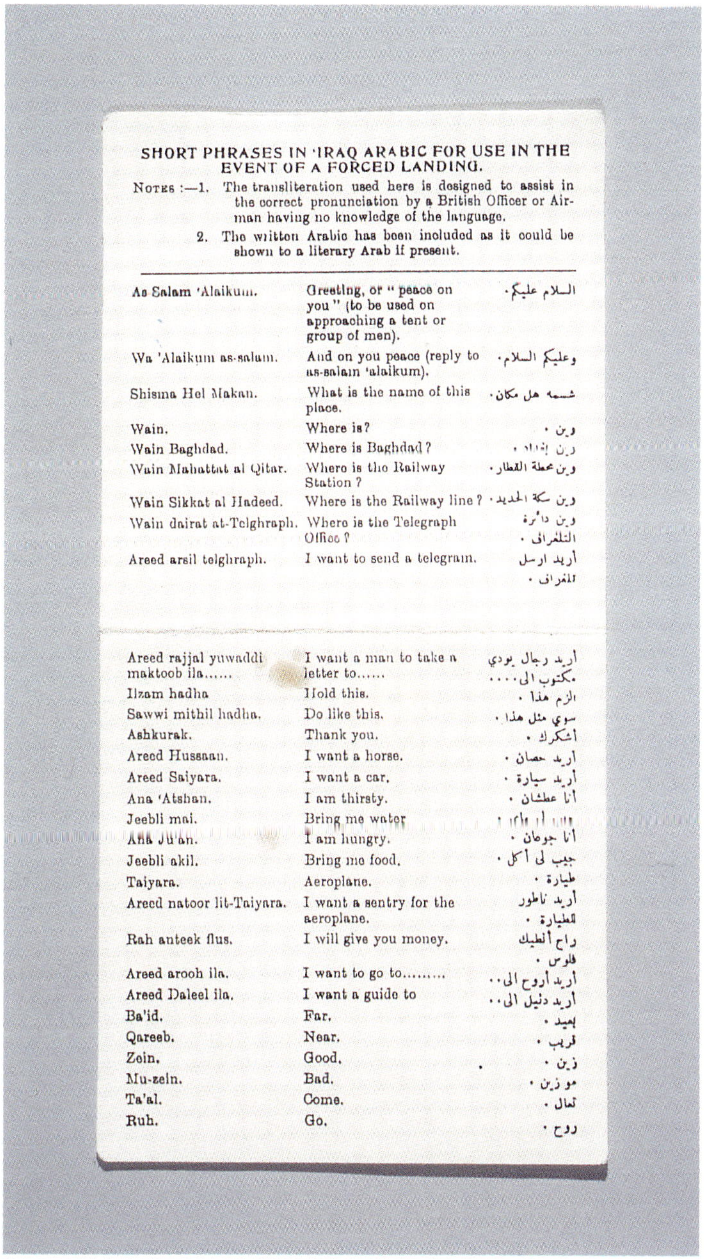

SHORT PHRASES IN 'IRAQ ARABIC FOR USE IN THE EVENT OF A FORCED LANDING.

NOTES :—1. The transliteration used here is designed to assist in the correct pronunciation by a British Officer or Airman having no knowledge of the language.

2. The written Arabic has been included as it could be shown to a literary Arab if present.

As Salam 'Alaikum.	Greeting, or "peace on you" (to be used on approaching a tent or group of men).	السلام عليكم.
Wa 'Alaikum as-salum.	And on you peace (reply to as-salam 'alaikum).	وعليكم السلام.
Shisma Hel Makan.	What is the name of this place.	شسمه هل مكان.
Wain.	Where is?	وين .
Wain Baghdad.	Where is Baghdad?	وين بغداد .
Wain Mahattat al Qitar.	Where is the Railway Station?	وين محطة القطار.
Wain Sikkat al Hadeed.	Where is the Railway line?	وين سكة الحديد.
Wain dairat at-Telghraph.	Where is the Telegraph Office?	وين دائرة التلغراف .
Areed arsil telghraph.	I want to send a telegram.	أريد ارسل تلغراف .
Areed rajjal yuwaddi maktoob ila......	I want a man to take a letter to......	أريد رجال يودي مكتوب الى....
Ilzam hadha	Hold this.	الزم هذا .
Sawwi mithil hadha.	Do like this.	سوي مثل هذا .
Ashkurak.	Thank you.	أشكرك .
Areed Hussaan.	I want a horse.	اريد حصان .
Areed Saiyara.	I want a car.	أريد سيارة .
Ana 'Atshan.	I am thirsty.	أنا عطشان .
Jeebli mai.	Bring me water	[illegible]
Ana Ju'an.	I am hungry.	أنا جوعان .
Jeebli akil.	Bring me food.	جيب لى أكل .
Taiyara.	Aeroplane.	طيارة .
Areed natoor lit-Taiyara.	I want a sentry for the aeroplane.	أريد ناطور للطيارة .
Rah anteek flus.	I will give you money.	راح أنطيك فلوس .
Areed arooh ila.	I want to go to.........	أريد أروح الى..
Areed Daleel ila.	I want a guide to	أريد دليل الى..
Ba'id.	Far.	بعيد .
Qareeb.	Near.	قريب .
Zein.	Good.	زين .
Mu-zein.	Bad.	مو زين .
Ta'al.	Come.	تعال .
Ruh.	Go.	روح .

Fig. 3.8: This language aid for RAF air crews serving in the Middle East is thought to be the first Phrase Card issued to airmen anywhere. It is designed for easy use in the field, and contains 30 phrases in both phonetic and written Arabic. Later, during World War II, Phrase Cards in numerous languages would be prepared by the British for issue to Allied airmen. (REB 1124) Courtesy of David Vincent.

CHAPTER 4

East Meets West
Foreign Airmen in the Chinese Air Force

The first half of the 20th Century saw the disastrous culmination of some 350 years of European and American contact with China. The vast distances and difficult geography that once protected the Chinese were overcome by European political and commercial desires and by changes in marine technology. By the early 1920s China had lost control of its own destiny, falling into regional chaos as organized crime, military leaders, and warlords took control of portions of the country, while the British, French, Germans, Americans, Russians, Japanese, and Portuguese had partitioned and annexed China's ports, major rivers, and cities.

To further complicate matters, a civil war between the Chinese Communist Party (CCP) and the Kuomintang (KMT), or Nationalist Party, under Chiang Kai-shek began with the Nationalist massacre of Communists in Shanghai in April 1927. At the same time the Nationalists and Communists were fighting each other, they faced a common regional enemy in the Japanese.

Japan and the United States were the only two nations to come out of World War I without major devastation and with comparatively few casualties. At the end of the war, these two countries were the strongest military powers in the world, a situation better realized by the Japanese than by the isolationist Americans. Led by an expansionist military, Japan had regional if not global ambitions. With most European nations and the United States dominated by isolationists and preoccupied with their depressed economies, Japan invaded and occupied Manchuria in 1931, establishing a protectorate known as Manchukuo. This was accomplished without resistance from Nationalist forces, who withdrew in the vain hope that assistance from western nations would be forthcoming.

Less than a year later, in February 1932, a large-scale battle erupted between the Nationalist Chinese army and Japanese troops in Shanghai's Japanese settlement. This battle included the use of aircraft launched from Japan's first two aircraft carriers: *Hoshu*, which launched the first strike on 5 February, and *Kaga*.

One of the earliest air battles in China was fought on 22 February 1932. This brief dogfight over Shanghai featured fighters from the Imperial Japanese Navy and a Boeing Model 218, the export version of the American Navy's Boeing F4B-4 and the Army's P-12E, flown for the Chinese Air Force by an American, Robert Short. While attacking three Japanese Mitsubishi B1M Type 13 bombers during an air raid on Shanghai, Short was killed by gunfire from a Nakajima Type 3 carrier-based fighter. Despite the outcome of the engagement, Short's attack established a precedent for what would later become a common aspect of the air war: American pilots flying American aircraft in defense of China.[1]

The Chinese Air Force facing the Japanese in 1932 consisted of about 270 aircraft of mixed foreign types belonging to the national government, several provincial governments, a number of airlines, and several flying schools. T. V. Soong, given the task of rebuilding the Nationalist air force, recruited fifteen American pilots and technicians led by retired U.S. Army officer Colonel John J. Jouett, a West Point graduate and a supporter of General Billy Mitchell. The unofficial air mission from the United States arrived in Shanghai on 9 July 1932 and established the Central Chinese Aviation School at Shien Chiao, near Hangchow.[2] During their three year contract the group trained 375 Chinese pilots and directed the establishment of a Curtiss-Wright Corporation aircraft repair and assembly plant. Most of Jouett's group left China in June 1935 at the end of their three year contract, possibly due to animosity between Jouett and Chiang Kai-shek over the Americans' refusal to become involved in China's internal affairs by flying missions in Fukien Province against a warlord who was opposed to Chiang.

In June 1937 Claire Lee Chennault arrived in Shanghai to undertake a confidential survey of the Chinese Air Force at the invitation of Madame Chiang Kai-shek, who had become the head of China's Commission for Aeronautical Affairs. The Chinese Air Force in 1937 consisted of 31 squadrons, each with nine aircraft authorized, if not present or flyable. A British air attaché reported that the squadrons had 284 combat aircraft, and that there were another 400 aircraft attached to various flying schools. Chennault indicated, however, that only 91 of these aircraft were suitable for combat.[3]

On 7 July 1937, using a fabricated attack on its forces by Chinese troops at the Lukouchiao (Marco Polo) Bridge just outside of Peking as an excuse, the Japanese invaded China by land from the north and by sea from the south. Chennault wired Chiang Kai-shek at once, offering his services in whatever capacity they might be required. On 9 July the Generalissimo replied, accepting Chennault's offer.

As the conflict between the Japanese and Chinese intensified, there was an influx of international pilots and aircraft into China for service with the Nationalist government. This force of German,

Russian, American, Dutch, and Italian pilots included veterans of World War I and of the Spanish Civil War.

Chinese-American pilots from the United States began arriving in China as early as the late 1920s simply to pursue a career in aviation, something they were denied at home because of their ethnicity. Some of these pilots, born and raised in the United States, spoke no Chinese. When war came many of these men made the transition to military aviation while others remained in civil aviation, often flying supplies and troops wherever they were needed.

Other Chinese-Americans made their way to China once the war had begun. One of these pilots was Art Chin (Chen in some spellings). Of all the American pilots who flew in China, Chin, who arrived in China in 1932, doubtless had the most experience. Although the aircraft he flew were often overmatched by those of his Japanese adversaries, he became a successful fighter pilot. Former Oregon State Senator Ken Jernstedt, a member of the AVG 3rd Pursuit Squadron, remarked, "Art was an 'ace' long before the AVG arrived in China."[4] Chin remained active in the Chinese Air Force following the end of World War II. In 1949, as the Nationalist government was driven from China by the Communists, it was Chin who flew many of China's national treasures to Formosa.

For nearly two years, much of the air defense that was needed to keep China in the war was provided by the Soviet Union. Fearing the expansionist policies of the Japanese and still smarting from their defeat in the Russo-Japanese War, the Russians signed a non-aggression pact with China in September 1937. By November, the first of more than 2000 Russian pilots, crew members, maintenance personnel, and officers were in China. The Russians replaced the German and Italian missions, which had been expelled on 21 March 1937 by Chiang Kai-shek after the fall of Nanking because of their governments' ties to the Japanese.

The Red Air Force sent four fighter and two bomber squadrons to China. Many of the first Russian pilots in China had gained combat experience fighting the fascists during the Spanish Civil War, and the Russians developed a system of rotating their personnel back to the Soviet Union after three months of service in China so that additional crews could gain combat experience. The Soviets were reported to be top-notch fighter pilots and completely without fear in combat.[5] The Russian air mission in China was a costly one, and more than 200 Russian airmen were buried in China by the time the Russians were called home in the summer of 1939 to defend the Soviet Union against an expected German invasion.[6]

Chinese Air Force pilots also flew Russian aircraft. One of these pilots, Mai Lam, a relative of R. E. Baldwin's wife Shirlaine, ferried Russian aircraft from the Soviet Union to China and also flew them in combat. He was killed when his Polikarpov I-16 was shot down by Japanese aircraft.[7]

In October 1937, a squadron of foreign volunteers was formed by a decision of Madame Chiang Kai-shek, President of the Aero Commission, under the guidance of her advisor, Colonel Claire L. Chennault.[8] The 14th Squadron, also known as the 14th International Squadron, the Volunteer Bombardment Squadron (VBS) and the 14th Volunteer Bombardment Squadron (14th VBS), existed for only about five months, from November 1937 to March 1938. Vincent Schmidt, an American who had flown in both World War I and the Spanish Civil War, commanded the 14th Squadron. The squadron had about 30 members including 11 pilots – American, Dutch, Australian, French, and one Chinese American – and 16 Chinese air gunners and mechanics.

Service by European aviators in China presented many of the same problems RAF crews had faced in their colonial assignments. The aircraft involved were not entirely reliable, the distances covered in even non-combat flights could be significant, and the flights were often over rugged, desolate terrain and sparsely populated areas without established communications or even frequently traveled roads. The local populations could be friendly, hostile, or indifferent to a foreigner, and the reception by the Chinese population of a downed Western aviator was uncertain at best.

Clearly foreign pilots needed assistance in communicating their needs in a country such as China with its strong traditions of both isolationism and a dislike of foreigners. Moreover, China is a vast country with populations speaking different dialects. Even the few foreigners who spoke one of the dialects were likely to encounter difficulties in communicating with those Chinese who spoke another. Written Chinese, however, was a constant in that sea of diverse spoken dialects, and was universally understood by educated Chinese in all parts of the country.

From photographs taken as early as February 1938, it is clear that the American and European pilots of the 14th Squadron were issued what they referred to as "Rescue Patches". These were known to the Chinese as "Hu Chao", from "hu" meaning protection and

Fig. 4.1: The first Rescue Patches were issued in the winter of 1937-38 to members of the 14th Volunteer Bombardment Squadron (VBS). The group, commanded by Claire Chennault and stationed in Hankow, was composed of European and American pilots with Chinese air gunners, and supported by Chinese mechanics and technicians. American pilot George H. Weigle, left, poses in this photo with fellow American pilot Elwyn H. Gibbon. Weigle's Rescue Patch is worn on the front of his flight suit, as was typical for members of the 14th VBS. *Photo courtesy of the San Diego Aerospace Museum*

Fig. 4.2: An unidentified 14th VBS pilot, left, poses with American pilots George Weigle and Elwyn Gibbon on the airfield in Hankow. Rescue Patches can be clearly seen on the front of the two flight suits. *Photo courtesy of the San Diego Aerospace Museum*

Fig. 4.3: William Labussière, left, and André Boulingre, two French pilots of the 14th VBS, relax on the wing of a Chinese Air Force Vultee V11. Labussière's Rescue Patch is illustrated in Fig. 4.4. *Photo Courtesy of the San Diego Aerospace Museum*

"chao" referring to a type of silk fabric. The translation is "a silk patch for protection", hence the term "Rescue Patch".[9]

Shown in Fig. 4.1 are American pilots George H. Weigle, left, and Elwyn H. Gibbon. Taken in early 1938, the photo clearly shows Weigle's Rescue Patch attached to the front of his Chinese flying suit. Weigle is reported to have flown in the Spanish Civil War before coming to China to fly for the 14th VBS,[10] and he later was killed when the Curtiss Hawk he was flying crashed on takeoff in May of 1939.[11] Gibbon was briefly taken prisoner by the Japanese in 1938 when the ship on which he was returning to the United States docked in Yokohama. Hong Kong newspapers, read by the Japanese, had printed a story that he had flown for the Chinese. Eventually released, he went on to fly for TWA and Lockheed before returning to Asia to assemble and ferry aircraft from the American base at Karachi, India. He was killed in August 1942 when the P-66 he was testing crashed.[12]

Wiegle and Gibbon pose in Fig. 4.2 with an unidentified 14th VBS pilot on the airfield in Hangkow. Rescue Patches are seen on the front of two flight suits. This is of obvious interest, and may be related to the fact that, as illustrated on page 93 of Malcolm Rosholt's excellent and comprehensive work *Flight in the China Air Space 1910-1950*, some Chinese pilots wore the Nationalist flag on the front of their flying suits. Considering the many dialects spoken in China, this was perhaps done to identify downed pilots as Chinese rather than Japanese. A logical extension of this practice may have been to attach the foreigners' Rescue Patches to the fronts of their flight suits.

Two of the French pilots of the 14th VBS, William Labussière, left, and André Boulingre, are shown in Fig. 4.3 relaxing on the wing of a Vultee V11. Before joining the International Squadron, Labussière had flown in Africa in 1935-36, and until 25 September 1937 he flew fighters in Spain for the elected Republican government during the Spanish Civil War. He and Boulingre remained in China after the 14th VBS was disbanded, and in July 1938 they were posted to a fighter squadron in Kunming. Labussière returned to France in 1939 to help defend the nation against Germany.[13] A photo of William Labussière's Rescue Patch, provided by Canadian author G. Beauchamp who visited Labussière in France, is shown in Fig. 4.4.

Labussière's patch is printed on silk sewn to a cotton backing. The flag appears to be printed, while the text may have been printed or written by hand. The flag is that of the Republic of China, with the white sun and twelve rays (not a star, not a flower) on a blue field in the upper left corner of a red field. It bears two red seals, called "chops" by the Chinese, a large one in the lower left corner, and a smaller one in the upper left corner of the text area. Its serial number, 0242, is somewhat hidden beneath the large chop.

While a literal word-for-word translation from Chinese to English is difficult at best, it will be attempted because so much erroneous information has been published regarding the message given on Rescue Patches. The worst of these misconceptions stated that Rescue Patches gave the pilot's name and blood type. It has also been wrongly stated that the patches offer a reward for the pilot's safe return and that they bear the chop and/or signature of Chiang Kai-shek.

The Chinese text is read from the top of the column to the bottom, and the columns are read from from right to left. The English equivalents of the large characters in the four major columns are as follows:

Col. 4	Col. 3	Col. 2	Col. 1
Deliver	Soldier	Foreign	Come
Protect	Civilian	Person	China
	One		Help
	Body		Fight

Note that there is no plural in Chinese, that "fight" refers to the war effort against Japan, that "one body" is the Chinese way of

Fig. 4.4 : This 14th VBS Rescue Patch was issued to William Labussière in the winter of 1937-38. The four columns of large characters beneath the flag of the Republic of China, read, "This foreign person has come to China to help in the war effort. Soldiers and civilians, one and all, should rescue, protect and provide him medical care." The single column of eight small characters appears to be unique to patches issued to members of the 14th VBS and identify it as being issued by the Nationalist government's Air Force Front Line Command Headquarters. It bears two red seals, called chops, the larger being that of the Air Force Front Line Command Headquarters that issued it, and the smaller being Labussière's name transliterated into Chinese characters. Two characters accompany the serial number, 0242, which is overstamped by the Headquarters chop. This chit can be considered the prototype for all Rescue Patches issued to American airmen by the Kuomintang government during World War II. (REB 2110) *Photo courtesy of Labussière/Beauchamp*

saying "together" or "one and all", that "deliver" is used in the biblical sense which might best be regarded as "save" or "rescue", and that the the full meaning of the character translated as "protect" also includes providing medical assistance. Incorporating all of the above qualifications into the word-for-word translation, one obtains the following message:

> This foreign person has come to China to help in the war effort. Soldiers and civilians, one and all, should rescue, protect, and provide him medical care.[14]

The column of eight small characters to the left of the larger ones tells under what authority the chit was issued. It reads as follows: the first five characters are "Air Force Front Enemy Head", the next two characters taken together mean "Command", and the last character is "Quarter". Noting that the term "Front Enemy" means the combat front or the front lines of combat, this column can be translated as: "Air Force Front Line Command Headquarters".[15] This was the entity issuing Rescue Patches to foreign airmen in early 1938.

The large chop consists of the same eight characters as in the above-translated column, being the chop of the Air Force Front Line Command Headquarters. The value of having the chop on Rescue Patches should not be underestimated, for it was recognized as an important seal of a government agency and as such indicated that the documents were officially issued and backed by the government. The smaller chop is a Chinese name consisting of three characters, one on the entire right side of the chop and two, one above the other, on the left side. The name is Lai Bu Sieu.[16] This is, of course, the name Labussière transliterated into Chinese characters. Labussière had a personal chop made, as is the custom even today among travellers to China, and stamped it onto his Rescue Patch.

The serial number, 0242, is somewhat hidden on the left side beneath the large chop. The two small characters, one above and one below the serial number, indicate that this is the number of the document. The top character may be taken as the equivalent to our "st" as in 1st, "nd" as in 2nd, "th" as in 10th, 25th, 100th, and so on. The bottom character simply means "number". Together they mean "ordinal number". The translation of the column on this chit is therefore "the 242nd number" or "ordinal number 242".[17]

The same language and cultural barriers faced by 14th VBS pilots were also faced by the Russians and by some of the American-born Chinese who were flying against the Japanese in China. There is some well-reasoned speculation that Russian air crews may have used some type of chit to assist them in communicating with the Chinese in an emergency, but no evidence has surfaced to support this theory.[18]

Although the International Squadron lasted only a few months and their Rescue Patches were little noted by the public, their successors in 1941, the American Volunteer Group (AVG), and the cloth chits that they carried, became immortalized forever in the collective mind of America. The AVG, the famous *Flying Tigers*, have become American legends and symbolize for all time the American character: a special blend of compassion, self-sacrifice, and willingness to help defend others from aggression and tyranny.

It is outside the scope of this work to repeat the history of the AVG which has been exhaustively covered by numerous other writers and historians. Many members of the group itself, including Charles R. Bond, R.T. Smith, Erik Shilling, Charles Baisden, Joe Rosbert, Robert M. Smith, Frank Losonsky, and Claire Chennault, have written their own histories and autobiographies. At this writing other members, including Leo Schramm, are also in the process of writing books.

It is important for the sake of continuity to note that there was a significant period of time between the disbanding of the 14th Squadron on 22 March 1938, and the authorization by President Franklin Roosevelt on 15 April 1941 allowing China to recruit American volunteers for the AVG. The formal constitution of the First American Volunteer Group in Chungking, China, occurred on 1 August 1941; the first volunteers of the AVG reached Rangoon, Burma, in July 1941; and the last group of the original members were in Asia by November 1941.

AVG flyers were clearly concerned about their fate should they fall into the hands of the Japanese. Upon his arrival in Singapore on 6 November 1941, Charles R. Bond noted in his diary: "Read the local English paper with great interest . . . the Japanese notified the U.S. State Department and told them that American pilots flying for China and captured by the Japanese would be shot . . . I wonder how true this is. Some of the bunch are pretty pessimistic."[19]

Given the serious nature of the Japanese threats to shoot American flyers if captured, the fact that both the British and Chinese had prior experience with Blood Chits and Rescue Patches, and knowing Chennault's role as an advisor to China's Commission for Aeronautical Affairs, it is not surprising that written documents requesting assistance from the local populace were created for the AVG.

AVG pilots received their first chits from the British while in Burma. J.R. "Dick" Rossi writes, "We did have a British issued chit of thin cardboard (or heavy paper) in English and Burmese. These were hand numbered . . . mine . . . was numbered 90. These were folded once and fit into a shirt pocket."[20] Shown in Fig. 4.5,[21] it reads, "I am a British/American Airman. I am in want of assistance. Please send me to the nearest Police Station." There is no offer of a reward for assisting the airman.

From Burma the AVG moved into China. Rossi writes, "The main body of the AVG moved up to Kunming in December, 1941. The movement had already started before Pearl Harbor. That event disrupted the movement to some extent. The 3rd Pursuit Squadron was sent to Rangoon, and the rest of the Group went on to Kunming. Most of the planes and pilots arrived there on 18 December 1941. A few people were left in Toungoo to finish up things there. The road convoys arrived at different intervals."[22]

Many authors have expressed the sentiment that the Blood Chit was a result of Erik Shilling's experience on 23-24 December 1942. During the Third Squadron's move from Burma to Kunming, China, Shilling flew one of the unit's CW aircraft filled with his personal belongings. Engine trouble forced him down short of Kunming,

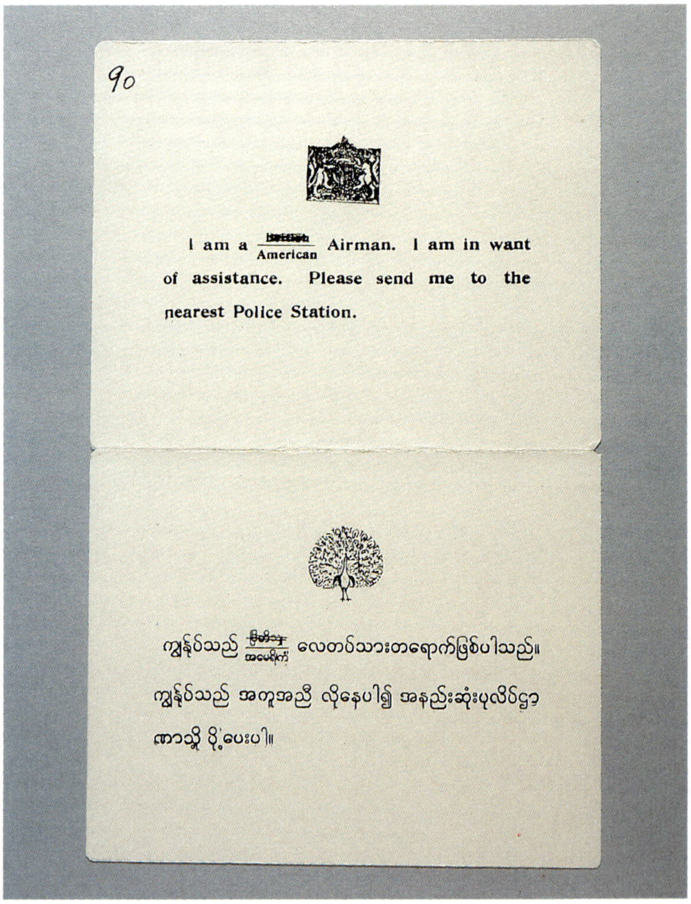

Fig. 4.5: The first chit received by American Volunteer Group (AVG) flyers was a cardstock item issued in Burma by the British. It is possibly the simplest chit ever made: there is no offer of a reward. the chit simply identifies the bearer as either British or American, and it asks that he be directed to the nearest police station. On the chit belonging to J.R. "Dick" Rossi of the AVGs 1st Pursuit "Adam & Eve" Squadron, the word "British" is struck out. Rossi's chit is hand numbered "90" in the upper left corner. These chits were folded once and kept in a shirt pocket. (REB 1212) *Courtesy of J.R. "Dick" Rossi*

and he was kept prisoner by members of a mountain tribe who built a stockade around him and his aircraft, and covered the aircraft so it could not be seen from the air.

When Shilling realized that one of the villagers was inciting the the rest of them to harm him, he attempted to convince them that he was a friend by by showing them his American passport. The instigator inspected it carefully, as if reading it, but Shilling deduced from his manner that the person could not actually read. He grabbed his passport, and with a great flourish opened it to the correct page and offered it back to the villager. His quick thinking may have saved his life, for the villager "lost face" and he was not bothered by this individual again.

Word had gone out through the radio warning network that he was missing, and that message apparently reached the villagers because their attitude toward him changed, although they would not let him leave. He kept the villagers entertained with records played on the wind-up phonograph that he had stowed in the aircraft, while they constructed a sedan chair. In it they carried him to a radio warning net station where, with the help of an English/Chinese dictionary, he learned that the villagers had originally suspected him of being Japanese.[23]

Coming as it did on the eve of the issuance of the Chinese government's Rescue Patches to AVG pilots, it is easy to understand how the incident might be interpreted by those unfamiliar with the International Squadron's Rescue Patches and Chennault's involvement with them as inspiring their creation. Schilling's experience did clearly illustrate the need for some way for AVG pilots to communicate with the Chinese.

Other problems faced by American pilots were also made apparent by Shilling's experience. One of these is that many Chinese did not know what a Japanese person looked like, and assumed that any downed airman might be the enemy. The isolation of many parts of China contributed to this lack of knowledge, and as an example of this isolation Ken Jernstedt noted that the residents of the next valley over the mountains from Kunming were not even aware of the existence of oceans, let alone what an American or Japanese person looked like.[24] Another problem that Shilling's adventure highlighted was the lack of literacy among much of China's population. Jernstedt also confirmed this, noting that the Chinese who were literate generally did not live in the remote areas that the AVG flew over.[25]

The first Chinese Aero Commission Rescue Patches were apparently issued to the AVG pilots in Kunming on Christmas Day, 1941. Charles R. Bond, Jr., Vice Squadron Leader of the 1st Pursuit, noted in the 25 December entry to his *A Flying Tiger's Diary* that "We were issued the Chinese 'blood chit' emblems to sew on the back of our jacket today. It is about six by ten inches and has the Chinese flag on it. Also Chinese writing that directs the people to assist us in any way possible – that we are their friends."[26]

Ken Jernstedt recalls that he and the other members of the 3rd Pursuit Squadron who had stayed behind in Burma when the balance of the group relocated to China did not receive their Rescue Patches until they arrived in Kunming in January 1942.[27]

AVG flyers first referred to their chits as "patches", "rescue patches", or "identification patches". R.T. Smith writes, "Cloth patches . . . were sewn to the backs of our flight suits and leather jackets . . . Later these things became known as 'blood chits,' though I don't recall hearing them called that while in the AVG."[28] J.R. "Dick" Rossi confirms this: "It was not until much later that I heard them referred to as 'Blood Chits'."[29]

Like the 14th VBS Rescue Patches, those issued to AVG members featured printed Republic of China flags over the text. Although similar in design, there are varying styles of calligraphy on these patches. Illustrated in Figs. 4.6 through 4.9 are four AVG silk Rescue Patches. Three of these, those in Figs. 4.6 through 4.8, belong to Dick Rossi and are on display at the San Diego Aerospace Museum. These photos highlight some of the similarities and differences in the AVG silk Rescue Patches.

Shown in Fig. 4.6 is an AVG identification patch which belonged to Jasper Harrington of the 1st Pursuit "Adam and Eve" Squadron. The main text is identical to that on the 14th VBS Rescue Patches:

> This foreign person has come to China to help in the war effort. Soldiers and civilians, one and all, should rescue, protect, and provide him medical care.[30]

While the International Squadron Rescue Patches were issued by the Nationalist government's Air Force Front Line Command Headquarters, the column of five small characters on this and all AVG patches indicates that this was not the agency issuing AVG chits. The first two characters mean "aviation" or "aeronautical", and the last three characters mean a "committee" or "commission", indicating that the agency issuing AVG patches was China's Commission for Aeronautical Affairs, best known as the Aero Commis-

Fig. 4.6: Shown here is the Rescue Patch of AVG 1st Squadron Chief Jasper J. Harrington. The main text is identical to that of the earlier 14th International Squadron's Rescue Patches. The smaller column of five characters identifies the patch as being issued by the Nationalist government's Commission for Aeronautical Affairs, headed by Madame Chiang Kai-shek. The serial number, 0467, is accompanied two characters, and the Aero Commission chop is centered in the text area as is typical of all AVG patches. (REB 2212) *Courtesy of J.R. "Dick" Rossi and the San Diego Aerospace Museum.*

sion.[31] The five characters on the chop are identical to those in the column of smaller characters, indicating that the chop is that of the Aero Commission.

The same two characters accompanying the serial number on the 14th VBS patches also accompany the serial number, 0467, on this patch.

Many AVG chits featured a new addition to the old 14th VBS text: two new characters enclosed in vertical parentheses in the second column (from the right). They are transliterated into English as "may guo", or "may gwo" in some spellings. These characters translate as "Beautiful Country" which is one Chinese name for the United States. Other Chinese terms for the United States include "flowery flag", because the stars on the American flag reminded the Chinese of flowers, and "gum san" meaning "gold mountain", a term referring to California and dating from the nineteenth century when the lure of work and wealth brought thousands of Cantonese to the western United States.

The term "may guo" does not literally mean that America is a "beautiful country" to the Chinese. Rather, the term is a much used contraction of the longer "may lay gin guo", in which "may lay gin" *sounds* something like the word "America" or "American" (a-may-ree-kin = may-lay-gin). In the same vein, the term for the United Kingdom is "ying guo", a contraction of "ying gut lay guo", where "ying gut" *sounds* something like "England" (eng-lund = ying-gut).[32] Later in the war, the "may guo" characters would be stamped onto the Aero Commission chits issued to U.S. air crew members, but on the AVG Rescue Patches they are written by hand. The Rescue Patches illustrated in Figs. 4.7 through 4.9 contain these characters.

Fig. 4.7: The text on this AVG Rescue Patch seems identical to that on the patch shown in Fig. 4.6, but the fabrics are slightly different. The addition of two handwritten characters enclosed in vertical parentheses at the bottom of the second column from the right, transliterated as "may guo", identify the bearer as an American. The text, as revised by the inclusion of these characters, reads, "This foreign person (American) has come to China to help in the war effort. Soldiers and civilians, one and all, should rescue, protect and provide him medical care." (REB 2214) Two characters accompany the serial number, 0383. *Courtesy of J.R. "Dick" Rossi and the San Diego Aerospace Museum.*

J.R. "Dick" Rossi notes these two variations in the text when he writes, "regarding our cloth chits, we had two versions. One had the parenthesis with . . . the American Identification, while the others merely said foreigner."[33] The translation of this new text is identical to that of the old, with the addition of the nationality of the pilot:

> This foreign person (an American) has come to China to help in the war effort. Soldiers and civilians, one and all, should rescue, protect, and provide him medical care.[34]

The printing of the text on the Rescue Patch in Fig. 4.7 appears to be identical to that on the patch shown in Fig. 4.6, but the fabrics appear to be slightly different. This is not a trick of the photos, as this difference is evident when the chits are seen firsthand. Also, like the patch in Fig. 4.6, only two characters accompany the serial number, 0383.

The text on the Rescue Patch illustrated in Fig. 4.8 was written by a different hand than the previous two, and the "may guo" characters on this patch are written in a similar but not identical hand to those in Fig. 4.7. The serial number, 0297, is accompanied by three Chinese characters. The top character translates simply as "character". Modern, or vernacular, Chinese is more like spoken Chinese and might require this third character to read "ordinal number", although it only requires two characters in classical Chinese. Since the chits are official government documents they are written in classical Chinese, not in the vernacular, and a more likely reason for the inclusion of the third character is that it may have been anticipated that several series of chits were to be issued. The third character would have facilitated this.[35]

Perhaps the easiest way to visualize this system of series numbering is to note that in the west items are sometimes numbered using A-1, A-2, A-3, etc. for one series, and B-1, B-2, B-3, etc. for another. This is not uncommon in China, but where one culture uses the Roman alphabet the other uses names such as "heaven character", "earth character", etc. The first series of chits thus identified would have been the "heaven character" series, the second would have been the "earth character" series, and so on. This would

Fig. 4.8: The text on this Rescue Patch is written in a different hand than the text on the patches illustrated in Figs. 4.6 and 4.7. It also features the "may guo" characters, but the serial number, 0297, is accompanied by three characters. There is no single "pattern" AVG Rescue Patch, and many small variations exist. (REB 2216) *Courtesy of J.R. "Dick" Rossi and the San Diego Aerospace Museum*

have been accomplished by simply adding "heaven", "earth" and so on above this new "character" character. The first chit of the first series thus numbered would have then been "heaven character 1st number". From the differing styles of calligraphy it seems that several series of chits may have indeed been made, but they were never actually identified by series even though the provision to do so was made.[36]

The red in the flag on the Rescue Patch shown in Fig. 4.9 is light in color when compared to those in Figs. 4.6 through 4.8. It is strikingly like the Rescue Patch of Claire Chennault displayed at the National Air and Space Museum in Washington, D.C. The red in the flag on Chennault's chit is the same light color and the text is likewise identical. The "may guo" characters on both appear to have been added in the same hand, and particularly noticeable on both are the large parentheses surrounding these characters. Chennault's patch, shown in a black-and-white photo in *China-Burma-India* of the Time-Life Books World War II series (page 70)[37] is serial numbered 0001, while that shown here is 0043.

The back of the chit shown in Fig. 4.9 is illustrated in Fig. 4.10. It has been signed by 36 members of the original American Volunteer Group, plus Henry C. Y. Lee, the Chinese Air Force Communications Officer who met Claire Chennault when he first arrived in China in 1937. Lee served as Chennault's translator throughout his tenure there, organized the AVG's Radio Warning Net, and later organized the Chinese-American Composite Wing for Chennault. He went on to become a General in the Nationalist Chinese Air Force in Taiwan. These are post-war signatures.

While the illustrations show that AVG Rescue Patches had a wide range of small differences, one feature common to all of them is the position of the Aero Commission chop. Dick Rossi writes, "On all of the chits we were issued, the Chinese chops were pretty much centered in the text below the flag."[38] As can be seen in the photos, this is a distinguishing characteristic of all the AVG identification patches.

While minor differences in the translation of these Rescue Patches are to be expected, there is no documentation that they ever contained either the signature or the chop of Chiang Kai-shek. The notion that the Aero Commission chop on AVG Rescue Patches represented the signature of Chiang Kai-shek appears to have come from the AVG itself. Shown in Fig. 4.11 is the back of a Rescue Patch, number 0387, reported to have been sent by the AVG to the family of John T. Donovan, killed 12 May 1942 over Hanoi. It reads, "This represents a permanent identification worn by all our pilots. The red seal indicates the Generalissimo's signature. Translation below: Come China Help Fight, Foreigner (American), Soldiers People All Together, Help Protect! Aeronautical Commission, No. 387." The chop, although not that of Chiang as reported above, was an important element of all Chinese Blood Chits, for it was recognized as the official seal of a government agency, in effect verifying the value of the document.

Chennault's Rescue Patch is appropriately numbered 0001. Presumably Jasper Harrington was issued his patch, numbered 0467, in December 1941 along with other members of the 1st Squadron including Charles R. Bond who was issued number 0212. Ken Jernstedt of the 3rd Pursuit was issued number 0250 in January 1942. A chit similar to that shown in Fig. 4.9 on display at the Air Force Museum numbered 0155 was, according to information in the display, kept on hand for use by visitors to the AVG. A chit numbered 0337 was sent home to the family of pilot Pete Atkinson following his death,[39] and one numbered 0387 reportedly went to the family of John T. Donnovan. Without a more comprehensive listing of numbers and issue dates it is difficult to speculate about the order in which chits were issued or the range of numbers actually issued.

As the message on the back of Rescue Patch numbered 0387 (Fig. 4.11) indicates, many of the Rescue Patches were worn on flight gear. Photos from the period, appearing in a number of the books on the AVG familiar to many readers, show many AVG members with chits sewn to the back their flight suits and leather jack-

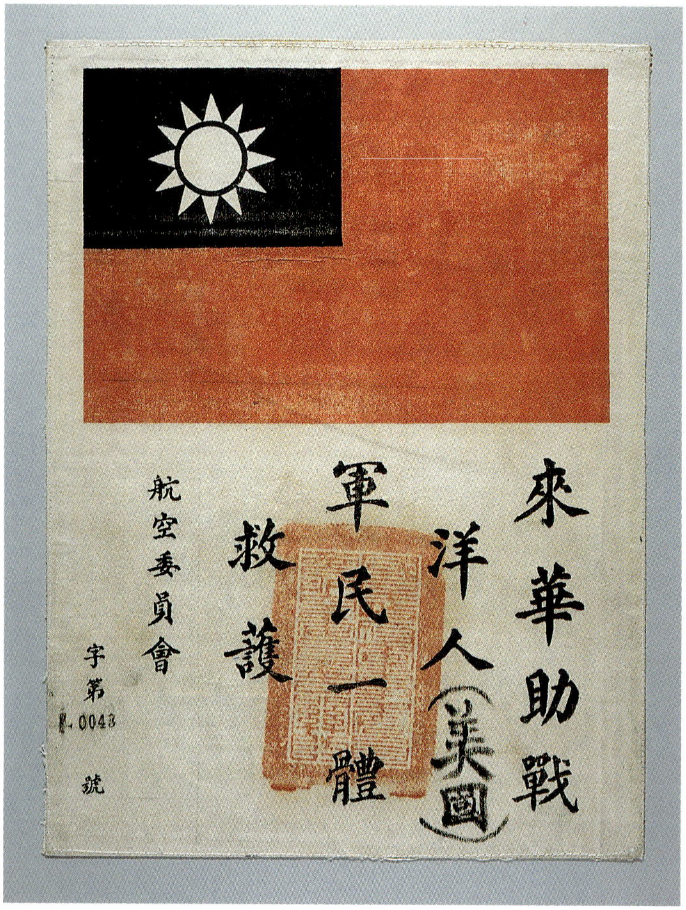

Fig. 4.9: This Rescue Patch is nearly identical to that of Claire Chennault, which is displayed at the National Air and Space Museum in Washington, D.C. Both feature flags with a light shade of red and identical styles of text, including the distinctive large parentheses surrounding the "may guo" characters. Some sources have misidentified the large parentheses on these early chits as a circle or oval around the two characters. Chennault's chit is numbered 0001, while that illustrated here is numbered 0043. (REB 2210) *International Blood Chit Museum*

ets. One such photo, Fig. 4.12, shows AVG pilot Charles R. Bond wearing his A-2 jacket, to which is sewn his Rescue Patch. This was a marked departure from the earlier Rescue Patches of the 14th VBS, which were worn on the front of flight suits.

Sewing a chit onto a garment ensured that it would not be lost and could be easily read if the wearer were unconscious, but it also restricted the clothing available to the pilot if he wanted to benefit from the chit, when needed. Ken Jernstedt wore a variety of flight gear, from his Navy issue flight suit, with or without his leather flight jacket, to just a shirt. For this reason he never attached his identification patch to his flight gear, but instead carried it in a convenient pocket.[40] Wearing the chit on a jacket or flight suit also increased wear and tear on the chit. D.L. "Tex" Hill indicated that his Rescue Patch eventually disintegrated because he wore it on the back of his flight jacket.[41]

Whether or not the AVG identification patches were effective in soliciting the assistance of the Chinese is open to debate. Ken Jernstedt doubts that chits were of much benefit because most Chinese in the areas over which the AVG operated could not read. He added, however, that the flag on the chit might have been useful because it was well recognized and clearly indicated that the bearer was not Japanese. He also noted that the Chinese were so thankful that Americans were protecting their skies that they would do anything for a downed AVG pilot once they knew who he was.[42]

There are several stories told by AVG pilots who were forced down. Not all of these stories mention Rescue Patches, nor does it seem that the patches were essential to rescue in many cases, although some thought it helpful.

Ed Rector recounted his first combat mission during which he set one Japanese bomber on fire, and attacked a second until his guns stopped firing. Low on fuel, he force landed the aircraft in a plowed field below a village. He took out his pistol, but put it back in its holster after seeing the inhabitants of the village coming down the hill toward him. He credits the Aero Commission Rescue Patch sewn to the back of his jacket and the fact that the Nationalist government had informed the Chinese people that American flyers were fighting against the Japanese, plus the Chinese Air Force insignia on his aircraft, for the friendly reception that followed.[43]

Charlie Bond notes that on 6 May 1942 he flew a combat mission without wearing or carrying the Chinese chit. After intercepting a flight of Japanese bombers during a raid on their airfield, Bond was jumped by three Japanese fighters while flying over Kunming to view the damage. He force-landed his burning fighter

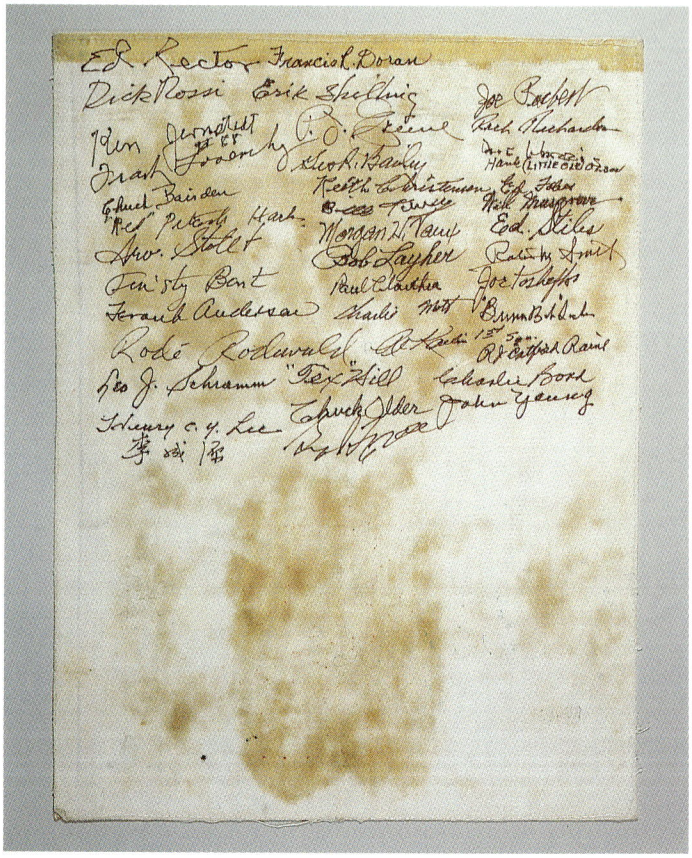

Fig. 4.10: Shown here is the back of the Rescue Patch illustrated in Fig. 4.9. It has been signed by 36 members of the original American Volunteer Group, plus Henry C.Y. Lee, the Chinese Air Force Communications Officer who served as Chennault's personal translator throughout the war. Lee also organized the AVG's Radio Warning Net for Chennault. These are postwar signatures *International Blood Chit Museum*

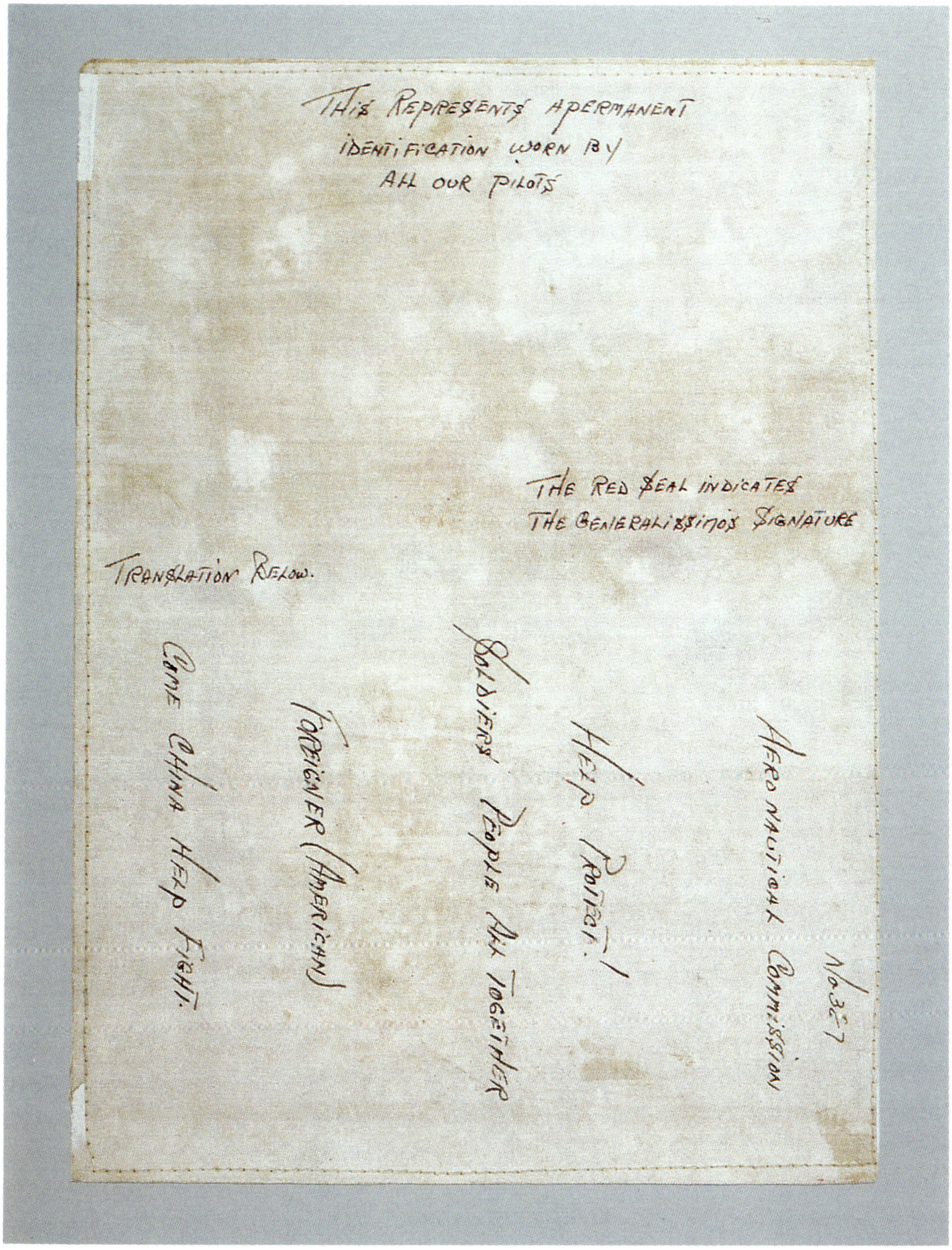

Fig. 4.11: The back of Rescue Patch number 0387, reported to have been sent by the AVG to the family of John T. Donovan, is shown here. It reads "This represents a permanent identification worn by all our pilots. The read seal indicates the Generalissimo's signature. Translation below: Come China Help Fight, Foreigner (American), Soldiers People All Together, Help Protect! Aeronautical Commission, No. 387." The translation is quite good, but the red seal, or chop, is that of the Aero Commission, not of Chiang. *Courtesy of Robert E. Hunter, M.D.*

in a Chinese cemetery, and, after dousing his burning scarf and flight suit in a small stream, faced the reality of being a downed airman in China:

> There were many Chinese in the cemetery crouching behind the burial mounds. I motioned for one of them to come over to me. He had one hand behind his back. I went through the gestures of picking up a telephone and calling, while uttering the usual Chinese greetings, "Wau. wao. whoo. Wa." I did not have my flying jacket, on the back of which was sewn the Chinese flag and directives written in Chinese to help us.
>
> Nevertheless, he seemed to get the message, smiled lightly, and then brought his hand from behind his back, revealing a large rock. He was ready to give me the works, thinking I was the enemy . . . It is amazing what one can do in an emergency with sign language.[44]

Fig. 4.12: AVG pilot Charles R. Bond looks skyward in this photo. Sewn to the back of his A-2 jacket is his Chinese Aero Commision Rescue Patch, number 0212, issued to him in Kunming in December 1941. Although attaching the Patch to a garment insured that it would be available even if the pilot were found unconscious, it also meant that it would only be available if that particular farment was worn. For a variety of reasons, some AVG pilots had their Patches sewn to their flight suits or jackets, while others simply carried them folded in apocket. *From* A Flying Tiger's Diary *by Charles R. Bond, Jr. and Terry H. Anderson, reprinted by permission of Texas A&M University Press.*

In analyzing the his experiences, Bond relates, "Had I been wearing my chit the two times I was shot down, I doubt if it would have helped much more since I was in friendly territory – actually near my base of operations in both cases. I was pretty convincing with my sign language!!"[45]

In the mind of the American public, the beautiful silk Rescue Patches issued by the Nationalist government are forever associated with the American Volunteer Group of the Chinese Air Force. They are a small but colorful sidelight to this band of modern American heroes who took it upon themselves to defend an entire nation against what appeared to be impossible odds.

One of the key questions in the history of Blood Chits is how they spread from use by the RAF in India and Mesopotamia to other military forces and areas of operations. Of particular interest is the origin of the Aero Commission Rescue Patches in China.

British pilots and air crews who rotated from colonial policing duties in India and the Middle East back to England and then on to other assignments, including posting to China, certainly took their knowledge of chits with them. In 1927 the Royal Air Force's No. II (AC) Squadron [AC denotes Army Cooperation] was posted to service in the international settlement in Shanghai. Several of its members had already completed assignments in India and the Middle East, and would have been familiar with the standard practice in those areas of carrying Goolie Chits. When the squadron returned to England later that same year many of its members were reassigned to the Fleet Air Arm and remained on duty in China.[46] The British use of chits in India and the Middle East was, therefore, known to British pilots in China and in neighboring Burma during the 1930s.

Ken Jernstedt related that "The Old Man", as he calls Chennault with obvious affection, "had been around" and could well have had knowledge of the British use of Goolie Chits. This was not said to endorse any theory that this was actually the case, but to point out that Chennault's wealth of experience and knowledge was vast, and certainly could have included this information.[47]

Jernstedt also noted that the use of chits in China was most likely Chennault's idea.[48] The fact that there is no evidence that Chinese Rescue Patches were ever issued to pilots other than those commanded by Chennault during period ending 4 July 1942 gives strong credibility to this view. Chennault certainly understood the value of his small number of pilots and the problems they faced when forced down over China. There may have been some uncertainty about whether the Rescue Patches would be effective, but the risks of the innovation's failure were limited, and if successful the benefits would be significant.

It may never be known whether the Chinese government's Rescue Patches were developed from the British Goolie Chits in use during the preceding two decades, or were the product of an independent, original idea – both possibilities clearly exist. Regardless of their origins, however, by providing these items stamped with an official government seal to pilots, and by communicating to the Chinese population that foreign airmen were helping their country, Claire Chennault and the Nationalist government began a system for the rescue of foreign airmen in China.

As the war escalated, the Allies would develop increasingly sophisticated methods and organizations to ensure the safe return of their air crew members in China and elsewhere.

CHAPTER 5

The Mercy of Allah World War II in Africa and the Middle East

Air crew members from the United States and the British Empire forced down during World War II had a "secret weapon" unlike anything possessed by their adversaries. This weapon was not a new technological marvel, but an organizational one. The governments of these allies established secret evasion and escape (E&E) organizations that provided everything from training, to kits and aids, to organized evasion networks in enemy-occupied territory, in an effort to effect the safe return downed air crewmen. The Allied effort was unparalleled in military history, and established strong traditions that continue today within the British and American armed services.

Whether or not the World War II Allied airman downed in enemy-occupied territory was truly prepared for what lay ahead, the training and aids provided to him by E&E organizations gave him the confidence to take proactive steps to evade capture. They also provided important psychological support because the airman knew that many individuals and organizations were working to ensure his safe return.

On 23 December 1939, less than three months after World War II in Europe began, Military Intelligence 9 (MI9) became Britain's official agency responsible for E&E matters, under the command of Brigadier Norman Crockatt. MI9's creation was the result of a slow but significant change in thought regarding prisoners of war that began years earlier when it was discovered that evaders and escapers during the 1914-18 War returned with valuable intelligence. The British were encouraged in their E&E efforts by the knowledge that during World War I some 107,000 Allied servicemen with no E&E training had managed to evade or escape from German hands and return to Allied control, and that the German military was tied up recapturing another 750,000 to 800,000 escapers at loose at one time or another among their civilian population.[1]

Preventive Training, or P Branch of MI9, gave training "concerning escape methods and MI9 operations" to "all air crews and people on special operations who might be in greater danger of being captured than the soldier in a line regiment." Escape, or E Branch, operated patrols "behind the enemy lines or in occupied countries for the purpose of rescuing escaped prisoners or people cut off behind the lines or forced landed in occupied territory" and also handled "the enlistment of local aid and the laying out and maintenance of escape routes."[2]

In 1940, Col. Dudley Clarke was posted to British headquarters in Cairo, Egypt, where he organized a unit known as "A-Force". The primary mission of A-Force was to convince the Germans and Italians that British strength was far greater than it actually was. In August 1940, N-Section of A-Force was established to conduct British E&E operations in Africa and the Mediterranean. N-Section was headed by Lieutenant Colonel A. C. Simonds, a veteran of intelligence operations in Ethiopia, Greece and Palestine. In the field, N-Section was known as Advance A-Force, and the Long Range Desert Group (LRDG) routinely transported its personnel across miles of inhospitable desert to set up caches of water and other supplies for use by evaders, and to organize evasion networks among the Arab populace.[3]

The war ranged over the African continent from Morocco, Algeria, Egypt, Libya, Tunisia, and the Sudan in the north to Iran and Iraq, to British Somaliland (now Aden) on the east coast and to Chad, Cameroon, French West Africa, Senegal, Kenya, and Madagascar. The Mediterranean campaign extended some 2200 miles from Gibraltar to the Suez Canal. Although eventually overshadowed by the war in Europe, the war in Africa and the Mediterranean was no less fierce even though it lacked notoriety in the public's mind following the Normandy invasion.

The North African Campaign was fought in a difficult climate over vast distances of desert, mountain, and ocean. The geography of the continent ranged from the trackless expanses of the Sahara to the rugged, snow-covered Atlas Mountains. The survival and evasion requirements of air crews were wide-ranging not only because of the harshness of the climate and terrain, but because their missions also often took them out to sea to attack German and Italian supply lines.

The desert climate was particularly harsh and unforgiving. Temperatures ranged "from 125 degrees Fahrenheit midday in the Summer to 25 degrees Fahrenheit in the early mornings of the Winter" and "the baking sun beats down unhampered by clouds and mist, raising the temperature of the ground, tents, planes and everything on which it strikes to such a heat that metal objects can scarcely be touched; while at night with no insulation by moisture, the ground and air cool rapidly."[4] What respite Winter brought was not much better, as aircraft sank in the mud so deep they could not be moved onto runways.[5] In addition, sand ingested into engines caused severe problems.

Desert campaigns brought with them a new set of priorities. In both operations and evasion, water was the prime commodity. The military was blunt about the need for water for air crew survival, advising crews to "make sure you have your escape aids with you. Above all, have a filled water bottle."[6] That water was valued above other survival and evasion aids is illustrated by instructions given to Intelligence Officers lecturing crews before missions, which stated that "if the weight allowance is small (10 lbs. or less) put all of it into water. Only if they are carrying adequate water and have weight allowance left, should they pack food, emergency equipment, etc. If they can be accommodated, Emergency Kits that seem to be the most suitable for any given type of mission should be packed. These should be revised and adapted to the crews' purpose."[7] This last quote is significant, for in addition to its comment regarding water it explains the many variations found in E&E kits. Collectors are sometimes perplexed by the variety found in what they think should be standardized kits, when in reality the kits were adapted at the unit level to meet the needs of the moment.

Along with the importance of carrying water and emergency kits, the application of common sense and a positive mental attitude were stressed while reassuring crews that they had a good chance of returning if forced down. Air crews in North Africa were told that "if stranded persons follow the practical suggestions given in this section, and add to them common sense and ingenuity, they have every chance to come through safely,"[8] and that "air crews who come down in enemy occupied territory have a good chance of evading capture and of regaining their own forces. Experience has shown that this chance is greatly increased if the proper equipment is carried and if the information given by your Intelligence Officer . . . is remembered and used."[9]

The cooperation and generosity of the local inhabitants was critical in the rescue and safe return of Allied air crew members. British and American evasion agencies provided air crews with information about the local people they would meet if forced down, and on how to seek out and recognize potential helpers:

> Choose carefully which native to speak to: a farmer working in a field is perhaps best, or a country Arab going along the road. The country Arab (who may be any color from white to black) is simple but shrewd; he should be given a little money and promised more if he helps you. The Moroccan Arabs have a long tradition of blockade running, and it seems to give them pleasure to crawl through the French lines at night; you can probably get them to deliver you to the nearest American post if you give them a good talk and have any luck.[10]
>
> Friendly Arabs in Zaghouan area . . . will make the offer to evaders or escapers of cigarettes even though they may not actually possess cigarettes.[11]
>
> Ferry pilots and others whose duties or training take them near the borders of Spanish Morocco should be specifically warned that Spanish Moroccan Moors cannot be trusted.[12]

Of immeasurable help to the rescue of airmen was the fact that there existed in the desert a strong cultural tradition of hospitality. Air crews were given very specific instructions about their behavior should they be guests of the local population:

> Do not refuse their food if it is offered to you. This signifies a great deal in the Moslem culture; in fact if you have eaten out of the same pot with an Arab you are his guest and it is his religious duty to take care of you.
>
> After eating, if you are in a tribal camp or a country family, you will find that the Arabs like to talk at length and tell stories. It is their custom to exaggerate everything in these stories, not with an idea of being believed, but merely as entertainment after eating. If you are asked questions (as you will be endlessly) try to make a good story even by gestures.
>
> They are as a rule extremely brave people, and even if they are not, they pretend to be. Therefore it is important NEVER to show any fear of them. Treat them with confidence – say "Salaam" when you speak, and when they pray before eating, bow your head.[13]

North African residents helped Allied air crews at great personal risk. This risk was drilled home to airmen, who were told, "Anyone who helps you risks Death. If caught you face nothing worse than a German prison camp, bad as that may be. They face certain Death for themselves and families. You must protect them with the greatest care. Never write or tell anything about receiving help to anyone."[14]

In Africa and the Middle East, N-Section of A-Force undertook the jobs of distributing Blood Chits to air crew members and paying helpers:

> A separate branch of "A" Force was concerned with the recovery of prisoners of war and stragglers, largely through Arab agents who were paid per capita for the men they brought back to the British lines. RAF crews and members of both the Special Air Service [SAS] and the Long Range Desert Group [LRDG] carried "blood chits" which were encashable by the Arabs who helped them in this way.[15]

War first came to Africa on 3 October 1935 when Italy invaded Ethiopia without the formality of a declaration of war, following border clashes with Ethiopian troops which began as early as December 1934. The war lasted until 5 May 1936, when the capital, Addis Ababa, was captured by the Italians and the country became part of Mussolini's new Roman Empire.

There was no major intervention by the League of Nations or other outside powers to defend the Ethiopian government of Haile Selassie. Of note, however, is the service of an American pilot, John C. Robinson, an African-American graduate of the Tuskegee Institute, who volunteered to fly for the Ethiopians. He was commissioned as a Colonel, and given command of the 19 aircraft and 50 pilots of the Ethiopian Air Force.[16]

The British counter-offensive to drive the Italians from Ethiopia began in the Fall of 1940 with South African units accompanied by colonial troops including the King's African Rifles and two Indian infantry divisions, as well as irregular forces including the Sudan Defence Force and the "Gideon Force" led by Orde Wingate. The air war included South African Air Force (SAAF) and Royal Air Force (RAF) units equipped with contemporary biplanes.

The RAF had been issuing Goolie Chits to flight crews on service in the Middle East and North Africa since the 1920s, and the employment of Blood Chits in the African campaigns was a natural extension of their previous use. The first of these Goolie Chits appear to have been those issued to RAF crews assisting the government of Ethiopia to resist the Italian invasion and occupation.

One of the more interesting Goolie Chits issued for this campaign was a pocket-sized green leather-covered folder with a paper chit bound into it, apparently issued by the government of Ethiopia. The example shown in Fig. 5.1 was issued to Squadron Leader R.A. Collis, DFC, the RAF pilot who flew Orde Wingate into the Abyssinian Highlands during the campaign to stop an Italian advance in 1940.

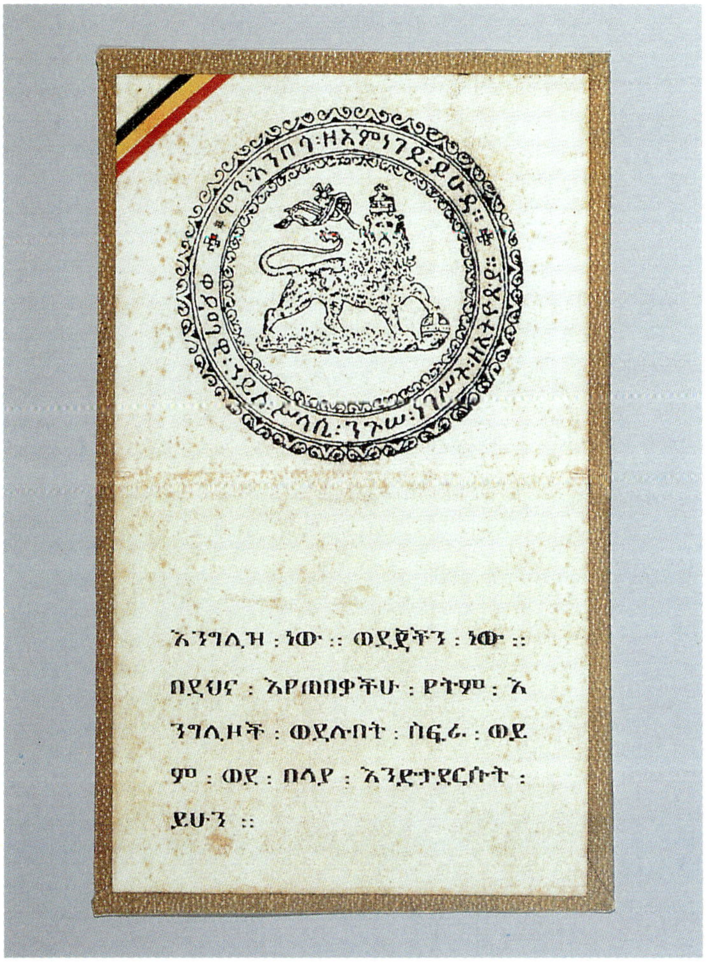

Fig. 5.1 : This leather covered folder contains an Ethiopian government paper chit in Amharic, the language of Ethiopia, and features the seal of Emperor Haile Selassie. It was issued to S/Ldr R. A. Collis, DFC, in 1940. The words at the bottom of the seal read, "The King of Ethiopia, Haile Selassie," and the message is "The British are favored by us. Protect the bearer and take him wherever the British are." There is no reward offered, but the chit would have been effective because it was issued by the Emperor. S/Ldr Collis was the RAF pilot who flew Orde Wingate into the Abyssinian Highlands during the campaign to stop an Italian advance in 1940. (REB 4210) *Courtesy of R. A. Collis, DFC*

At the top of the chit is the seal of Haile Selassie, Emperor of Ethiopia. The text at the top of the seal is written in Gize, the ancient language of Ethiopia, and the remainder of the text is written in Amharic, Ethiopia's contemporary language. The words at the bottom of the seal read, "The King of Ethiopia, Haile Selassie." The literal translation of the message is "British are our favorite. Protect us and take us wherever they go." As previously noted, word-for-word translations are often misleading, and the correct interpretation of this message is "The British are favored by us. Protect the bearer and take him wherever the British are."[17] A reward is not offered for providing assistance, but the chit was issued by the Emperor and under his seal, and it bears the statement that the British are favored by "us" meaning by the Emperor, so the chit would have been effective.

S/Ldr Collis was issued a second chit, illustrated in Fig. 5.2, printed on a tightly woven, high quality linen. The chit is written in two languages, Amharic above the English Government seal and English below it. The chit also gives advice on techniques of identifying oneself and on the best, safest way to approach Ethiopians. The commanding officer of No. 47 Squadron, Squadron Leader James Pelly-Frey, was issued one of these chits while serving in Ethiopia in 1940. On it, to either side of the British Government seal, is stamped the seal of Emperor Haile Selassie.[18] Another chit of this type, collected by R.E. Baldwin, is printed on thin, coated linen and is stamped with the Emperor's seal.

A third type of chit for RAF air crews in Ethiopia, illustrated in Fig. 5.3, was produced late in 1941 and issued in 1942. Written in Amharic and English, it requests assistance for downed flyers but does not promise a reward. The text is unlike the typical British chits of the day, leading to speculation that it may have been issued by Haile Selassi, or may have been prepared by the British and issued in the Emperor's name. It states that the instructions are "our recommendation", meaning those of the Emperor, and the seal of the Haile Selassie has been stamped on the chit to give it the required authority.

The introduction of the Ethiopian chits signaled the beginning of an era in which Britain, and later the United States, produced and issued a multitude of different Blood Chits, each designed for a specific area of operations. Some of the Goolie Chits produced by Britain for use in Africa and the Middle East are illustrated in Fig. 5.4, displayed on an English-made cloth evasion map, K3/H2, covering North Africa and Spain. These items, discussed individually later in this chapter, illustrate the variety of chits issued to British and Commonwealth air crew members in this theater. Also shown is cloth map K5/K6 covering portions of Sudan, Nigeria, Cameroon, Spanish Guinea, French Equatorial Africa, and the Belgian Congo.

One of the early British chits was designed for use in Iraq, a former Turkish colony placed under British mandate after World War I, and becoming an independent state in 1927. The treaty granting independence guaranteed Iraqi assistance to Britain in case of war and allowed British forces to pass through the nation. In April and May 1941, pro-German elements of the Iraqi government and

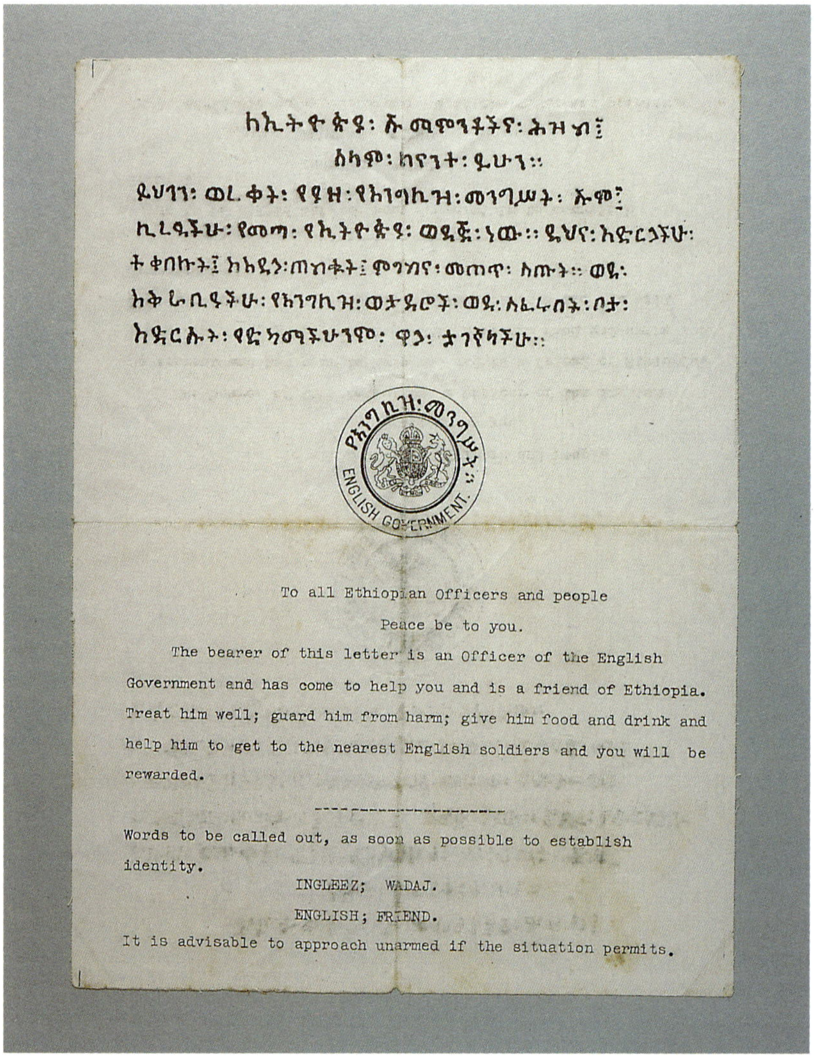

ENGLISH GOVERNMENT

To all Ethiopian Officers and people

Peace be to you.

The bearer of this letter is an Officer of the English Government and has come to help you and is a friend of Ethiopia. Treat him well; guard him from harm; give him food and drink and help him to get to the nearest English soldiers and you will be rewarded.

Words to be called out, as soon as possible to establish identity.

INGLEEZ; WADAJ.

ENGLISH; FRIEND.

It is advisable to approach unarmed if the situation permits.

Fig. 5.2: This RAF Goolie Chit was issued to S/Ldr. R. A. Collis, DFC, when he was on operations in Ethiopia in 1940. It is printed on a thick, closely woven, high quality linen. It bears an English Government seal and messages in Amharic and English, plus words to be called out to identify oneself and tips on the safest way to approach local people. This was the first of many British chits to include useful words and phrases in phonetic equivalents of native languages. (REB 1210) *Courtesy of R. A. Collis, DFC*

military led by Raschid Ali began a month-long siege of a major Royal Air Force training base at Habbaniyah, Iraq. The Iraqi government requested assistance from Germany, and the first Luftwaffe aircraft, assigned to Fleigerfuhrer Irak, arrived in Iraq on 7 May. Although the Iraqis and Germans were defeated by the end of May in well-coordinated air and ground attacks, the siege of Habbaniyah revealed the complexities of alliances in the region and the importance of Allied air bases in colonial territories or areas of influence.

In September 1941 Squadron Leader A. C. Rawlinson, Royal Australian Air Force (RAAF) on service in Iraq, was issued a Goolie Chit, illustrated in Fig. 5.5, printed on a linen base.[19] The Arabic text appears above an English Government seal, and below it the message is repeated in English. Note that this chit offers a reward for helping the bearer. In the lower right corner is the notation "PME 1554-9/41." The last numbers, 9/41, represent the date the chit was produced, September 1941. The RAF Museum was not able to determine the meaning of the acronym "PME", although it may represent the intended area of use for the chit, Persia - Middle East.

S/Ldr Rawlinson, later a Group Captain, commanded RAAF No. 3 Squadron, and recalled that there were several earlier issues of Goolie Chits in the Western Desert in 1940 and 1941, and in the June-July 1941 Syrian Campaign.[20] Unfortunately no examples of these earlier chits were available for illustration.

Americans on operations in the Middle East were also provided chits by the British. As an example, a chit of the type shown in Fig. 5.5, on display at the U.S. Air Force Museum, was carried by Col. Caleb Haynes on the first flight by the USAAF Ferrying Command's Asia Corps to Cairo, Egypt.

Air crew members sometimes cut down their Goolie Chits in size so that they could be folded into a smaller package, facilitating movement from pocket to pocket when changing flight gear. One such item is illustrated in Fig. 5.6. Although similar to Gp/Capt

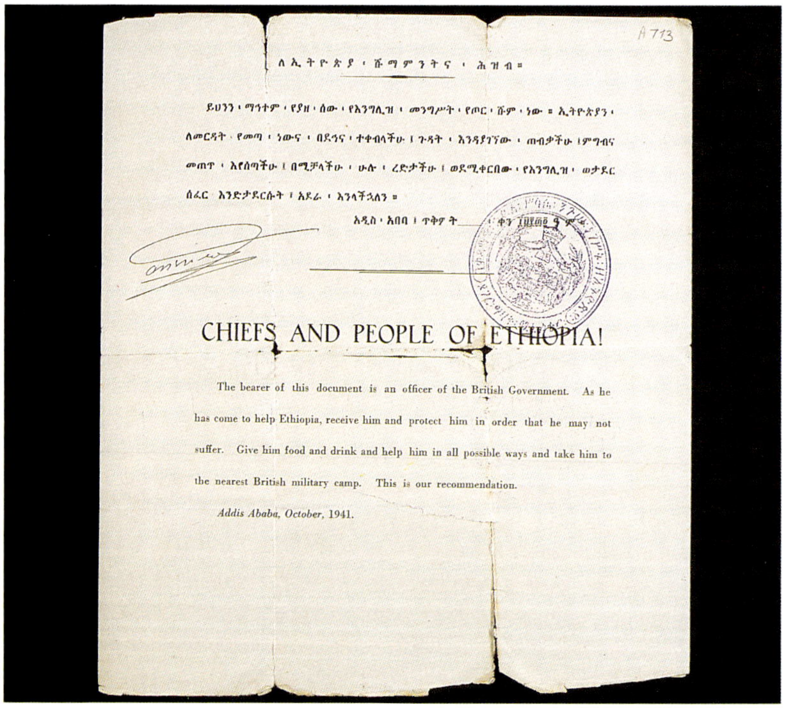

A 713

CHIEFS AND PEOPLE OF ETHIOPIA!

The bearer of this document is an officer of the British Government. As he has come to help Ethiopia, receive him and protect him in order that he may not suffer. Give him food and drink and help him in all possible ways and take him to the nearest British military camp. This is our recommendation.

Addis Ababa, October, 1941.

Fig. 5.3: This chit, issued by Haile Selassi or possibly prepared by the British and issued in his name, was printed in Addis Ababa in October 1941 and issued to RAF air crew members in 1942. It bears messages in English and Amharic. It does not offer a reward, but states that the instructions are "our recommendation", meaning those of Emperor Haile Selassie, and it is stamped with his seal to give it the required authority. (REB 4212) © *British Crown copyright/MOD. RAF Museum Photo, printed by permission of the Ministry of Defence (MOD).*

Fig. 5.4: Some of the great variety of Goolie Chits issued to British personnel in Africa and the Middle East are shown here, displayed with two English evasion maps, K3/H2 and K5/K6, covering areas in Africa. Some chits were single sheets, others were folded cards, and yet others were booklets of tear-out chits for various regions with hints on dealing with local peoples. *International Blood Chit Museum*

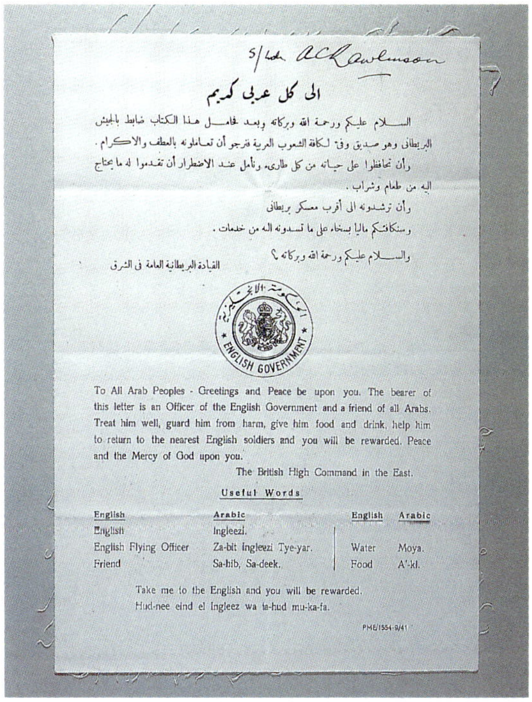
To All Arab Peoples - Greetings and Peace be upon you. The bearer of this letter is an Officer of the English Government and a friend of all Arabs. Treat him well, guard him from harm, give him food and drink, help him to return to the nearest English soldiers and you will be rewarded. Peace and the Mercy of God upon you.

The British High Command in the East.

Useful Words.

English	Arabic	English	Arabic
English	Ingleezi.	Water	Moya.
English Flying Officer	Za-bit Ingleezi Tye-yar.	Food	A'-kl.
Friend	Sa-hib, Sa-deek.		

Take me to the English and you will be rewarded.
Hud-nee eind el Ingleez wa ta-hud mu-ka-fa.

PME/1554-9/41

Fig. 5.5 : S/Ldr A. C. Rawlinson, later Gp/Capt, was issued his linen Goolie Chit printed on a linen base in September 1941 while on service with RAAF No. 3 Squadron in Iraq. It bears an English Government seal and messages in Arabic and English. A reward is offered for assisting the bearer. The notation "PME 1554-9/41" appears in the lower right corner, the last two numbers representing the date the chit was produced, September 1941. The meaning of the letters "PME" are unknown, but may represent the chit's intended area of use, Persia-Middle East. (REB 1214) *Courtesy of A. C. Rawlinson*

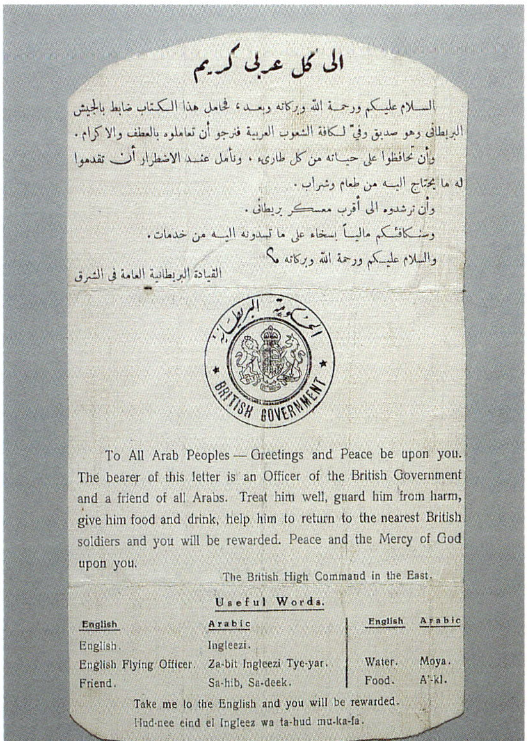
To All Arab Peoples — Greetings and Peace be upon you. The bearer of this letter is an Officer of the British Government and a friend of all Arabs. Treat him well, guard him from harm, give him food and drink, help him to return to the nearest British soldiers and you will be rewarded. Peace and the Mercy of God upon you.

The British High Command in the East.

Useful Words.

English	Arabic	English	Arabic
English.	Ingleezi.	Water.	Moya.
English Flying Officer.	Za-bit Ingleezi Tye-yar.	Food.	A'-kl.
Friend.	Sa-hib, Sa-deek.		

Take me to the English and you will be rewarded.
Hud-nee eind el Ingleez wa ta-hud mu-ka-fa.

Fig. 5.6 : This Goolie Chit on paper backed with linen is written in Arabic and English. It was cut down from its original size to make it easier to transfer from pocket to pocket when changing flight gear. It is similar to Gp/Capt Rawlinson's chit (Fig. 5.5), with minor differences in the layout of the English text. Written in pencil on the back appears to be "H. E. Kirby, P202941V, SAAF." The notation "SAAF" is the acronym for the South African Air Force. (REB 1216) *International Blood Chit Museum*

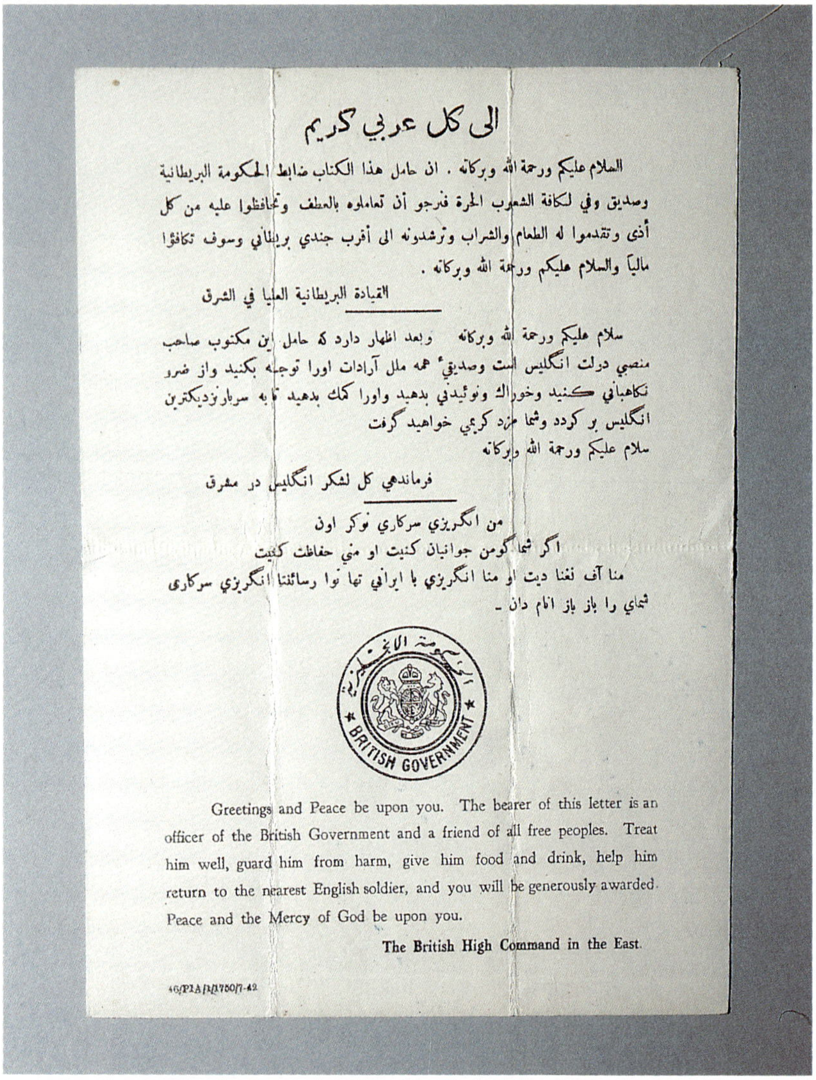
الى كل عربي كريم

BRITISH GOVERNMENT

Greetings and Peace be upon you. The bearer of this letter is an officer of the British Government and a friend of all free peoples. Treat him well, guard him from harm, give him food and drink, help him return to the nearest English soldier, and you will be generously awarded. Peace and the Mercy of God be upon you.

The British High Command in the East.

46/PIA/1/1750/7-42

Fig. 5.7 : Printed on paper and backed with a linen cheesecloth, this Goolie Chit was issued to RCAF F/Lt Ian Glen, DFC, at RAF Station Habbaniyah, Iraq, in November 1944. The notation "46/PIA/1/1750/7-42" appears in the lower left corner. "PIA" is the acronym for the RAF's Persia-Iraq Air Command, and the numbers "7-42" represent the printing date, July 1942. This type of chit was issued between 1942 and 1945 at RAF Stations Habbaniyah and Shaibi, in Iraq. (REB 1218) *Courtesy of Ian Glen, DFC, RCAF (Retd)*

Rawlinson's chit, there are several obvious small differences in the layout of the English text. This chit is made of paper backed with linen cheesecloth, and written in pencil on the back is what appears to read "H. E. Kirby, P202941V, SAAF."

Royal Canadian Air Force (RCAF) Flight Lt. Ian Glen, DFC, was issued his Goolie Chit, illustrated in Fig. 5.7, at RAF Station Habbaniyah, Iraq, in November 1944. Glen flew a Hurricane as Officer Commanding of RAF No. 1415 Meteorological Flight after completing his "ops" in Italy as a Spitfire pilot with the RCAF. The chit is printed on paper backed with a linen gauze and in the lower left corner is the notation "46/PIA/1/1750/7-42". The last two numbers represent the printing date, July 1942. Glen reports that "PIA" is the acronym for the RAF's Persia-Iraq Air Command. The chit's message is printed in three languages: Arabic at the top, Persian-Arabic in the middle and in another variation of Persian-Arabic at the bottom. In Africa the original Persian language was modified by the introduction of Arabic characters and sounds, while it remained relatively unchanged in isolated areas such as Afghanistan. This type of chit was presumably issued to all air crews flying over Persia (Iran) and Iraq. Although Glen was an RCAF officer, the Canadian military did not issue the Goolie Chit to him. These items were provided by the British:

> The only RCAF air crew who operated in areas that were provided chits were personnel in RAF units. There were, I think, two [RCAF] bomber squadrons operating out of North Africa and one Spitfire squadron, No. 417 RCAF, in North Africa and Italy and they all operated in RAF Wings. While the local native tribesmen and nomads were supposedly friendly, we always carried our "Goolie Chits" in case we were forced down in the desert.[21]

Former RAF Flight Lieutenant Jack Butterfield obtained a chit of this type from another former RAF member who told him they were issued from 1942 to

الى كل عربي كريم

BRITISH GOVERNMENT

Greetings and Peace be upon you. The bearer of this letter is an officer of the British Government and a friend of all free peoples. Treat him well, guard him from harm, give him food and drink, help him return to the nearest English soldier, and you will be generously awarded. Peace and the Mercy of God be upon you.

The British High Command in the East.

Fig. 5.8: This linen backed paper chit's message is written in Arabic, two versions of Persian-Arabic, and English. The notation "46/PIB/2000/8-42" in the lower left corner indicates that it was printed in August 1942. The meaning of "PIB" is unknown, but extrapolating from the meaning of "PIA" on the chit shown in fig. 5.7, it may represent a Bomber presence in the Persia-Iraq area. This chit was issued to Flt/Sgt Don Townend who flew Handley Page Halifax bombers on special operations to India and Middle East with No. 269 Squadron, 38 Group, RAF. (REB 1220) *Photo Courtesy of Don Townend*

1945, to RAF members serving at Habbaniyah and Shaibi, in Iraq.[22]

While serving as a Flight Sergeant with No. 296 Squadron, 38 Group, Royal Air Force, Don Townend received the chit shown in Fig. 5.8.[23] Stationed at Earls Colne, Essex, during 1944 and 1945 he flew Handley Page Halifax bombers on special operations to India and Middle East. He was issued chits appropriate to each area, including this one for Persia and Iraq.[24]

The chit, on linen-backed paper, carries a text in Arabic and two versions of Persian-Arabic above the British Government seal, and in English below it. In the lower left corner is the notation "46/PIB/2000/8-42", with 8-42 being the production date. The meaning of the acronym "PIB" is unknown, although extrapolating from the meaning of "PIA" given above and the type of aircraft flown by Townend, it may denote a bomber presence in the Persia-Iraq area.

One British chit seems to have been issued in greater numbers than any other used in this theater. Sometimes referred to as a "common card chit", it is illustrated in Fig. 5.9. It was issued to air crews in quantity, and was reportedly also carried by some ground forces. On the front of the chit is the Royal Arms, printed in red. The words "British Government" appear to the right of this, and the equivalent in Arabic appears to the left. Unfolding the card reveals an Arabic text on the left panel and the equivalent message in English on the right panel. On the back is a list of "Useful Words" given in English and their phonetic Arabic equivalents. Col. Gabriel Taschereau, who flew with Royal Canadian Air Force (RCAF) No. 425 Squadron of 331 Wing was issued one of these chits at Dishforth, Yorkshire, in May 1943 before proceeding to Kairouan, Tunisia. The same chit was also issued to the RAF's No. 420 Squadron and No. 422 Squadron.[25]

Perhaps because of its large scale distribution, the Germans created and distributed a decent forgery of this chit, shown in Fig. 5.10.[26] Its message was certain to insult and enrage any Arab to whom it was presented:

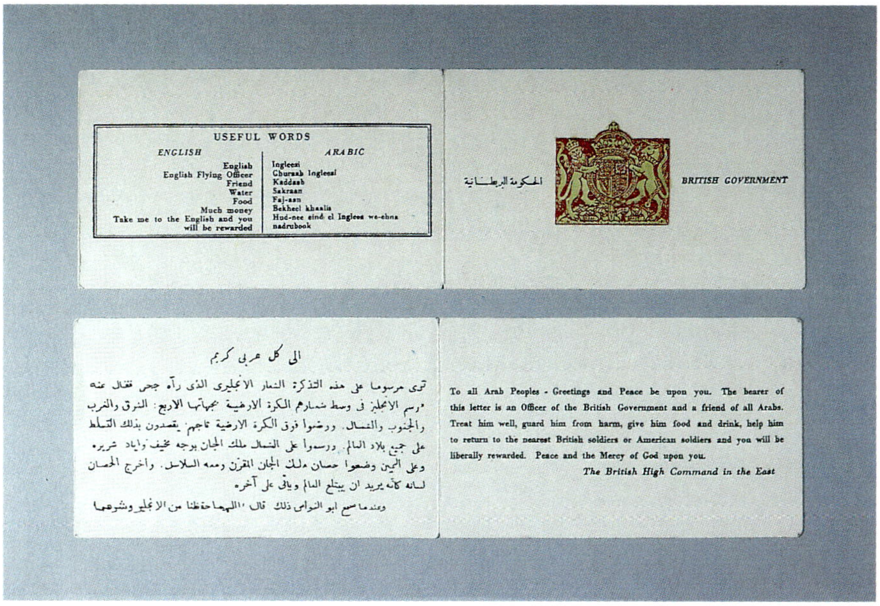

Fig. 5.10: The Germans distributed forgeries of the card chit shown in Fig. 5.9 in an attempt to create ill-will against the British. The English text was identical to the authentic chit, but the Arabic message depicts the components Royal Arms as evidence of British desire to colonize the Arab world. The phonetic "Useful Words" section is designed so that anyone using them would have received an unexpectedly hostile reception. For example, the phrase for "water" actually refers to intoxicants, which are forbidden in Islam. (REB 4208) *Reprinted from* Falling Leaf, *quarterly magazine of the Psywar Society, by permission of R. G. Auckland, Hon. Editor.*

Fig. 5.9: Sometimes called a "common card chit", this type of chit was issued in great numbers to both air crew members and ground forces. Printed on cardstock and folded in half, the chit contains messages on the inside in Arabic and English, and a list of useful words appears on the back. Probably because this chit was widely distributed, the Germans made a forgery of the card, shown in Fig. 5.10. (REB 1222) *International Blood Chit Museum*

the Germans made excellent replicas of these cards with the British coat-of-arms on the front and seemingly exactly the same. But the Arabic text, instead of being the same as the English version, was entirely different: "To the respected Arab people. Greetings and Peace be upon you. Here on this card you can see the emblem of the British which was once seen by Joha (a legendary humorous character). Joha said that the British had drawn the earth with the four directions (the shield); they put the crown of their monarchy at the top to reflect the fact that they would subdue the whole world. On the left-hand side is the King of Demons (Jinn) (the lion) having a frightful face and evil hands. On the right is the horse of the Demon King (the unicorn) with horns and chains and its tongue pointing outwards to confirm their intention of colonising the whole universe. Another great Arab poet, Abu Nowas, hated the British and their evil operations.

The "Useful Words" panel was also mis-translated with considerable effect. The "English flying officer" turns into an "English raven" and "water" is turned into "intoxicants" (so repulsive to the Moslems); "friend" becomes "liar," whilst "Take me to the English and you will be rewarded" reads "Take me to the English and you will be flogged."

Note: words in parentheses are *not* in the original text.

These cards were handed out to British troops arriving in Algiers by people purporting to be British officers and it is believed that others were distributed by hand and air over areas where they were likely to be effective in the same way as the fake currency and other propaganda material . . .

So far no record seems to be known of these forgeries' actual employment in the field, but any British serviceman who attempted to use them is unlikely to have had the welcome he expected.[27]

One of the best known and most comprehensive single-sheet Blood Chits used in North Africa was that shown in Fig. 5.11, a paper chit with a linen cheesecloth backing. It offers a text in Arabic and English, and includes hints for both desert travel and recommended conduct when meeting Arabs, together with a list of useful words and phrases. At the top of the chit is the Royal Arms flanked by the words "British Government" in Arabic to the left and in English to the right. Below this is a message in Arabic, followed by an English version of the message. Below the message is a section titled "Some Points on Conduct When Meeting the Arabs in the Desert", a caution regarding desert travel, a section titled "A few useful words", and a paragraph of useful information regarding how the older Arabs interpreted distance and time. One of the most important yet intangible evasion considerations is acknowledged by the last two words on the chit: "Good Luck".

Flt./Sgt. Don Townend was issued a chit of this variety at Earls Colne, Essex, England, while attached to No. 296 Squadron. Townend's chit was numbered 1292, stamped in the upper right corner above the words "British Government" and again on the right side just above the English text. Townend states that this type of chit was issued to all squadrons of 38 Group.[28]

William Fulcher was also issued one of these chits, and in late 1942 had reason to use it:

> As a member of the RAF, Middle East, I was assigned to 148 Squadron, at the moment stationed on the Suez Canal at El Kabrit. One of the first things assigned to me was a "Goolie Chit" to carry on our operations. It was explained to me to carry this, for if you went down behind enemy lines, the local

الحكومة البريطانية | BRITISH GOVERNMENT

الى كل عربي كريم

السلام عليكم ورحمة الله وبركاته وبعد ، فحامل هذا الكتاب ضابط بالجيش البريطاني وهو صديق وفى لكافة الشعوب العربية فنرجو أن تعاملوه بالعطف والاكرام . وأن تحافظوا على حياته من كل طارىء ، ونأمل عند الاضطرار أن تقدموا له مايحتاج اليه من طعام وشراب . وأن ترشدوه الى أقرب معسكر بريطاني . وسنكافئكم مالياً بسخاء على ماتسدونه اليه من خدمات .

والسلام عليكم ورحمة الله وبركاته — القيادة البريطانية العامة في الشرق الاوسط

To All Arab Peoples — Greetings and Peace be upon you. The bearer of this letter is an Officer of the British Government and a friend of all Arabs. Treat him well, guard him from harm, give him food and drink, help him to return to the nearest British soldiers and you will be rewarded. Peace and the Mercy of God upon you. ***The British High Command in the East.***

SOME POINTS ON CONDUCT WHEN MEETING THE ARABS IN THE DESERT.

Remove footwear on entering their tents. Completely ignore their women. If thirsty drink the water they offer, but DO NOT fill your waterbottle from their personal supply. Go to their well and fetch what you want. Never neglect any puddle or other water supply for topping up your bottle. Use the Halazone included in your Aid Box. Do not expect breakfast if you sleep the night. Arabs will give you a mid-day or evening meal.

REMEMBER, NEVER TRY AND HURRY IN THE DESERT, SLOW AND SURE DOES IT.

A few useful words

English	Arabic	English	Arabic
English	Ingleezi	Day	Nahaar or Yom
Friend	Sa-hib, Sa-deck.	Night	Layl
Water	Moya	Half	Nuss
Food	Akl	Half a day	Nuss il Nahaar
Village	Balaad	Near	Gareeb
Tired	Ta-eban	Far	Baeed

Take me to the English and you will be rewarded.	Hud nee eind el Ingleez wa tahud Mu-ka-fa.
English Flying Officer	Za-bit Ingleezi Tye-yara
How far (how many kilos?)	Kam kilo ?
Enemy	Germani, Taliani, Sizliani

Distance and time: Remember, Slow & Sure does it

The older Arabs cannot read, write or tell the time. They measure distance by the number of days journey. "Near" may mean 10 minutes or 10 hours. Far probably means over a days journey. A days journey is probably about 30 miles. The younger Arabs are more accurate.

GOOD LUCK

Fig. 5.11: The chit shown here is one of the most complete single-sheet chits issued in the North African theater. It was widely distributed, and in addition to the request for aid in Arabic and English it includes hints for desert travel, recommended conduct when meeting Arabs, and a list of useful words and phrases. One of the more important yet intangible evasion considerations is acknowledged by the last two words on the chit: "Good Luck." (REB 1226) *International Blood Chit Museum*

natives would take care of you and get you back behind your own lines.

On the night of November 12, 1942, our plane, starboard engine and wing on fire, finished our drop then turned away, and we parachuted into enemy territory near Derna, Libya.

I left my chit in my shirt that I did not wear that night. One of the boys had his chit with him and it covered all of us . . . To the Arabs it was a piece of money. Bedouins, knowing there was a reward for them to get us back safe, hid and looked after us until the British Eighth Army caught up with us.

Some ten days later we finally found our squadron and became members of the RAF Late Arrivals Club and also received a ten day leave to Cairo. Then it was back to the old grind.[29]

Perhaps the most famous chit of this type was carried by Warrant Officer Mickey Vertican, DGC, CGM, on a bombing raid against the remains of the Africa Corps lining the highways converging on Tunis from Bizerta. As related in *Down in The Drink*, the flight ended with his Halifax ditching in the Mediterranean. Vertican and his crew spent eleven days in a dinghy before making landfall on the North African coast, and his Goolie Chit was used as a diary to record the log of their voyage.

On the eleventh morning Vertican and the surviving members of his crew, six other men, awoke to find a strong wind blowing them toward shore. One member thought he saw a person on shore, and another wondered if their two Goolie Chits would cover all seven of them.[30]

Once on shore, two Arabs approached them. Vertican removed his Goolie Chit from his pocket and greeted the Arab as instructed on the chit. The Arabs brought water and fresh eggs for the airmen, and then "partly helped by the Arabs they made their way to a small hutted village about four hundred yards inshore . . . They were taken to the biggest hut in the village where they were received with great friendliness by the chief. There was no mistaking the kindness of the Arabs . . . tea was made . . . Then came a meal of bread, dates, eggs and water. Vertican understood from one of the Arabs that the local Sudanese Defence Force had been sent for, and when they came he delivered up his goolie chit and received assurance that the nearest British post would be informed within a few hours."[31]

Several booklets with hints on conduct when meeting native peoples plus tear-out Goolie Chits covering many areas in the Middle East were produced by the British. Three of these booklets, open to illustrate their typical contents, are illustrated in

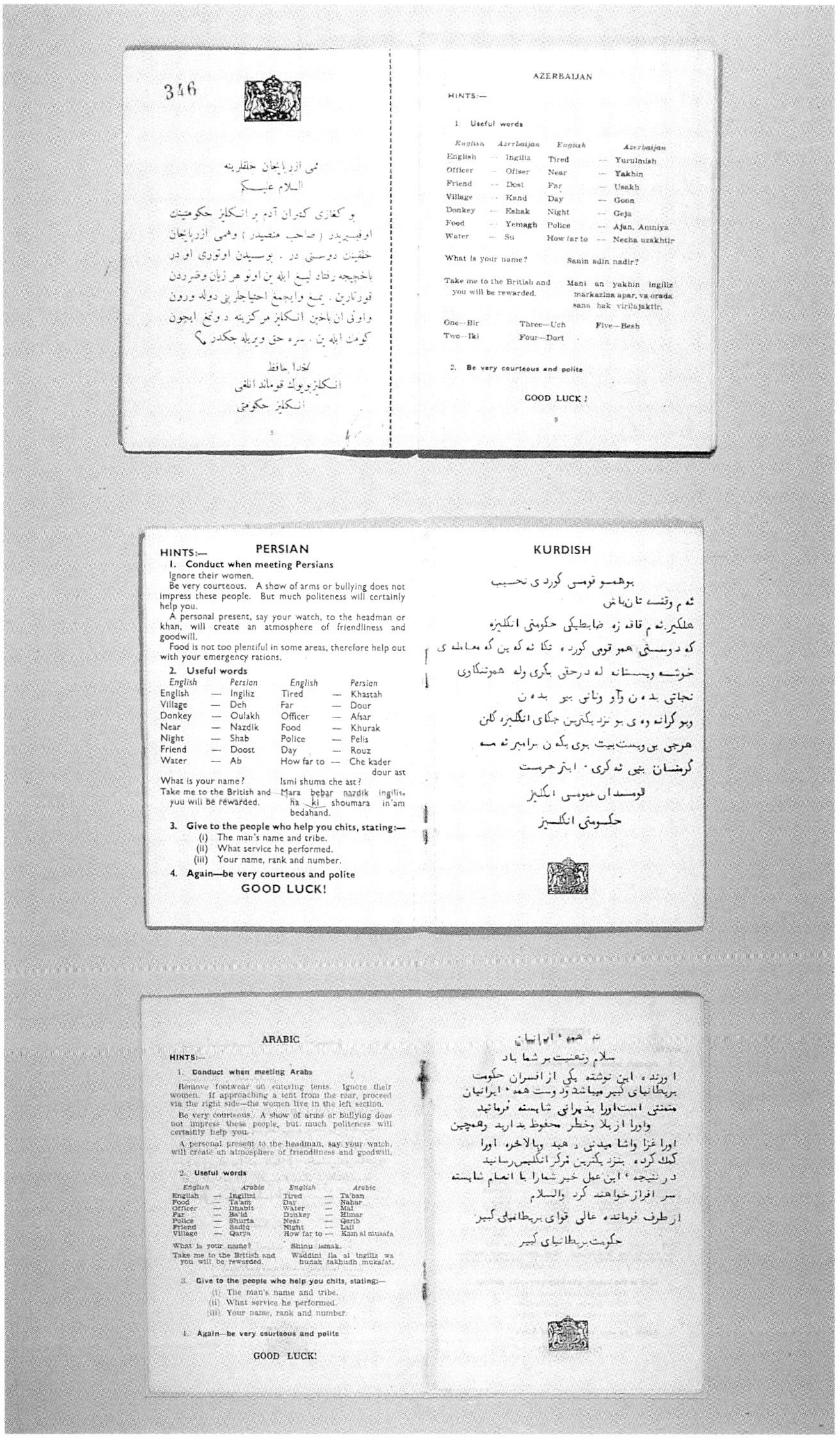

Fig. 5.12: Three booklets of tear-out chits and hints on conduct when meeting natives are shown here. The top book contains chits in six languages: Assyrian, Azerbaijan, Arabic, Persian, Kurdish and Turkish. The middle book contains chits in these languages, plus Russian, as does the bottom booklet. The top booklet is serial numbered 346, was printed by the Printing and Stationery Services of the Middle East Forces (MEF), and the faded writing on the cover indicates that it belonged to a member of the Middle East Air Force (MEAF). The middle and bottom booklets are not serial numbered and contain no information as to their manufacture. The booklets instruct the user to give "chits" — referring to Promissory Notes — to those people who have helped them. (Top: REB 1228, Center: REB 1230, Bottom: REB 1232) *International Blood Chit Museum*

Fig. 5.12. The covers of the booklets are of blue paper without any title or other information other than, on the top booklet, the serial number 346 stamped on it.

The top booklet contains tear-out chits in six languages: Assyrian, Azerbaijan, Arabic, Persian, Kurdish and Turkish, plus four blank tear-out pages for the airman to write out Promissory Notes. The Royal Coat-Of-Arms and the serial number 346 appear on each tear-out page. Page three of the booklet provides the following information regarding the chits:

> The following is a general translation of the messages contained in this booklet:-
>
> TO ALL ________ PEOPLES
>
> GREETINGS AND PEACE BE UPON YOU
>
> The bearer of this booklet is an officer of the British Government and a true friend of all ________ peoples. Treat him kindly, protect him from harm, give him food and drink, and help him to return to the nearest British Unit.
>
> For this good deed you will be rewarded.
>
> PEACE BE UPON YOU
> HEADQUARTERS
> MIDDLE EAST FORCES

The inside covers are unusual in that they are forms for the flyer to record which page he had torn out, to whom he had given it, what assistance was provided, and any other information he desired or felt necessary. This was to be done only after he had returned to British control, because if the book were recovered by the enemy, it would point directly to those persons who aided the airman and would mean death for both them and their families.

The first two pages give instructions for the use of the booklet:

> After contacting native, ask for help, and show him page written in his own language.
>
> After receiving help, write a brief description of help received on reverse side, adding your Number, Rank, Name and Signature.
>
> Ensure that the helper writes his name on the page also. Without some proof of identity, the reward will be paid to the person presenting the claim.
>
> Tear out the page, and give it to the helper.
>
> In the event of being passed from one native to another ... retain the page written in the native language and use a blank sheet instead. Until you are within easy reach of a British Unit, it is essential to retain the native script in case further help should be needed.
>
> Upon return to your unit, complete questionnaire on inside of cover, and give booklet to the unit I.O. or C.O. for return to Headquarters.
>
> A reward must be promised to the helper but NO DEFINITE AMOUNT STATED.

The booklet was printed by the Printing and Stationary Services of the Middle East Forces (MEF). The example pictured was damaged when removed from a scrapbook in which it had been glued, and a thin layer of the cover, on which the date of printing appeared, has peeled off. The name of the owner and other information written in pencil on the cover has faded but appears to read, "Jeffry, 214 Squadron, Bomber Command Detachment, Shullufa, M.E.A.F."

The booklet shown at the center of Fig. 5.12 is similar to the example at the top of the photo except that it has no instructions, lacks a serial number, is slightly smaller in overall size, and contains no information as to when or by whom it was printed. It contains hints on conduct when meeting local peoples plus tear-out chits in seven languages: Assyrian, Azerbaijan, Arabic, Persian, Kurdish, Turkish, and Russian.

Bob Gilroy, who flew Avro Yorks from England to Singapore via the Middle east, was issued one of these booklets toward the end of the war.[32]

David Underwood has a similar booklet that belonged to his uncle who "was air crew on Vickers Wellingtons with No. 70 Squadron RAF in the western desert. He did go missing for some time and he was looked after by Arabs for a while before returning back. I don't know whether he used a chit or not."[33] The printing in Underwood's booklet is of a different font, and it was printed by the Printing and Stationary Services of the MEF in 1942.

D. G. Upward, Director of the Southampton Hall of Aviation, writes that he was issued a similar booklet of tear-out chits "whilst serving with a Royal Air Force Transport Squadron based in India just after the war." Presumably the he would have flown missions taking him over areas where the languages in the booklet were spoken.[34]

The booklet shown at the bottom of Fig. 5.12 has Blood Chits in the same seven languages as in the middle booklet. The chits are not perforated as they are in the other two booklets shown. The format of this booklet is slightly different than the others: the hints for conduct when meeting local peoples are on the back of each chit for use with those peoples, and there are no language labels at the top of the chits.

The three booklets tell airmen to "give the people who help you chits."

The first American airmen, belonging to the 9th Bomb Squadron, a B-17 unit, arrived in North Africa from their former base in India. Then came a force of B-24s, and by June 1942, the month Tobruk fell, the 57th Fighter Group with P-40s, the 98th Bomb Group with more B-24s, and a B-25 unit had arrived. These formed the core of the Middle East Air Force (MEAF) which became operational in time to participate in the El Alamein offensive in October 1942, the same month that an evasion and escape agency was being established in the United States.

When the United States entered World War II it had no evasion and escape organizations, but fortunately found MI9's well-functioning program. When Major General Carl Spaatz went to England in February 1942 as head of the Allied Expeditionary Air Forces (AEAF), he became familiar with the workings of MI9. By that time MI9 had a program of evasion briefings for operational crews and was supplying them with evasion kits, maps and other aids. A school to train Intelligence Officers (IOs) in E&E matters, Intelligence School 9 (IS9), had also been established.[35] General Spaatz was allowed to detach an American officer, Col. Robely E. Winfrey, to MI9 for training.[36]

Following a visit by Britain's Air Vice-Marshall Charles Medhurst in March 1942 to brief U.S. officials on MI9's operation, the War Department formed the Military Intelligence Service (MIS), and on 6 October 1942, MIS-X was established to direct American evasion and escape activities. On 22 October 1942, the Captured Personnel and Materiel (CPM) branch was established under G-2 (Intelligence) to oversee MIS-X. CPM was commanded by Colonel Catesby ap C. Jones, who appointed Col. J. Edward Johnston as the War Department's Command Officer for MIS-X. Col. Winfrey returned to the United States after five months of training with MI9, and became commander of MIS-X in November 1942, at Fort Hunt, Virginia, outside of Washington, D.C. For reasons of security, Fort Hunt was referred to only as "Post Office Box 1142", its post office box number in Alexandria, Virginia.[37]

The directive establishing MIS-X indicated that two of its primary functions were to "coordinate provision of assistance and information pertaining to evasion of capture and escape methods to American Army personnel in theaters of operation" and to "supervise training of selected key personnel in accomplishing this purpose."[38] These specifically included the training of IOs who would serve as air crew "briefers", and the preparation, production and distribution of evasion and other emergency kits to American air crews.[39] Since the British had nearly three years experience in E&E matters at the time MIS-X was formed, their assistance was critical to the American effort, and they provided both training and supplies.

Following the establishment of MIS-X, an agreement was reached with MI9 to share resources and avoid unnecessary duplication. This agreement is notable because it laid the groundwork for the later, more publicized agreement between the British Special Operations Executive (SOE) and the American Office of Strategic Services (OSS). Under the agreement, made in May 1943, MI9 was given supreme authority in Europe, India and South East Asia, while MIS-X assumed responsibility for China, Indo-China, the South West Pacific Area, Japan, and Eastern Siberia. American liaison officers were attached to British units in areas of British authority, and British liaison officers were attached to American units in areas under American jurisdiction. Responsibilities for E&E on the African continent were shared by MI9 and MIS-X.[40]

Some of the first American Blood Chits for air crews operating in Africa appear to have been issued with the cooperation of the British. Several of these early American chits were printed on single sheets of paper.

One of these is shown in Fig. 5.13. The message in Arabic is given above the Great Seal. Below this is the equivalent message in English, and at the bottom the following useful words and phrases are given. While bearing the Great Seal of the United States and identifying the bearer as an American, this chit states that the bearer is "assisting the English Government", and that a reward will be given for taking the bearer to the English rather than to the Americans. The chit's serial number, 105, is written in the lower right corner in red indelible pencil.

Another single sheet Blood Chit, printed in black ink on plain paper by the U.S. Government Printing Office (GPO) in 1943, is shown in Fig. 5.14. It is very much like the chit illustrated in Fig. 5.13, with an Arabic message above the Great Seal, above an English message, above a list of useful words and phrases. This chit also specifies that the bearer is assisting the British, though the specific wording varies slightly. The useful words section is identical to the chit shown in Fig. 5.13 with the addition of one word. Although an American chit, it also identifies the bearer as "assisting the British Government". Moreover, it states that it is issued by a "permit" of the British. The notation "U.S. Government Printing Office: 1943 0-533957" appears at the bottom of the chit.

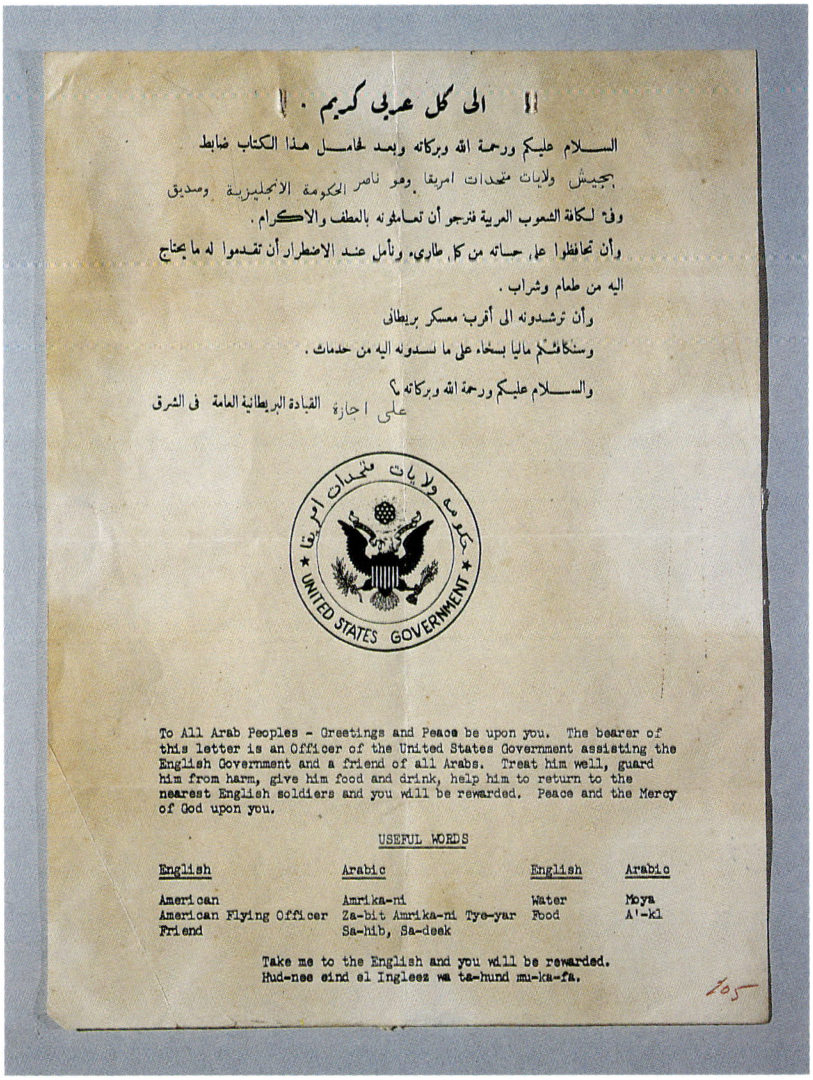
!! الى كل عربي كريم . !!

السلام عليكم ورحمة الله وبركاته وبعد لحامل هذا الكتاب ضابط بجيش ولايات متحدات امريقا. وهو ناصر الحكومة الانجليزية وصديق وفي لكافة الشعوب العربية فنرجو أن تعاملوه بالعطف والاكرام.

وأن تحافظوا على حياته من كل طارىء، ونأمل عند الاضطرار أن تقدموا له ما يحتاج اليه من طعام وشراب.

وأن ترشدوه الى أقرب معسكر بريطاني

وسنكافئكم ماليا بسخاء على ما تسدونه اليه من خدمات.

والسلام عليكم ورحمة الله وبركاته

على اجازة القيادة البريطانية العامة في الشرق

To All Arab Peoples - Greetings and Peace be upon you. The bearer of this letter is an Officer of the United States Government assisting the English Government and a friend of all Arabs. Treat him well, guard him from harm, give him food and drink, help him to return to the nearest English soldiers and you will be rewarded. Peace and the Mercy of God upon you.

USEFUL WORDS

English	Arabic	English	Arabic
American	Amrika-ni	Water	Moya
American Flying Officer	Za-bit Amrika-ni Tye-yar	Food	A'-kl
Friend	Sa-hib, Sa-deek		

Take me to the English and you will be rewarded.
Hud-nee eind el Ingleez wa ta-hund mu-ka-fa.

105

Fig. 5.13: Some chits were issued to American air crew members in cooperation with the British. One example is this paper chit, which indicates that the American bearer is assisting the English Government and requests that he be returned to English forces for a reward. A serial number, 105, is written on the chit in red pencil. (REB 1234) *Courtesy of Bob Gill*

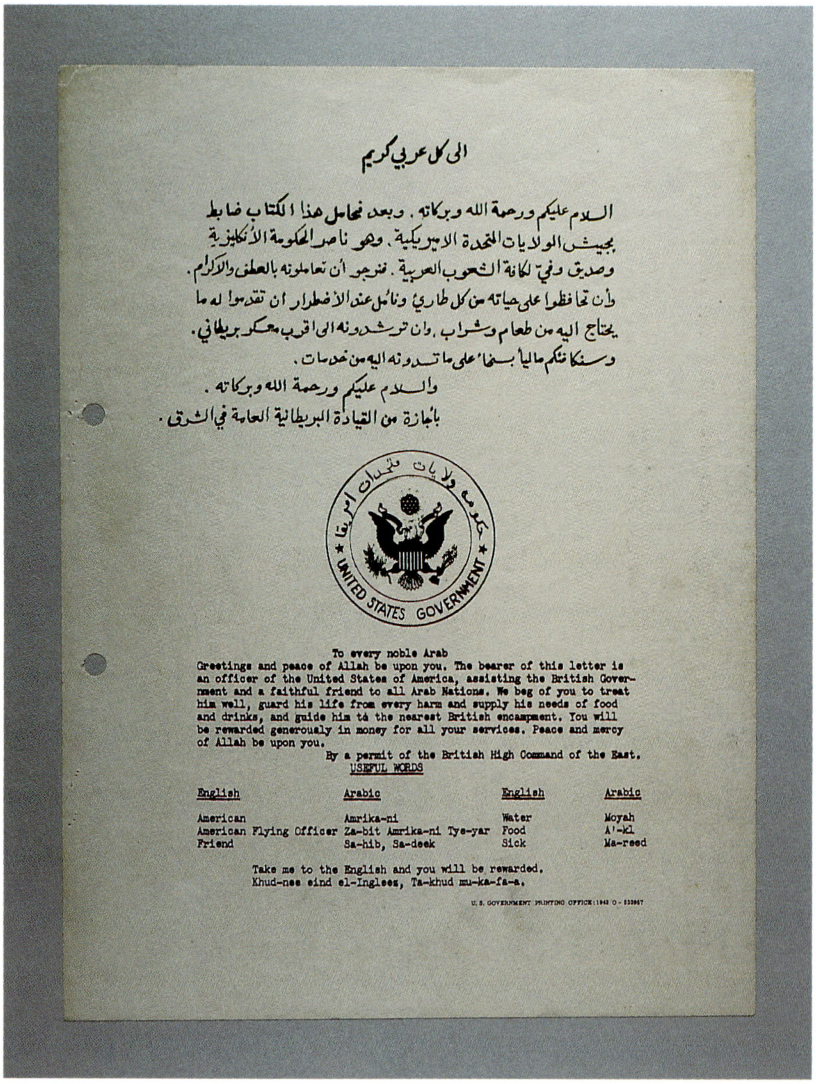

الى كل عربي كريم

السلام عليكم ورحمة الله وبركاته . وبعد فحامل هذا الكتاب ضابط
بجيش الولايات المتحدة الاميركية ، وهو ناصر الحكومة الأنكليزية
وصديق وفي لكافة الشعوب العربية . فنرجو أن تعاملونه بالعطف والاكرام .
وأن تحافظوا على حياته من كل طارئ ونأمل عند الاضطرار ان تقدموا له ما
يحتاج اليه من طعام وشراب . وان ترشدونه الى اقرب معسكر بريطاني .
وسنكافئكم مالياً بسخاء على ما تسدونه اليه من خدمات .
والسلام عليكم ورحمة الله وبركاته .
باجازة من القيادة البريطانية العامة في الشرق .

UNITED STATES GOVERNMENT

To every noble Arab

Greetings and peace of Allah be upon you. The bearer of this letter is an officer of the United States of America, assisting the British Government and a faithful friend to all Arab Nations. We beg of you to treat him well, guard his life from every harm and supply his needs of food and drinks, and guide him tá the nearest British encampment. You will be rewarded generously in money for all your services. Peace and mercy of Allah be upon you.

By a permit of the British High Command of the East.

USEFUL WORDS

English	Arabic	English	Arabic
American	Amrika-ni	Water	Moyah
American Flying Officer	Za-bit Amrika-ni Tye-yar	Food	A'-kl
Friend	Sa-hib, Sa-deek	Sick	Ma-reed

Take me to the English and you will be rewarded.
Khud-nee eind el-Ingleez, Ta-khud mu-ka-fa-a.

U.S. GOVERNMENT PRINTING OFFICE: 1943 O - 533957

Fig. 5.14: Another chit issued to American forces in cooperation with the British is this paper chit, printed in 1943 by the U.S. Government Printing Office. It promises a reward for returning the American airman, whom it says is assisting the British Government, to the nearest British encampment, and states that it is issued under a permit of the British. (REB 3210) *International Blood Chit Museum*

One style of American Blood Chit containing a stamp that dates it to one week before the invasion is referred to as a "passport chit". It was designed to impress any bureaucrat, regardless of nationality, who loved official-looking documents. The examples illustrated in Fig. 5.15 were issued to S/Sgt John B. Heinecke and T/Sgt. Joseph Donato. The chits are printed on cardstock and folded once so that they open like a book. The top chit, opened to show the front and back of the outside, was folded twice more by the owner, S/Sgt. Heinecke, making it easier to carry. On the cover is the Great Seal of the United States above the American flag, and at the bottom are lines for the airman to enter his name and rank. Sgt. Heinecke has also entered his serial number. A list of useful words and phrases in English with their phonetic Arabic equivalents is printed on the back page.

T/Sgt. Joseph Donato, a radio operator/gunner on a Martin B-26 *Marauder* of the 37 Bomb Squadron (Medium), 17th Bomb Group, was issued the chit shown at the bottom of the photo in March 1943 while based at Telegrma Air Base near Constantine, Algeria. It is opened to show the inside of the chit. The message is given in Arabic, English, and French. The Great Seal of the United States and a stamp reading "Official U.S. Army, October 1st, 1942" are stamped in red ink, giving the chit the appearance of a passport.

It is possible that these Passport Chits were printed by the British, or that they were designed with British cooperation, because as of 1 October 1942, MIS-X had not yet been established.

Donato indicated that "the chit . . . was carried in a pocket of either my A-2 jacket or my coveralls, that we wore over everything else. After the African campaign was over in May of 1943, I no longer kept my chit for flying purposes, as our targets were over Italian areas and I could speak the language."[41]

Operation Torch, the invasion of North Africa by British and American Troops, began on 8 October 1942, the same month that Col. Winfrey took command of MIS-X. The largest amphibious operation yet in the war, the invasion had been opposed by the American Joint Chiefs of Staff. They were overruled by the President who, for a variety of reasons including urging from the British, wanted American forces in action. While actually a joint operation, it was made as American as possible to reduce the vicious antagonism among the English, the Vichy French and the Free French. Nonetheless, the invasion created bitter and long-lasting ill-will between the British and French, and prompted the Germans to occupy the rest of France and to add some 250,000 soldiers to their forces in Africa.

The North African landings featured a rarity for the war against the Germans: the use of American Navy carrier-based aircraft. The evasion and survival needs of the Navy pilots, many of whom who were to experience their first combat, were not forgotten and a "safe area" for emergency landings was designated where they might obtain help.

Later in the African campaign, the United States produced a booklet of chits illustrated in Fig. 5.16. Clearly influenced by earlier British chits, the booklet contains sections on conduct when meeting Arabs in the desert, information on how Arabs reckon time and distance, and useful words. This is nearly identical to the information given on the British chit illustrated in Fig. 5.11. On the cover of the booklet is the Great Seal of the United States. The booklet contains chits and useful words in the four languages used in the western desert: Arabic, French, Spanish and Portuguese. The Arabic chit and its translation are illustrated in Fig. 5.17. The translation provided for the French, Spanish and Portuguese chits is, "The bearer of this letter is an officer of the United States of America and

Fig. 5.15: These cardstock "passport chits" were issued to S/Sgt John B. Heinecke (top) and T/Sgt Joseph Donato (bottom). On the cover is a place for the owner to write his name and rank. A message from President Franklin D. Roosevelt is given in English, Arabic and French on the inside, and on the back cover is a list of useful words and phrases. Two stamps appear in red, the Great Seal of the United States and a smaller one reading, "Official U.S. Army, October 1st, 1942." These stamps give it a more formal and official appearance than some of the other chits used in this theater. (REB 3212) *Top: International Blood Chit Museum; Bottom: Courtesy of Joe Donato, 37th B.S., 17th B.G. (B-26 Marauders)*

is a friend of the French (Spanish, Portuguese) people. Please treat him well, supply him with necessary food and drink, and guide him to the nearest American encampment. You will be generously rewarded for all your services."

North Africa served as a stopping point for a fleet of aircraft bound for the Soviet Union, Burma, India and later the Pacific. The Air Transport Command ferry crews faced some of the same risks as did combat crews, and they were issued aids to assist them should they be forced down. Some of the materials available to Americans who ferried aircraft along a route which ran from the United States to South America, across the South Atlantic from Brazil to Central Africa, and then turned north and east, are illustrated in Fig. 5.18.

This material included booklets such as the language guide, *TM 30-301 Portuguese as spoken in Brazil* and the *Pocket Guide to North Africa* shown here. Pocket Guides to various areas and individual countries gave servicemen a basic understanding of the geography, peoples and cultures they would encounter while on duty in those locations, and also included useful phrases in phonetic versions of local languages. The small *Pocket Guide of Rio de Janeiro and List of Useful Phrases* was published by the U.S.A. War Emergency Committee in Rio, and distributed free to servicemen at the U.S. Servicemen's Center there. It contains maps of Rio, tourist information, and 23 pages of phrases in Portuguese as spoken in Brazil.

Special banknotes with blue serial numbers and yellow seals, detailed in Fig. 5.19, were used by American forces in North Africa in an effort to foil any German attempts at forgery, and so that they could be

Fig. 5.16: A booklet of chits and useful words in Arabic, French, Spanish and Portuguese, the four major languages used in the western desert, was provided to American air crew members. Some of its contents were copied almost verbatim from the British chit shown in Fig. 5.11. The cover, featuring the Great Seal of the United States, is shown here. (REB 3214) *Courtesy of John A. Scott*

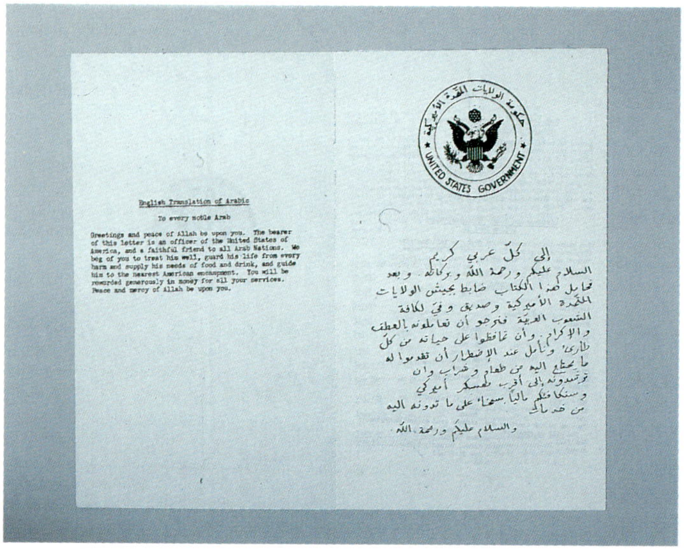

Fig. 5.17: Shown here is the Arabic chit that is contained in the booklet of chits illustrated in Fig. 5.14. The Great Seal appears above this chit, but not above the French, Spanish or Portuguese chits in the booklet. *Courtesy of John A. Scott*

In order to avoid flying over Saudi Arabia on a direct, shorter route, the route led over Khartoum to Cairo, around the top of Arabia and then down to Abadan. At that time flying over Saudi Arabian territory was forbidden. They did not permit it and in case of forced landing, the safety of "infidel" airmen was in great danger from the natives. We were warned of the possible dire consequences of a forced landing in Arabia and of falling into the hands of fanatical natives, hence the chit.[43]

At the top of the chit is printed "Army Air Forces, Headquarters Air Transport Command, Washington, D.C., U.S.A." Beneath this is an Arabic text above the Great Seal, which is offset to the right side of the chit, and beneath this is the English message. The useful words and phrases section appears below the English text. At the bottom is are the notations "U.S. Government Printing Office: 1944 0-576399" and "3-7730, AF".

quickly demonetized if a substantial amount of them fell into enemy hands. Known as "gold seal" certificates, these special banknotes could be used to reward helpers in North Africa, and later in the war were included in evasion kits issued to American crews flying missions over Yugoslavia and eastern Europe from Italy. Three denominations of these notes are shown here.

The cloth map shown is the only one produced by the United States for use in North Africa during World War II. It is printed on a cotton fabric known as "balloon cloth" and is simply a copy of the Road Map of West Africa produced by the French company Michelin, recognized then, as today, as one of the world's premier makers of maps for travellers. One of the earliest American cloth maps, it appears in several illustrations in the June 1943 edition of *National Geographic Magazine*.[42]

The paper Blood Chit, detailed in Fig. 5.20, was issued in 1944 to William F. Beck, a pilot assigned to the Air Transport Command's 20th Ferrying Group based at Nashville, Tennessee. Beck ferried A-20 light bombers from the United States to Abadan, Iran, for Lend-Lease delivery to the Soviet Union, and ferried North American B-25s to American units in the Middle East. The chit was printed on paper in brown ink by the U.S. Government Printing Office (GPO) in 1944, and although similar to two other paper chits (Figs. 5.13 and 5.14) it does not state that the American aviator bearing the chit is assisting the British. Beck recalls:

Fig. 5.18: Shown here are some of the materials available to Americans who ferried aircraft from the United States to Africa via South America. These include language guides, guides to specific countries and areas, a Blood Chit, a cloth map, and special "gold seal" currency. *International Blood Chit Museum*

Fig. 5.19: Three denominations of the special "gold seal" bank notes that were used in North Africa are shown here. These notes featured yellow seals and blue serial numbers, and they could be quickly demonetized if a substantial amount of them fell into enemy hands. These notes could be used to reward helpers in North Africa, and later were included in evasion kits issued to airmen flying missions over Yugoslavia from Italy. *International Blood Chit Museum*

It is impossible to verify the text without an example of the vest, however, it is doubtful that a $10,000 reward was ever offered during World War II. If true, the reward must have been offered by the airline or another private party, rather than the government or the military. Rumors that fabulous rewards were offered for returned air crewmen appear to have been common during the war, and persist today in spite of the fact that the official documents give the nominal reward amount for North Africa as US $100. Refer to Appendix III for more information concerning the payment of rewards.

Royal Air Force flyers who successfully returned to friendly control became members of the *Late Arrivals Club*. They were entitled to wear a special insignia officially known as the "Winged Foot", designed by Wing Commander George W. Houghton, the Senior RAF Public Relations Officer in the Middle East. The badge was first issued in June 1941 in the western desert. It is reported

A small group of American civilian pilots also received an item that may have been a chit. Max Welton-Stell, a retired Pan American World Airways Captain, recalls:

> When our company negotiated a ferry contract with the Air Force Division of the U.S. Army, I volunteered, for extra pay, of course, to ferry military aircraft to the Russians in Tehran, Iran; the British in Eritrea and Egypt; and the Dutch in Ceylon, now Sri Lanka.
>
> We took delivery from factory pilots from Lockheed, Douglas and North American factories respectively, at Miami, and by way of Brazil, the west coast of Africa, Chad, Khartoum and Sudan to the final destinations, which always required flights across Arab nations.
>
> We were advised that all non-Moslems were considered "infidels" by those natives and that a forced landing in the interior areas might create complications for us. Each captain was given a vest of cheap material, cheesecloth as I recall, dyed red, with yellow lettering in Arabic. The gist of this message was "Take these airmen safely to the nearest United States consulate or embassy office and you will be paid $10,000 in gold."
>
> To my knowledge, none of these offers were put to the test.[44]

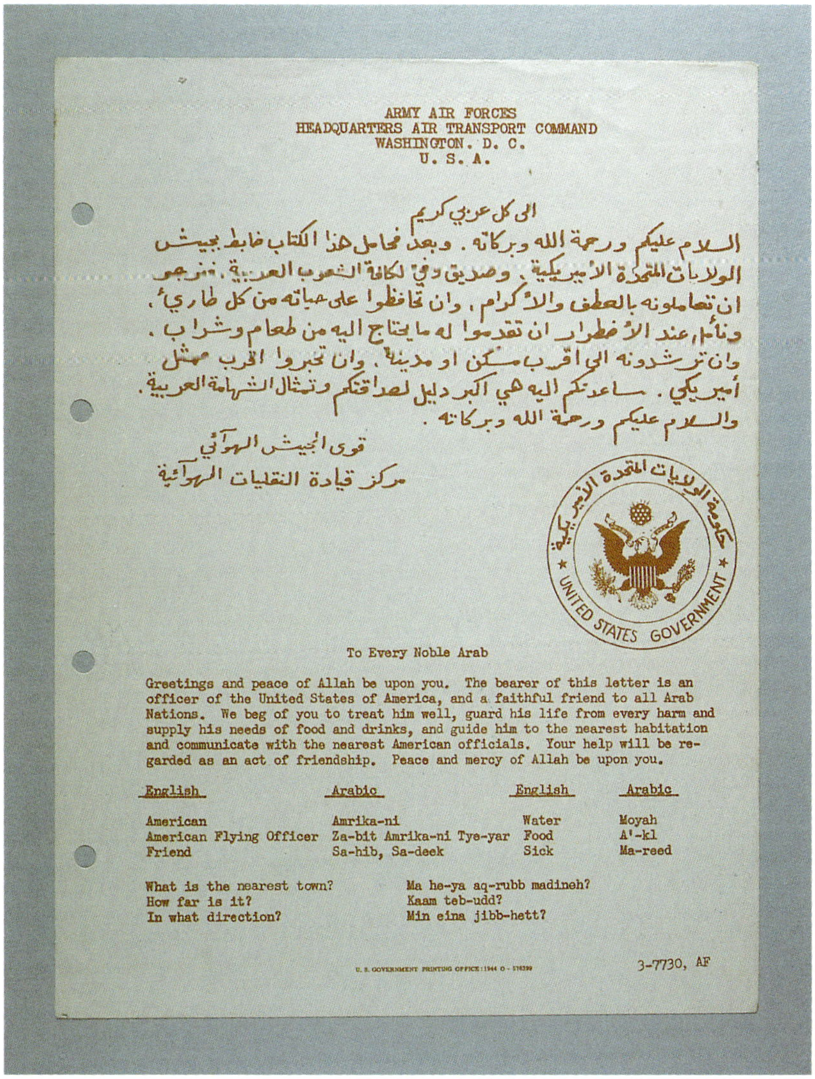

ARMY AIR FORCES
HEADQUARTERS AIR TRANSPORT COMMAND
WASHINGTON. D. C.
U. S. A.

الى كل عربي كريم
السلام عليكم ورحمة الله وبركاته . وبعد فحامل هذا الكتاب ضابط بجيش
الولايات المتحدة الأميركية ، وصديق وفي لكافة الشعوب العربية . نرجو
ان تعاملونه بالعطف والاكرام ، وان تحافظوا على حياته من كل طاريء ،
ونأمل عند الاضطرار ان تقدموا له ما يحتاج اليه من طعام وشراب ،
وان ترشدونه الى اقرب مسكن او مدينة ، وان تخبروا اقرب ممثل
أميركي . مساعدتكم اليه هي اكبر دليل لصداقتكم وتمثال الشهامة العربية .
والسلام عليكم ورحمة الله وبركاته .

قوى الجيش الهوائي
مركز قيادة النقليات الهوائية

حكومة الولايات المتحدة الأميركية
UNITED STATES GOVERNMENT

To Every Noble Arab

Greetings and peace of Allah be upon you. The bearer of this letter is an officer of the United States of America, and a faithful friend to all Arab Nations. We beg of you to treat him well, guard his life from every harm and supply his needs of food and drinks, and guide him to the nearest habitation and communicate with the nearest American officials. Your help will be regarded as an act of friendship. Peace and mercy of Allah be upon you.

English	Arabic	English	Arabic
American	Amrika-ni	Water	Moyah
American Flying Officer	Za-bit Amrika-ni Tye-yar	Food	A'-kl
Friend	Sa-hib, Sa-deek	Sick	Ma-reed

What is the nearest town? — Ma he-ya aq-rubb madineh?
How far is it? — Kaam teb-udd?
In what direction? — Min eina jibb-hett?

U. S. GOVERNMENT PRINTING OFFICE : 1944 O - 578399 3-7730, AF

Fig. 5.20: This paper Blood Chit, printed in 1944 by the U.S. Government Printing Office, belonged to William F. Beck, a pilot with the 20th Ferrying Group. It contains messages in Arabic and English together with the Great Seal, and was issued by Headquarters, Air Transport Command. (REB 3216) *International Blood Chit Museum, a gift of William F. Beck*

that the club existed only in the Middle East, and that there were no American members.[45] Other clubs also existed, such as the *Caterpillar Club* sponsored by several parachute manufacturers for men who parachuted from aircraft, the *Sea Squatters Club*, and the *Goldfish Club* for airmen who went down at sea. Illustrated in Fig. 5.21 are examples the insignia for the *Late Arrivals Club* and the *Goldfish Club*.

Over the Mediterranean, the Atlas Mountains, and the North African desert, the Allies learned a lesson in the application of airpower that they would improve upon during the remainder of the war. The Americans and British also learned valuable lessons about air crew survival and evasion that they would apply in other theaters, regardless of climate or geography. In the conflict on the European continent these lessons, and others, would be put to good use.

Fig. 5.21: British flyers who successfully returned to Allied lines became members of exclusive clubs, such as the *Late Arrivals Club* for those who walked out, and the *Goldfish Club* for those who were downed at sea. Shown here are the insignia that members of these two clubs were entitled to wear on their uniforms. On the left is a wire bullion badge officially known as a "Winged Foot", first awarded in the western desert in June 1941, and to its right is a cloth Flying Goldfish badge. The two waves beneath the goldfish are said to indicate that their owner twice returned from being downed at sea. *International Blood Chit Museum*

CHAPTER 6

Steadfast Friends and Dangerous Allies World War II in Europe

Shortly after Major General Carl Spaatz's introduction to MI9 in February 1942, an American staff branch known as P/W&X was established, commanded by Lieutenant Colonel W. Stull Holt, a former World War I pilot. Holt served as the 8th Air Force's liaison with MI9, and P/W&X remained under command of the European Theater of Operations (ETO) Headquarters for the duration of the war. P/W&X received evasion kits and other aids from MI9 for distribution to American crews, and supervised the training and debriefing of American personnel in the ETO. Another of Holt's duties was to cooperate with the American counterpart of MI9, the secret organization known as MIS-X.[1]

Pre-capture evasion aids in the European theater, including Blood Chits, Evasion Maps, Phrase Cards, and Escape Compasses, were produced in Britain and provided to air crews through regional E&E organizations. The organization responsible for western Europe was IS9(WEA). In August 1944, the central Mediterranean area plus Albania and Yugoslavia became a separate area under a new organization known as IS9(CMF) headed by Royal Navy Commander Rodd. He and Major Philip Holder, an American who had served in P/W&X, controlled most of its operations from Bari, Italy. The remainder of A-Force, under Colonel A.C. Simonds, became IS9(ME) and was responsible for the Aegean, the Balkans, and the remainder of the Middle East.[2] WEA was the acronym for Western European Area, CMF was that for Central Mediterranean Forces, and ME was the acronym for Middle East.

The air war in Europe was the ultimate testing ground for theories of airpower and tactics, for each side's leadership, and for new weapons. It also tested the courage, endurance, and resourcefulness of the participants. For the pilots and crews of Allied aircraft, the parachute jump from their fatally stricken aircraft was also a leap into the unknown. Crewmen could be killed or injured while under their parachute's canopy or on landing. Their fate once safely on the ground in occupied Europe was equally uncertain.

Despite being a signatory to the Geneva Conventions on the treatment of prisoners of war, the Nazis introduced a political element to a prisoner's fate. Although rarely enforced, Hitler ordered in May 1944 that Allied air crews who attacked civilians or downed German airmen should be summarily executed as Terrorfliegers (Terror Fliers).[3] This category included all bomber crews since bombs often fell in civilian areas whether or not they were actually targeted. While most captured Allied personnel faced the traditional fates of any prisoner, for at least 82 unlucky Jewish and gentile American airmen and for 86 other Allied flyers, capture meant not a German Luftstalag but shipment to Buchenwald, a death camp, where they became victims of the Nazi's genocide against Jews, Gypsies, ethnic Slavs, and others.

The development and fielding of evasion aids, the training and briefing of crews, and the relationships with local resistance groups were important factual and psychological assists to crews facing the statistical probability of death or capture before completing their required number of combat missions.

Allied air crews were instructed clearly and directly that to benefit from their evasion aids, they had to be carried on missions. A British document noted that "The aids and devices are to be carried on every operational flight. They can be of no use to you if they are in your tent."[4] Americans were further advised, "Moreover, they should be securely hidden on your person, not just put into the aircraft. You might have to get out in a hurry and have no time to check whether you left anything behind. Guard against loss by sewing into inner pocket or putting zipper on knee pocket of coveralls."[5]

Airmen were briefed that once on the ground in occupied Europe, they did have a reasonable chance of returning to friendly control:

> If any AAF men land on the Continent unhurt, they have an even chance of returning to base without being captured. Statistics prove that this is true. 90% of the population bitterly hate the conqueror, and of these, 80% will help the evaders. Successful evasion depends to a large extent on initiative, coupled with a knowledge of evasion tactics, plus a certain amount of common sense.[6]

Help from the many resistance groups was the key to the successful repatriation of thousands of Allied flyers who went down in occupied Europe. American airmen were advised:

> There are many resistance organizations in all occupied countries, but they are highly secret groups because mere membership in such an organization liables [*sic*] the member to a firing squad . . . Once you have been taken in by the organization, you have nothing to do but obey orders, regardless of how strange they may seem to be. Everything will be arranged for you for your return journey.[7]

Resistance and guerilla organizations were not only active in western Europe, where highly organized evasion networks to whisk Allied air crews to freedom were given names such as "The Comet", they were also active in Yugoslavia and the Balkan nations. This was so in spite of bitter conflicts between nationalistic, ethnic, and religious groups such as the Croats and Serbs, between communist and non-communist factions such as the Partisans under Tito and the Chetniks under General Mihailovich, and in spite of local militias like the Ustashi that were allied with the Germans.

American crews were briefed regarding evasion in Yugoslavia that "the guerillas will be very friendly, once they have proof of evaders' identities, and will pass them on to one of the British officers who are working with them and who will be able to organize the rest of the journey."[8] The guerillas referred to are Tito's Partisans, with whom British agents were attached. The Chetniks were forced to ally themselves with the Germans because of American and British support of the Communists. In spite of this, the Chetniks assisted downed Allied airmen, subverted the Nazis, and continued to believe in the Allied cause in spite of what they considered their political betrayal by the west.

Allied crews in the ETO received stern advice that their helpers faced danger, including death, if the occupation forces should learn of their identities: "If a helper is caught, he is shot and his house burned . . . Never write down and have in your possession the names of Helpers. If you are caught and such names are found on you, it is an absolute certainty that your helpers will be shot as soon as the enemy can lay hands of them."[9]

American OSS teams and air rescue units were also heavily involved in returning Allied evaders. An OSS team went into the Balkan states in December 1943 and established escape routes among the Partisans, and the 15th AF established an Air Rescue Unit which returned some 1,600 Allied airmen to friendly control by the war's end.[10]

Although resistance groups aided downed airmen and the SOE and OSS provided liaison teams to the Partisans, the lack of knowledge of European languages used in areas of Allied operations com-

Fig. 6.1: Phrase Cards, such as that shown here with its slipcover, were issued to Allied air crews well before mid-1943. Communication with local people was accomplished by pointing to the phrases which appear in English and foreign languages, hence the instruction "Not To Be Produced In Public." The card is shown with a restricted booklet that accompanied it when collected, and a British evasion map printed on edible tissue paper. The booklet informs air crews of their rights and expected behavior if captured, and of tricks used by the enemy to obtain information. Also included is a paragraph of instructions for airmen stranded, detained, or interned in a neutral country. Unlike the Phrase Card and map, the booklet was not to be taken on missions. (REB 1910) *International Blood Chit Museum*

plicated evasion operations. Crews were briefed that "it is well to have memorized a number of stock phrases in order to express needs and for bluffing in emergencies."[11]

The British and Americans were not the only military forces to realize that downed air crews would encounter difficulties with language in territory controlled by their enemies. The Germans recognized this early in the war, and a compass and German/English Phrase Book and dictionary, the *Luftwaffen-Fachwürterbuch*, was carried by Luftwaffe crews during the Battle of Britain in the Summer of 1940.[12] A German evasion compass is illustrated in the Additional Photos section following Chapter 8.

Allied Intelligence Officers were urged to give specific examples during E&E briefings of how even a small amount of language ability could mean the difference between freedom and capture. One of these examples relates that an "escaper might have been successful in crossing the frontier from Italy into Switzerland if he had known a few Italian words. He entered a lavatory in the station marked 'SIGNORE' (Women) instead of the one marked 'SIGNORI' (Men) and the station guard promptly arrested him . . . Intelligence officers should emphasize the necessity of knowing a few words or phrases of the country over which their squadrons are operating. (See phrase cards: English, French, Spanish, German, Italian.)"[13]

In response to the need to know something of the local languages, Phrase Cards, referred to in the above quotation, were developed. These were accordion-fold paper or card-stock items with English phrases on the left side of a page and the corresponding phrase in another language on the right side. They were not organized specifically for questions and answers as were Pointie Talkies, but they were to be used in a like manner: by pointing to phrases.

As early as July 1943, a P/W&X memo listed "2500 phrase cards" among the evasion equipment issued during the preceding month.[14]

Phrase Cards were generally issued in a slipcover. A Phrase Card in French, German, Dutch, and Spanish is illustrated with its slipcover in Fig. 6.1 along with a restricted booklet that accompanied it when collected, and an evasion map of the Iberian Peninsula printed on edible tissue paper, known as a "Tissue Map". The booklet, printed in January 1943, informed crews of of their rights and expected behavior in the event of capture and imprisonment, and of tricks used by the enemy to obtain information from prisoners. Also included was one paragraph of instructions for airmen stranded, detained, or interned in a neutral country. The Phrase Card and map were intended to be carried on missions, unlike the booklet that warns "This Document Must Not be Taken Into a Plane or Aboard Ship".

Months before the 6 June 1944 invasion, Allied air units began a campaign to disrupt the French transportation systems in order to prevent German reinforcements from reaching the invasion beaches. These missions and the coming invasion prompted the inclusion of an increasing number of languages on Phrase Cards. In March 1944, three months prior to the invasion, it was reported:

> There will shortly be available for use of air crews, Phrase Books in most European languages. Undoubtedly many suggestions will be made that the lists should include phonetics. Before the books were printed, this question was carefully studied and it was decided to omit phonetics for the following reasons:
>
> 1. They are extremely difficult to pronounce naturally, especially in some languages such as SERBO-CROAT, RUMANIAN, etc.
> 2. The phonetics would require considerable more space and the books would consequently have to be much bulkier.
> 3. It was decided that it would be a boost for those using the books to point to the questions as it was considered impractical for potential users to learn in advance the pronunciation of the languages of the various countries they might have to pass through. For this reason, the books have been marked "Not to be produced in public."[15]

While the above quotation uses the term "phrase books", it clearly refers to Phrase Cards, which did not have phonetic spellings and were marked "Not to be produced in public". While actual Phrase Books did exist, they are not the items described above.

Among the many Phrase Cards produced is one printed on brown cardboard in red and purple ink in four languages: French, Norwegian, German and Spanish. The card is shown in Fig. 6.2 with its original slipcover. Another Phrase Card, printed in dark blue ink, contained fourteen languages: Albanian, Bulgarian, Croatian, Dutch, French, German, Greek, Italian, Norwegian, Polish, Rumanian, Russian, Serbo and Spanish. The front of this card is illustrated in Fig. 6.3.

To augment the information given on Phrase Cards, at least one British evasion map included language material. Map FGS-C/D of Eastern and Western Norway (not illustrated), contains map terms in Norwegian, Swedish, Finnish, and Russian, with English equivalents and a pronunciation glossary, in its border.

The months of pre-invasion interdiction missions culminated on the night of 5 June and the morning of 6 June 1944 when the

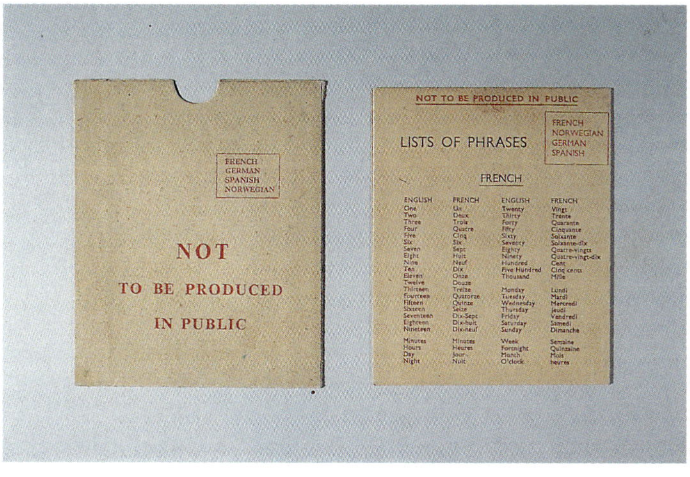

Fig. 6.2: Among the many varieties of Phrase Cards produced is this example in French, Norwegian, German and Spanish. It is printed on brown cardboard in red and purple ink, and is shown with its slipcover. (REB 1911) *Courtesy of John A. Scott*

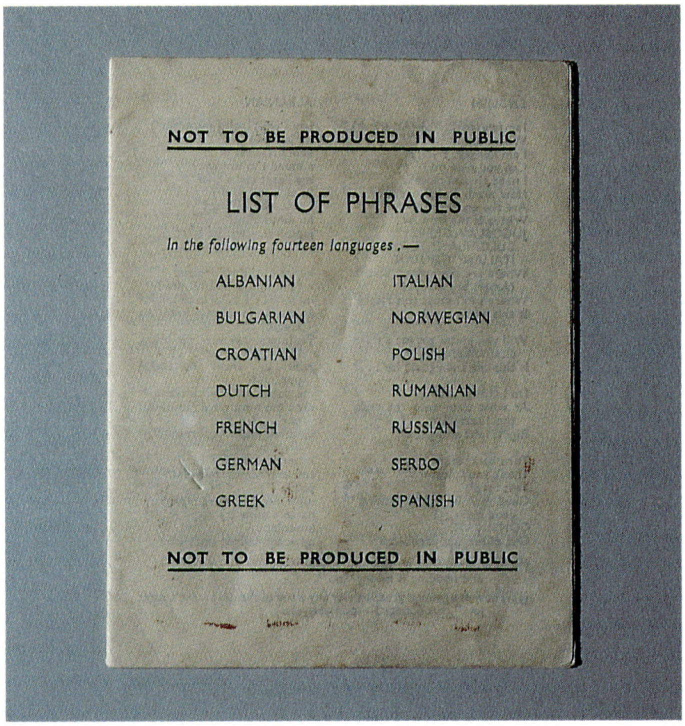

Fig. 6.3: Phrase Cards were made in many varieties and languages. This card is printed in fourteen languages: Albanian, Bulgarian, Croatian, Dutch, French, German, Greek, Italian, Norwegian, Polish, Rumanian, Russian, Serbo and Spanish. (REB 1912) *Courtesy of Bob Gill*

Fig. 6.5: Shown here are two types of brassards, or armbands, produced for the D-Day invasion. At the top is the British version, a plasticized flag sewn to burlap with snaps for closure, and below it is the U.S. version printed on oilcloth with pins for closure. *International Blood Chit Museum*

Allied air forces marshalled nearly 5500 fighters and fighter bombers, 1500 medium bombers, and 3500 heavy bombers to achieve air supremacy over the English Channel, the invasion beaches, and the forward areas as the landing troops moved inland.

Among the items issued to glider pilots immediately prior to D-Day were a Phrase Card in French, Dutch, German and Spanish, and the booklet *If You Should Be Captured*.[16] The booklet was printed in May 1944 specifically for the D-Day invasion and, in a departure from the norm for such publications, contained illustrations as well as text. Shown in Fig. 6.4 is one version of the D-Day Phrase Card, printed on thin blue paper in red and black ink, and two of the booklets, one open to show an illustration.

American and British troops participating in the D-Day invasion were also issued special identification aids bearing their respective national flags in the form of both sew-on items and armbands. Shown in Fig. 6.5 are the armbands, or brassards, produced for Overlord. The British version features a plasticized Union Jack sewn to a burlap strip with snap closures. The American version is made of a thin, rough-textured plastic-like material. The original pins issued for closure are attached to this item.

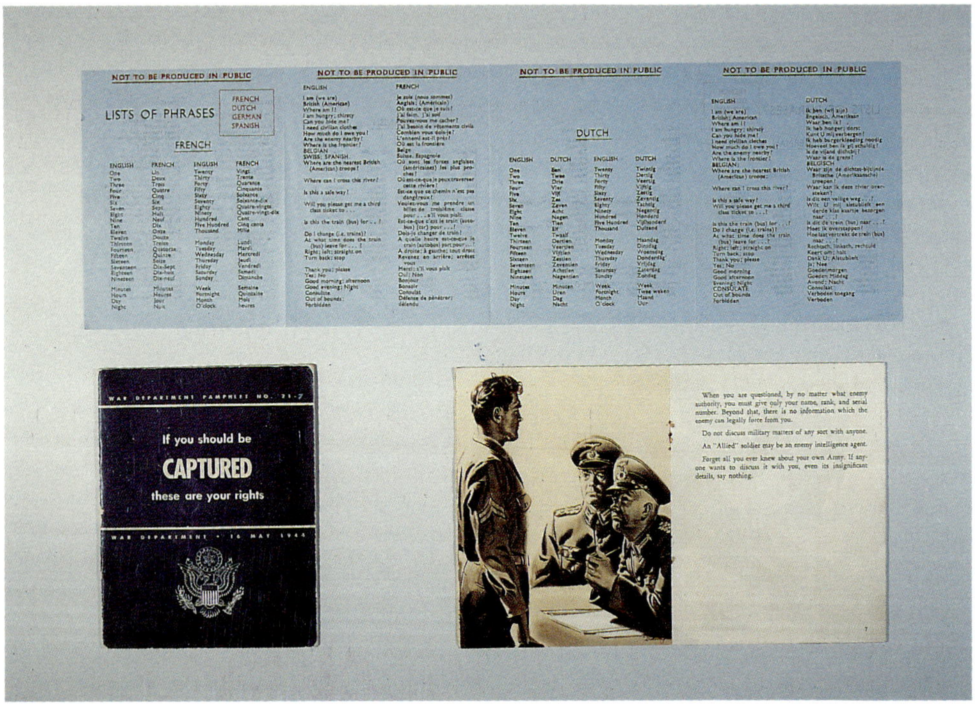

Fig. 6.4: Two items issued to air crews shortly before D-Day are the booklet "If You Should Be Captured" and a Phrase Card in French, Dutch, German and Spanish. The card, printed on very thin blue paper, would have been issued in a slip-cover, since lost. (REB 1910.1) *International Blood Chit Museum*

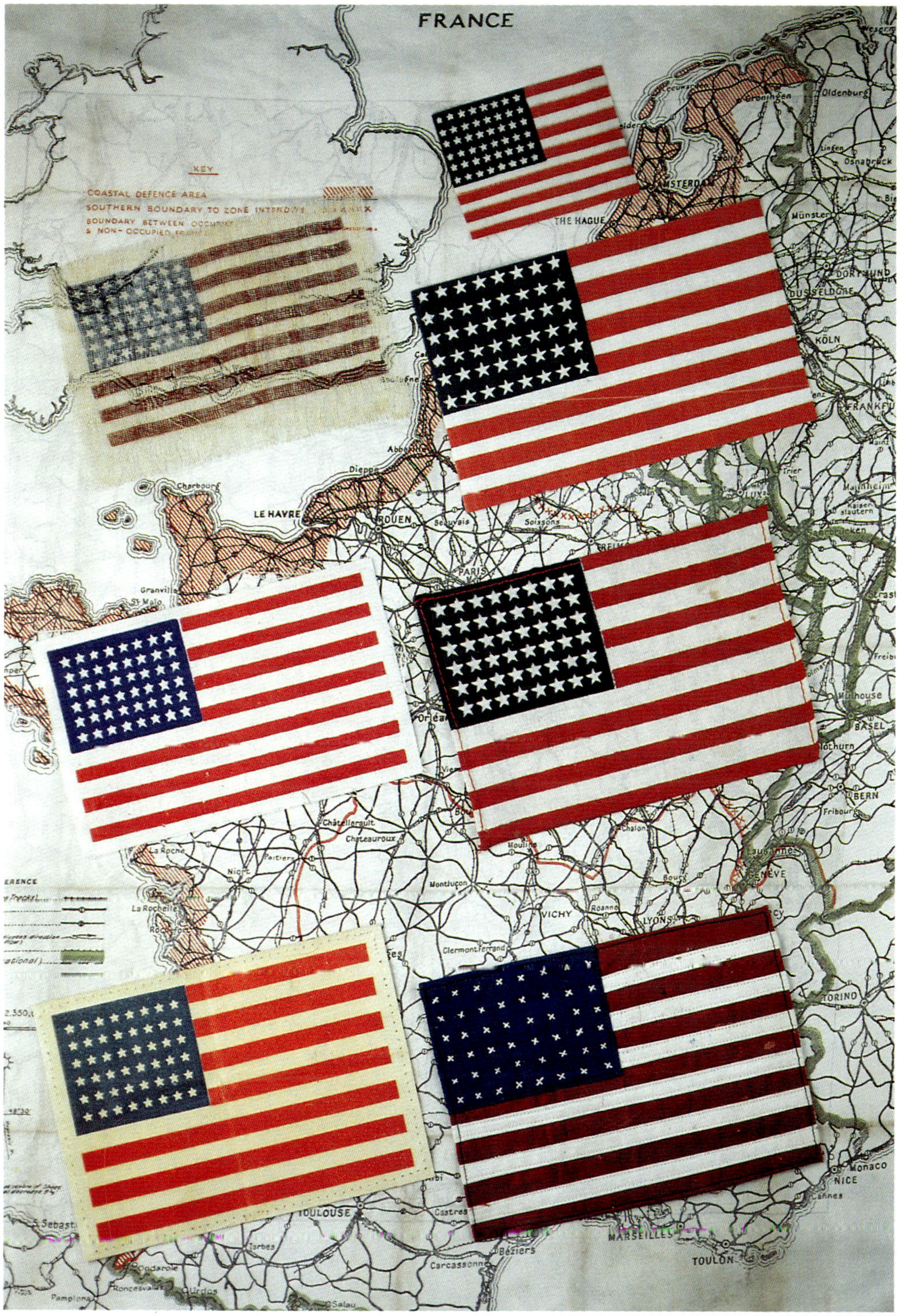

Fig. 6.6: Flags like these were sewn to uniforms for identification purposes during and following the Normandy invasion. The flag at the upper left is a printed on cheesecloth (gauze). Below it is a flag printed on cotton, with a white border around it. At the bottom of the left column is a flag which was apparently cut from an armband. It is yellowed and faded from wear. At the top of the right column are two unissued flags on linen which reportedly were printed for use following the invasion by the 5th Army. In the middle of the right column is a common U.S.-made cotton flag with sewn edges, reportedly removed from the back of a Paratrooper's jump jacket. At the bottom right is an English-made 8-piece appliqué flag. It was stitched on a sewing machine, and the stars, actually "X"s, were hand embroidered. The two items at the top of the left column are thought to be reproductions. *International Blood Chit Museum*

Illustrated in Fig. 6.6 are seven examples of sew-on flags used by American troops for identification purposes. These include items printed on linen, cheesecloth (gauze), and oilcloth.Fig. 6.7 shows British and American troops during Operation Overlord wearing some of the special identification armbands and sew-on flags.

In addition to Phrase Cards and identification materials, air crews were provided with a variety of pre-capture evasion and survival aids:

> This branch (E-Branch of M.I.9) furnished our crews and certain other operational personnel with sundry aids and devices designed to assist them in escaping or in making their way back to our own lines in the event they force land or are cut off behind enemy lines.
>
> These devices are usually in the following form:
>
> 1. An aid box containing concentrated food sufficient to sustain one for four or five days, a rubber water bottle, a compass, a book of matches, a roll of adhesive tape, a supply of benzedrine to help overcome excessive fatigue and a supply of halazone for purifying water.
> 2. A purse containing money for the country over which airmen operate, a map of the country, a compass and a hack-saw.
> 3. Compasses in various forms, rounds, wee's, swingers, bar compasses, trouser buttons, uniform buttons and pencil clips.
> 4. Blood chits.[17]

The Purse, mentioned in the above quotation, contained, as a rule, currency, a map or maps, a compass, and a hacksaw. It was called a "purse" because it contained currency. It was made of waterproof fabric and glued shut, to be opened only in the event of a forced landing or bail-out in occupied territory. Most Purses were marked with a band or bands of color in the upper left corner indicating the maps enclosed, a large letter denoting the purse's intended area of use and the currency enclosed, and a small letter in parentheses in the lower right corner. Some were also marked with

Fig. 6.7: British and American paratroops pause at a farmhouse near Le Muy in southern France on 15 August 1944. As an aid to identification of each nation's forces, the British paras wear armbands like that shown in Fig. 6.5 on their right arms, and the American airborne troops have flags like those shown in Fig. 6.6 sewn to their left sleeves. *Public domain photo, courtesy of Histoire et Collections, Paris*

Fig. 6.8: This Purse was for use in France as noted by the red "F" on the cover. The back of the purse is marked "Mark III" in red letters. It was known in official correspondence as a "red purse" in reference to the red stripe in the upper left corner. The white "43" over the red stripe indicates that it contains a 1943 edition map, sheet 43 C/D, so named because it was issued in 1943 and has sheet C on one side and sheet D on the other. The purse also contains a compass, and a hacksaw wrapped in stiff white paper. The meaning of the small "O" in parentheses in the lower right corner is not known. The term "purse" was used for this item because it contained currency. *International Blood Chit Museum*

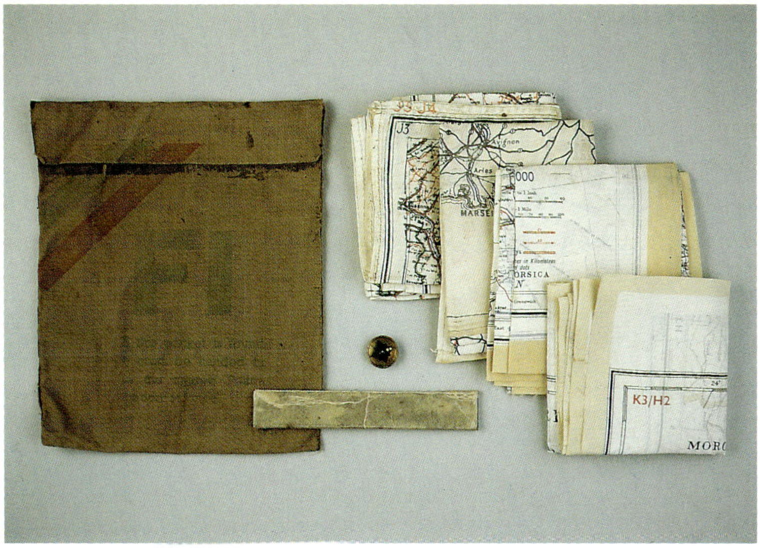

Fig. 6.9: The green letters "FI" on the cover of this Purse indicate that it was for use in France and Italy. The correct contents of the purse are in question, but it is shown with some of the maps with which it was collected. The maps shown are 9C(a)/9U covering France and the German-Swiss frontier, J3/J4 covering Northern and Southern Italy, D/H2 covering Southern France, Western Switzerland, Northwestern Italy, Spain and Portugal, and K3/H2 covering North Africa, Spain and Portugal. Bruce Magnuson, whose father flew from bases in North Africa, has a photo of a display in the unit's briefing room indicating the maps issued for France-Italy to be 9C(a)/9U, K3/H2, K1/K2, and C/D (not 43C/D). Two of those shown are identical to the maps identified by Magnuson, one shown contains sheet D but not C, and the remaining map shown was not listed on the board. Magnuson speculates that the contents of purses may have varied according to the time period during which missions were flown, a view supported by the *MIS-X Manual* which instructs Intelligence Officers to modify evasion kits as required for specific needs. The meaning of the small "F" in parentheses in the lower right corner is not known. The Purse is missing its currency, but the hacksaw enclosed in white paper and compass are shown. *International Blood Chit Museum*

a number across the colored bands denoting the year that the maps contained in the purse were produced.

In general, the Evasion Currency contained in Purses was local currency, so that Purses for France contained French Francs, those for Italy contained Italian Lire, and so forth. Two sources indicate that some evasion materials issued to American crews operating in the Balkans contained, variously, $48 [19] and $60 [20] in the blue serial number, gold seal U.S. banknotes also used in North Africa. It is rare to find a Purse containing its original currency today because at the war's end the currency was removed by E&E agencies, although collectors often complete Purses by adding the correct currency or scrip. One Purse in the U.S. Air Force Museum collection still contains its original 100-Franc notes.

Declassified documents refer to the colored stripes on the Purses as a means of identifying them. A 1943 memo indicates that "2250 Red Purses" and "750 Red/Green Purses" were issued during June 1943.[18] The Red Purse is that for

Fig. 6.10: The Maps Only "CE Purse" illustrated here contains map sheet 43 C/E in addition to a compass and hacksaw. The map sheets, indicated by the letters in the upper left corner of the Purse cover, were of a strip of land roughly from Brussels, Belgium, to Lublin, Poland. The map combining sheets 43 C and E was relatively uncommon, the more common sheets being 43 C/D and 43 E/F. No currency was included in Maps Only Purses. *International Blood Chit Museum*

Fig. 6.11: Official literature refers to this item as a "CDEF Purse" because it contains map sheets 43 C/D and 43 E/F in addition to the compass and a hacksaw. The maps covered most of Western and Eastern Europe and the Balkans. The CDEF Purse was issued in great numbers toward the end of the war. *International Blood Chit Museum*

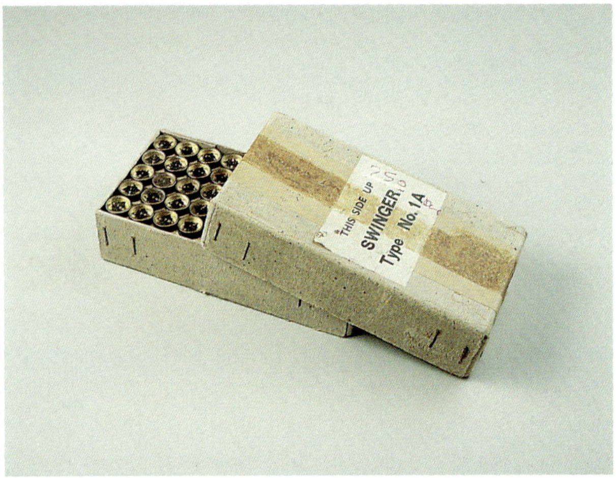

Fig. 6.12: A box of compasses as would have been provided to organizations assembling Evasion Purses is shown here. As the box cover indicates, the compass was known as "Swinger Type No. 1A." Each box contained 250 compasses, in five layers of fifty separated by sheets of cardboard. *International Blood Chit Museum*

France illustrated in Fig. 6.8, and the Red/Green Purse is that for France and Italy illustrated in Fig. 6.9.

Purses that did not contain currency were also distributed to air crews. These bore the words "Maps Only", even though they contained a compass and hacksaw in addition to maps. Some Purses, such as one obtained by the late John A. Scott from the estate of a Special Air Service (SAS) operative, were unmarked. Two varieties of "Maps Only" purses are illustrated in Figs. 6.10 and 6.11.

The saw contained in purses was, technically, a medical instrument known as a "gigli saw". It has teeth on both sides, has a hole in one end, and is tapered at the other end. The type of compass generally contained in purses was known as as a "Swinger Type No. 1A". Fig. 6.12 shows a box of these compasses as would have been provided to an organization assembling Purses.

At least one Purse, shown in Fig. 6.13, is of special interest because among its standard contents were two Phrase Cards. This was the CMF Pack, issued late in the war to air crews operating from bases in Italy. It was issued by IS9(CMF), the Allied E&E agency headquartered in Bari, Italy. The letters CMF are the acronym for Central Mediterranean Forces. Bruce Magnuson, whose father was an evader in Slovenia, reports that the CMF Purse contained U.S. banknote-sized scrip payable in Italian Lire. The two Phrase Cards issued in the CMF Pack are detailed in Fig. 6.14.

Some of the "gadget" compasses referred to in the above quote are illustrated in Fig. 6.15. These were often disguised as parts of the uniform and other everyday items so that they would go undetected by the enemy yet be readily available upon escape from a POW camp. Included are brass tunic buttons, a brass cap button, a blouse button, trouser fly buttons, a collar stud, a pencil clip, collar stays, and a belt buckle. Fig. 6.16 shows the four brass buttons opened to reveal the compasses within. Shown in Fig. 6.17 is the assembled buckle compass, the secret components of which were not visible when the buckle was attached to the belt. A "wee" compass (previously mentioned in a quotation) can be found in the Additional Photos section.

In addition to the issued aids, each air crew member was to have on his person small photos to be used if necessary in forging identity papers. The specifications for the photos were given as "Civilian clothes, correct size: (1.3 x 1.4) neutral gray background; should be full face, head and shoulders, no hat. Insist on having 3 copies along on every mission. Above all, keep clean and uncreased."[21] Three photos of this type are illustrated in the Additional Photos section.

Fig. 6.13: The Purse shown here contains two cloth maps, sheets 43 C/D and 43 E/F, a compass, and a hacksaw enclosed in a cellophane sleeve. The pack is missing its currency, reported to be scrip payable in Italian Lire. The words "CMF Packs" printed on the cover denote that it was for use by crews who operated from bases in Italy, where E&E operations were under IS9(CMF) headquartered in Bari. CMF is the acronym for Central Mediterranean Forces. The Purse, which was issued late in the war, is unusual because among its standard contains were two Phrase Cards, detailed in Fig. 6.14. *International Blood Chit Museum*

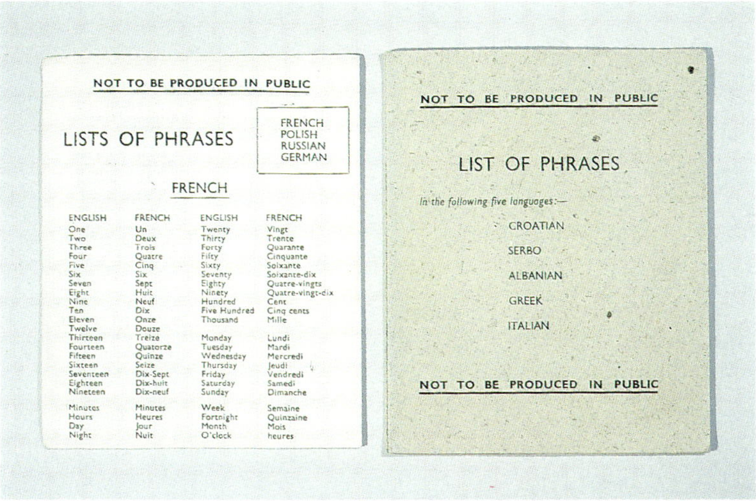

Fig. 6.14: The area of the CMF Pack's intended use can be deduced from the languages of its Phrase Cards, shown here. They are a white card in French, Polish, Russian, and German, and a green card in Croatian, Serbo, Albanian, Greek, and Italian. These cards are printed on simple although different papers, and in black ink only. This was perhaps an indication that the use of high-quality paper or cardstock, or the use of ink in several colors, was less important than mass-production for Phrase Cards produced late in the war. The white Phrase Card included in a CMF Pack collected by Walter Willett III was enclosed in a clear cellophane envelope. (Left: REB 1914, Right: REB 1915) *International Blood Chit Museum*

Fig. 6.15: Escape compasses were often disguised as everyday items or as parts of the uniform, so that the enemy would not discover them and they could be utilized upon escape from a POW camp. Those shown in the top row from left to right are an RAF tunic button, an RCAF tunic button, an RAF cap button, a U.S. tunic button, and an RAF collar stud. At the bottom, from left to right, are two collar stays above a pencil clip, an RAF buckle from the back of the Battle Dress tunic, two brass trouser fly buttons, a trouser fly button painted black, and an RAF blouse button. The blouse button has magnetic material embedded in the back, and could be suspended on a thread to find north. The collar stays could likewise be suspended from thread. One of the trouser fly buttons is a compass and the other is a needle upon which the compass would be balanced. The magnetized pencil clip was balanced upon the end of the sharpened pencil at the small dimple visible on its underside, and the long end pointed north. The collar stud had a cover over the glass to hide the compass within it, and was worn with the compass portion against the neck. The RAF tunic and cap buttons, plus the RCAF button, unscrew to reveal the compasses hidden within, while the American button is opened via a hinge on the back. Many of the compasses shown here were collected by Steve Griffith. *International Blood Chit Museum*

With knowledge from their early colonial experiences and from later operations in North Africa, the British developed a system of printed chits to be used by Allied airmen in the ETO. One report, written more than ten years following the close of World War II, has caused great confusion among contemporary researchers regarding the introduction of Blood Chits into the European theater. This document reads, in part:

Downed airmen in the Mediterranean Theater of Operation apparently employed an informal system of chits as a means of providing for the reimbursement of friendly nationals for help rendered. Downed airmen issued these chits on a voluntary basis, and, although the chits were eventually called 'blood chits', they differed from the more formal articles used in the Far East.

Fig. 6.16: Detailed here are the four escape compasses hidden within brass buttons shown in Fig. 6.15. They are, from left to right, the RAF cap button, the RCAF tunic button, the RAF tunic button, and the American tunic button. The compass in the RAF cap button is a black circle with a tiny white dot that points north. The American button compass is made by Hamilton Aero, and is so marked. The two RAF and RCAF tunic buttons have right-hand threads, while the RAF cap button has a left-hand thread. *International Blood Chit Museum*

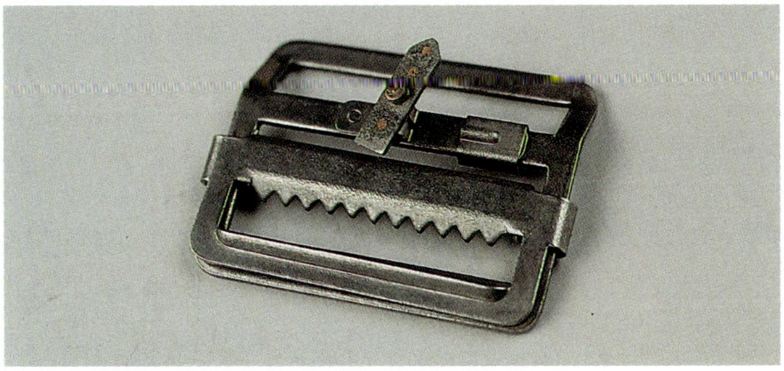

Fig. 6.17: The RAF tunic buckle contains a compass needle beneath a sliding sleeve. Moving the sleeve releases the needle and exposes a hinged pin on which the needle is then placed. The needle and pin were hidden beneath the cloth belt to which the buckle was sewn so that it would not be discovered by the enemy. The assembled compass is shown here. *International Blood Chit Museum*

Fig. 6.18: Among the many language-related evasion aids used in the European Theater were cloth Blood Chits in English and Russian, a combination ID card and phonetic Russian Phrase Card, a combination ID card and Blood Chit, and a small paper Blood Chit in eastern European languages. Armbands identical to those used during the Normandy invasion were also worn by some air crews flying into the Soviet Union to identify them as Americans. In addition to the items shown, Purses such as the CMF Pack contained Phrase Cards which could be used for more detailed communication after one's identity had been established through use of the Blood Chits and ID cards. *International Blood Chit Museum*

> An evasion report from an American shot down in Italy in September 1943 . . . states that helpers frequently asked the evaders to write down their names and serial numbers in order that the Italians could later present them to the British. This was in response to a radio promise by the British to reward all Italians who aided escaping PsW.[22]

The statement that the items referred to eventually developed into or became known as Blood Chits is most unfortunate. The items referred to are Promissory Notes, not Blood Chits, although the term "chit" was confusingly applied to both. Promissory notes were described as "Blank sheets of paper with the Coat of Arms printed at the top,"[23] and their use in the European theater was not as informal as implied in the post-war report quoted above. Detailed instructions were given to Allied air crews regarding how Promissory Notes were to be used, such as "An evader should write brief details of the help he has received from anyone whom he desires to reward, and should append to his signature his own name, rank and number only"[24], "Evaders may give chits to helpers and sign them with name, rank and number, stating the service performed and whether any payment was made to them out of the escaper's purse"[25], and "Give Chits to helpers signed with your name, rank and number only. Show the name of the helper and some indication of what has done for you and whether your have rewarded him in any way."[26] Promissory notes without the Royal Arms, simply blank pieces of paper, were included in evasion material issued to American personnel, and some of these are to be found in the collection of the United States Air Force Museum.

The first documented issue of actual Blood Chits to American air crews in the European Theater may be that in a memo from P/W&X to the I.O. of the 8th Air Force's 1st Bomb Wing that indicates that 1,160 Blood Chits were delivered to the Wing between 12 November and 18 December 1942:[27]

> The following equipment has been delivered to 1st Bomb Wing:
>
> 12/11/42- 350 Blood Chits (Receipt signed by Lt. Philip Haberman)
>
> 15/12/42- 350 Blood Chits (303rd Group) Signed by Lt. Claes E Johnson
>
> 15/12/42- 360 Blood Chits (305th Group) Signed by Capt Wm H Hitchcock
>
> 18/12/42- 100 Blood Chits (92nd Group) Signed by Capt Ira E Wright

The 1st Bomb Wing not only flew round trip missions over the continent but also participated in missions that originated from England, bombed German targets, and then flew on to bases in North Africa. It is likely the Blood Chits issued to the the 1st Bomb Wing crews in late 1942 were British chits for use in North Africa because, as documented in Chapter 5, Allied crews were routinely issued British chits for areas in Africa and the Middle East over which they operated.

It appears that the development of Blood Chits specifically designed for use in the European theater was a response to the severe problems encountered by evaders in Eastern Europe late in the war. These problems included airmen being shot at while in their parachutes, and being physically attacked once on the ground for lack of proper identification.[28]

There were a few isolated instances of Allied air units assisting the Soviets early in the war, including RAF No. 81 Squadron which went to Vaenga in northern Russia in July 1941, and the Free French 3rd Groupe de Chasse, that was sent in November 1942 from Lebanon to Ivanovo, 125 miles north-east of Moscow. There is no information to indicate that these groups used language aids.

An identification card was issued by the Soviets to one group of foreign airmen, indicating that they recognized the need for such an article by the foreigners in their midst. This card, issued in mid 1943 to French volunteers serving under Russian command, bore the airman's name and photo, and stated that he was a French Army officer and member of the 1st Fighter Air Regiment "Normandie" of the Red Air Force.[29]

Allied air crews were advised early in the war not to attempt to evade eastward unless absolutely necessary because of the many difficulties that would be encountered. American air crews were cautioned that it was "dangerous to cross battle lines, since Russians shoot first, and ask questions afterwards."[30] As the war continued, however, eastward evasions increased, and eventually these accounted for approximately 14% of all successful evasions in the European theater.

On 17 May 1944 the General Staff of the Soviet Union issued instructions to its troops to prepare to assist Allied airmen. This was in anticipation of Operation Frantic, the result of high-level marathon negotiating sessions between the Americans, English, and Russians to allow Allied aircraft to "shuttle bomb" targets in central and eastern Germany and Poland from bases in England and Italy. The aircraft would bomb their targets, fly on to Soviet terri-

Fig. 6.19: Identification cards containing phrases in phonetic Russian were among the first language aids developed for use in Operation Frantic, the shuttle bombing missions wherein Allied bombers bombed targets in Germany, refueled and re-armed at Russian airfields, and bombed a second set of targets when returning to their bases. Instructions were given to crews prior to the Frantic missions that the airman's name, rank and serial number were to be entered above the line on the card. The card shown was never issued. (REB 1236) *International Blood Chit Museum*

tory to be rearmed and refueled, and then bomb a second set of targets on the way back to their bases.

The Frantic missions began on 2 June 1944 when 104 B-17s and 64 P-51s from the 15th Air Force left their bases in Italy, bombed a rail yard in Hungary, and landed at an unpaved strip at Poltava in the Ukraine. They flew two missions from their Ukrainian base before returning to Italy.

Freed from pre-invasion tactical missions, the 8th Air Force flew its first Frantic mission on 21-22 June 1944. Flying from England, 114 B-17s and 71 P-51 fighter escorts bombed a German synthetic oil plant south of Berlin and flew eastward to bases at Mirgorod and Piryatin which, like Poltvava, were in the Soviet Ukraine near Kiev. Unknown to either the Russians or Americans, a Heinkel He 177 trailed the 8th Air Force formation to Poltava. That night the Lutfwaffe attacked the base with more than 110 tons of bombs and destroyed 43 B-17s, 15 P-51s, a number of Russian aircraft, and 450,000 gallons of aviation fuel. The surprise attack undoubtedly increased the Soviet's distrust of their Allies, and may have emphasized the need for language and identity aids for Allied airmen who might be downed in the midst of such distrust.

The Frantic operations and other missions flown over Soviet controlled territory presented grave difficulties for downed airmen seeking to evade capture by moving eastward. Blood Chits and other language aids developed for use in the Soviet Union and eastern Europe were an attempt by the Allies to mitigate the special dangers faced by Allied crews operating in these areas. Among these items, some of which are illustrated in Fig. 6.18, are chits in English and Russian, a combination ID card and Russian Phrase Card, a two-sided combination ID card and Blood Chit, and a small two-sided paper Blood Chit in eastern European languages. Phrase Cards, like those contained in the CMF Pack, augmented the Blood Chits. In addition, armbands of the type used by American troops in the Normandy invasion were issued to and worn by American air crews flying into the Soviet Union as an aid to their identification.

There was no "complete set" of these items as they were not developed at the same time, and they were issued singly and in various combinations by different units depending on availability and other factors.

One of these aids, shown in Fig. 6.19, is a combination ID card and a Russian language phonetic Phrase Card. The card seems to have been specifically developed for the Frantic missions and is listed along with other aids to be taken on these missions: "Before all take-offs each crew member should satisfy himself that he has on his person all necessary emergency equipment including . . . Escape kit and purse . . . Identification card printed in Russian, English and phonetics which will be provided to all personnel prior to departure."[31]

OPPOSITE: Fig. 6.20: Two-sided chits with the Union Jack and a message in Russian and English were issued to British air crews operating over Soviet controlled areas. They are made of two pieces of rayon sewn together back-to-back. The side with the large flag was to be held high while the bearer shouted "Ya Anglichahnin!" (I am English) when encountering Red Army troops. Detailed instructions for use are given in English on the back side, and were extremely important as Soviet troops were known to be trigger-happy. The chit on the left is sewn with a blanket stitch, while that on the right is sewn with straight stitching. The notation "F.R.1" appears on the bottom of the back side of one chit the chit. (Left: REB 1238, Right: REB 1238.1) *International Blood Chit Museum*

The card is also referred to in intelligence materials presented by briefers to air crews prior to their departure on Frantic missions:

> Caution crews to make sure they have escape kit and purse, stout G.I. shoes (extra socks recommended), knife, first aid pouch, toothbrush, soap. Keep identification card on person, also language cards.
>
> Crews should be familiar in a general way with the front lines.
>
> Carry identification card in safe pocket. Each crew member will print NAME, RANK AND SERIAL NUMBER on face of card above line at top. Try and memorize some phrases of Russian.
>
> If down in zone of operations of Red Army do not arouse suspicion of Red Army troops by any overt action, do no attempt concealment, do NOT bear arms in your hands, raise your hands on the approach of Red Army troops, indicate or display your identification card.
>
> In all cases, safeguard your identification card for proof to Soviet authorities that you are a bona fide American airman."[32]

The "identification card" referred to above is the combination ID card and Russian language Phrase Card, and the "language cards" referred to above are Phrase Cards. Although instructions directed airmen to fill in their name, rank and serial number, in practice these instructions were not always followed to the letter.

Jack Ilfrey was involved in Frantic missions, and writes, "On September 11, 1944, the 20th Fighter Group took off on Frantic VI – Shuttle Mission from England to Russia-Italy and back. We wore American flags on our arms and carried a card in English and Russian with our picture on the back."[33] The card Ilfrey refers to is the combination ID and phonetic Russian Phrase Card. As an additional aid to identification, photos were added to the back of the ID/Phrase Cards carried by the 20th Fighter Group pilots, although official instructions required only that the bearer's name, rank and serial number be filled out on the front of the card. The American flags referred to by Ilfrey were the invasion-type armbands.

OPPOSITE: Fig. 6.21: The American version of the two-sided rayon chit for use on missions over Soviet territory are nearly identical to the British version, changes being the use of the phrase "Ya Amerikanets" (I am American) and the American flag. There is no reward offered for assistance on these chits, but they are referred to in declassified ETOUSA documents as "Russian Blood Chits." Like the British versions, these chits are sewn with two types of stitching. The small notation "F.R.2" is printed at the bottom of the back of the chit with straight stitching, and implies that it was produced either in conjunction with, or later than, the British version which bears the notation "F.R.1." (Left: 3218, Right: 3218.1) *International Blood Chit Museum*

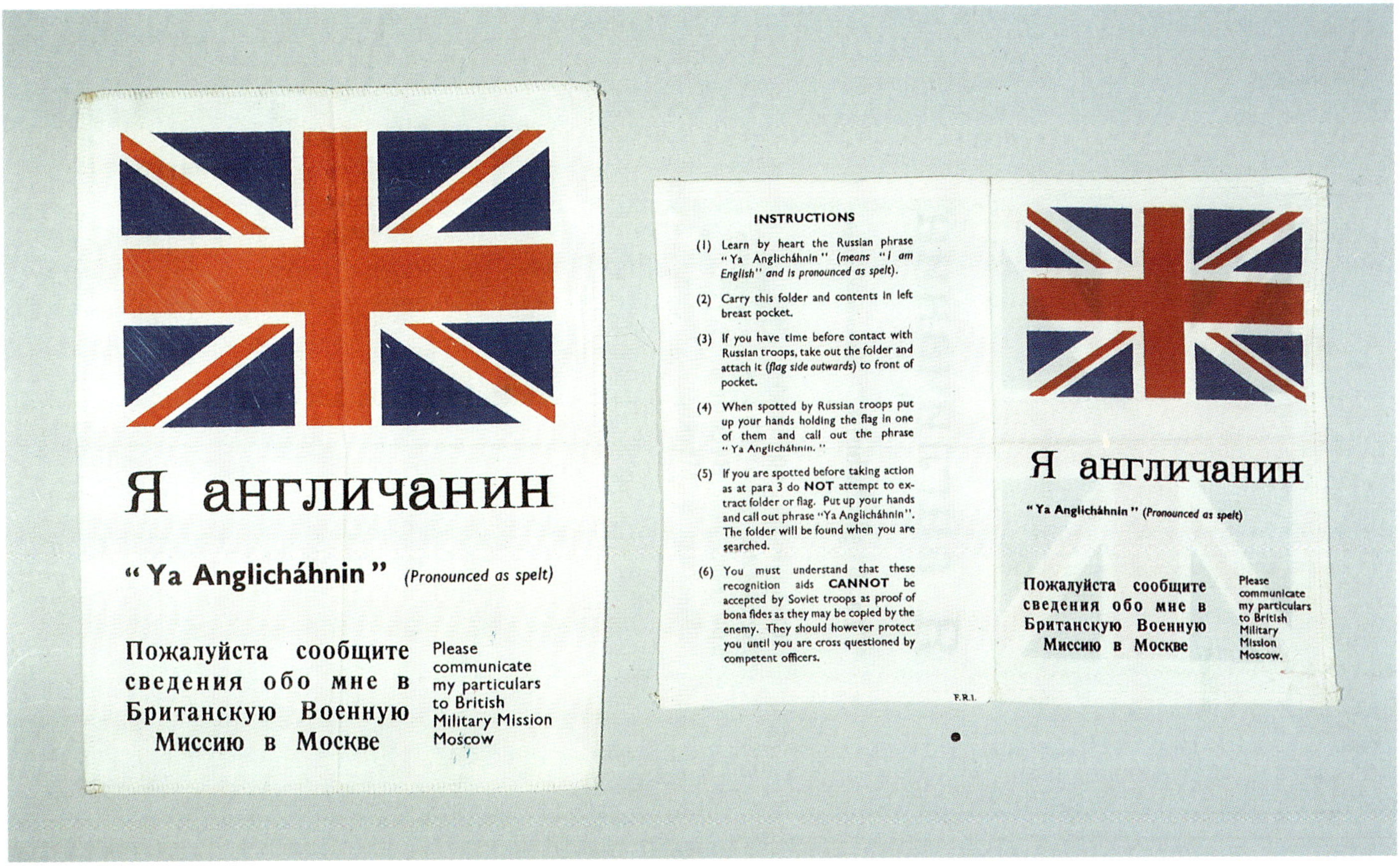
Я англичанин
" Ya Angliсháhnin " (Pronounced as spelt)
Пожалуйста сообщите сведения обо мне в Британскую Военную Миссию в Москве
Please communicate my particulars to British Military Mission Moscow
INSTRUCTIONS
(1) Learn by heart the Russian phrase "Ya Anglicháhnin" (means "I am English" and is pronounced as spelt).
(2) Carry this folder and contents in left breast pocket.
(3) If you have time before contact with Russian troops, take out the folder and attach it (flag side outwards) to front of pocket.
(4) When spotted by Russian troops put up your hands holding the flag in one of them and call out the phrase "Ya Anglicháhnin."
(5) If you are spotted before taking action as at para 3 do NOT attempt to extract folder or flag. Put up your hands and call out phrase "Ya Anglicháhnin". The folder will be found when you are searched.
(6) You must understand that these recognition aids CANNOT be accepted by Soviet troops as proof of bona fides as they may be copied by the enemy. They should however protect you until you are cross questioned by competent officers.
F.R.1.
Я англичанин
" Ya Anglicháhnin " (Pronounced as spelt)
Пожалуйста сообщите сведения обо мне в Британскую Военную Миссию в Москве
Please communicate my particulars to British Military Mission Moscow.

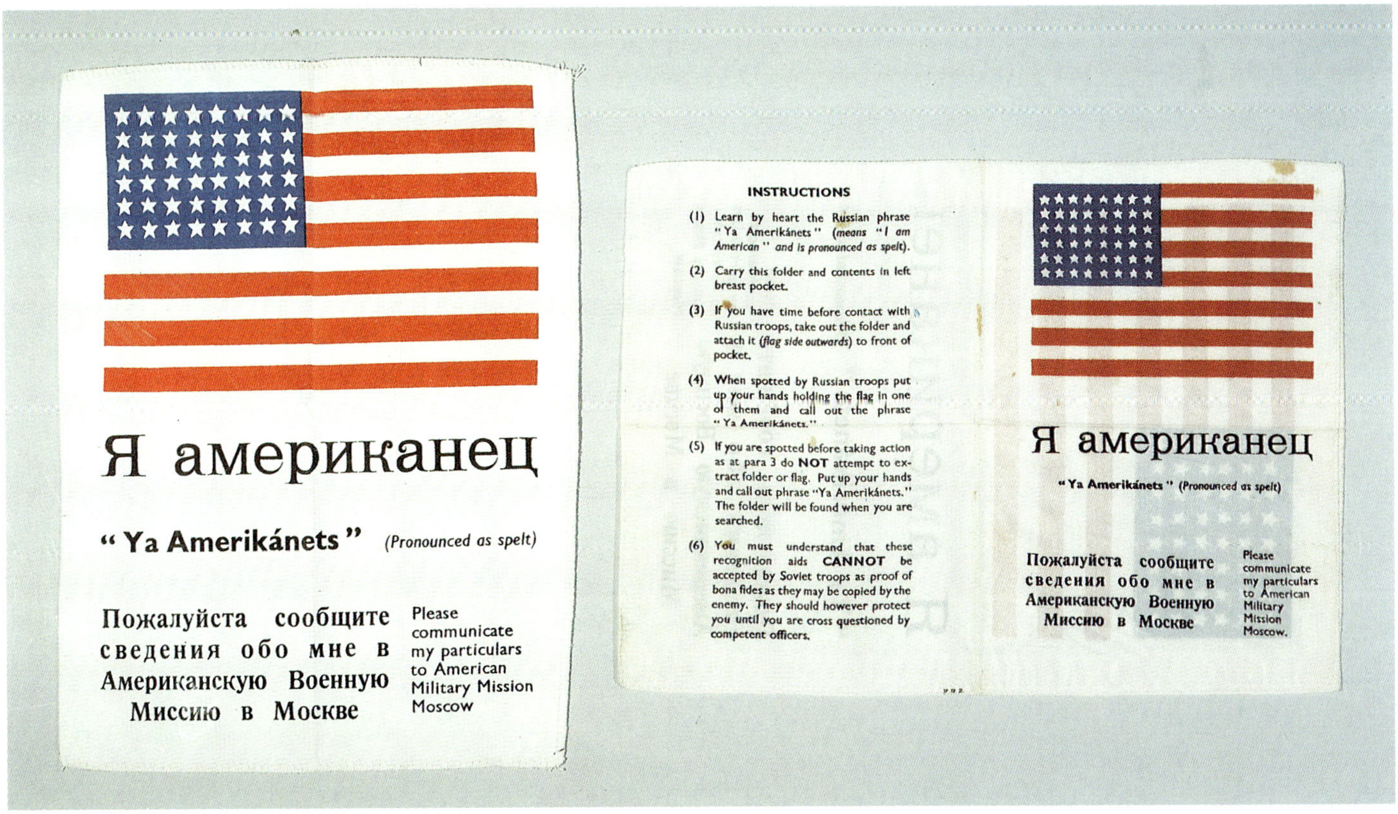
Я американец
" Ya Amerikánets " (Pronounced as spelt)
Пожалуйста сообщите сведения обо мне в Американскую Военную Миссию в Москве
Please communicate my particulars to American Military Mission Moscow
INSTRUCTIONS
(1) Learn by heart the Russian phrase "Ya Amerikánets" (means "I am American" and is pronounced as spelt).
(2) Carry this folder and contents in left breast pocket.
(3) If you have time before contact with Russian troops, take out the folder and attach it (flag side outwards) to front of pocket.
(4) When spotted by Russian troops put up your hands holding the flag in one of them and call out the phrase "Ya Amerikánets."
(5) If you are spotted before taking action as at para 3 do NOT attempt to extract folder or flag. Put up your hands and call out phrase "Ya Amerikánets." The folder will be found when you are searched.
(6) You must understand that these recognition aids CANNOT be accepted by Soviet troops as proof of bona fides as they may be copied by the enemy. They should however protect you until you are cross questioned by competent officers.
Я американец
" Ya Amerikánets " (Pronounced as spelt)
Пожалуйста сообщите сведения обо мне в Американскую Военную Миссию в Москве
Please communicate my particulars to American Military Mission Moscow.

Fig. 6.22: The notation "F.R.1" appeared on some British versions of chits made for use in Soviet-controlled territory, and the notation "F.R.2" appeared on some American versions. Sometimes these notations were folded under when the edges were sewn together and can be seen when chits are held up to the light. On some versions, these notations do not appear at all. No official explanation has been found for these notations. The periods after both the "F" and the "R" seem to rule out the notation representing "Frantic", the name for the shuttle bombing missions on which these chits were used, but the letters may represent something like "For Russia." *International Blood Chit Museum*

Fig. 6.23: Both the British and American versions of the rayon Russian chits shown in Figs. 6.20 and 6.21 were printed on two separate pieces of cloth which were then sewn together. Variations in the edge stitching are detailed here. The "F.R." notations shown in Fig. 6.22 seem to appear only on versions with straight stitching. Both versions are genuine. *International Blood Chit Museum*

Blood Chits are not mentioned in official documentation for the first Frantic missions, however an entire series of Blood Chits was developed for use by both British and American air crews in Soviet controlled territory. These chits continued to be produced and issued in quantity until the end of the war.

Examples of the two-sided rayon chits produced for use by British and Commonwealth air crews are illustrated in Fig. 6.20. These chits were printed on separate pieces of fabric which were then sewn together back-to-back. It was intended that the front side, with the large flag, could be held up and hopefully recognized from a distance while the airman shouted, "Ya Anglichahnin!" (I am English).

These chits do not offer a reward, but state in Russian and in English, "Please communicate my particulars to the British Military Mission Moscow." The right side of the back of the chit is identical to the front side, except that it is half-size.

Information on the left side of the back of the chit gives instructions in English to the user, an extremely important feature since the Russian troops were often trigger-happy.

Former Flying Officer E. J. Wesley Coe writes that he "carried one of these over the Balkans on operations from Amendola, on the Foggia Plain in Southern Italy during 1944-1945 when flying B-24J Liberators" with the RAF's 178 Squadron, 205 Group. Coe flew 40 missions, including supply drops to the Partisans in the Balkans, with Wing Commander J. Millar, Officer Commanding, 178 Squadron. He notes that the RAF did not have co-pilots, so the flight engineer sat in the right-hand seat.[34]

W. R. Wallace, a New Zealander who was a pilot with 297 Squadron of the RAF's 38 Group, recalls that they were issued these chits late in the war. This was in preparation for an eventually cancelled operation where transport aircraft would fly to Russian bases for refueling after dropping airborne troops west of Berlin.[35]

American versions of the two-sided rayon chits for use on missions over Soviet territory are shown in Fig. 6.21. They are nearly identical to the British versions except for the obvious changes: the American flag, the phrase "Ya Amerikanets" (I am American), and the request that the bearer's particulars be reported to the American mission.

The small notations "F.R.1" and "F.R.2", detailed in Fig. 6.22, are printed at the bottom of some British and American versions of these two-sided rayon chits. Fig. 6.23 illustrates two variations in edge stitching found on these chits.

Other variations of these British and American chits are illustrated in Fig. 6.24 and 6.25. These chits are identical in design and content to the chits to the rayon chits illustrated in Figs. 6.20 and 6.21, but on these versions the text is printed in blue rather than black ink and the user instructions are on the right side of the back instead of the left side. It is worthwhile to note that the on the U.S. version, the 48 white stars on the blue field are not of uniform shape or size. Produced and issued in large quantities, these chits are printed on a material that can best be described as "paper-like cloth", probably rayon. This unusual material was identified as cloth rather than paper by the late John A. Scott, curator of the Oregon Military Museum and an international expert on evasion and escape items, specifically cloth maps.[36] Scott noted that some British evasion maps were also printed on this type of fabric.

L. D. Volkmer carried the American version "whizzing around the Czech border in late March and April in P-47s so close to the war's end" while flying with the "1st Provisional Air Force, 358 FG, 365 FS, operating out of Toul, Lorraine and Sandhoffen, Germany."[37]

Paper versions of the American-flag chits have appeared on the collecting market in recent years, sometimes accompanied by reproduction paper Phrase Cards. While a number of the chits ap-

OPPOSITE: Fig 6.25: The reverse sides of the chits printed on paper-like cloth are shown here. The vast majority of chits for use in Soviet controlled territory were printed on this fabric. The stars on the American flag are not a uniform shape. *International Blood Chit Museum*

Fig 6.24: Many of the chits for use over Soviet territory were printed on a paper-like cloth, so thin that the printing on the reverse shows through the fabric. Their form and content is identical to the two-piece rayon versions except that the instructions on the backs of the chits, shown in Fig. 6.25, are on the right side of the page, and the text is printed in blue ink. The front sides of the British and American versions are shown here, and the reverse sides are illustrated in Fig. 6.25. (Left: REB 1240, Right: 3220) *International Blood Chit Museum*

Fig. 6.26: This Russian language blood chit card was issued to J. F. Notman, a Spitfire pilot with No. 93 Squadron, 326 Wing, RAF, around December of 1944 for missions over the Balkans. The text of the card reads, "Please communicate my particulars to the Allied Military Mission in Moscow." The first word on the card is a vernacular of "I beg your pardon", translated as "please." (REB 1242) *Courtesy of J. F. Notman*

pear to be genuine, readers are cautioned to be wary of paper artifacts because they are easily reproduced.

Small two-sided chits on cardstock, an example of which is shown in Fig. 6.26, were also produced for use by American and British airmen.[38] The chit illustrated was issued to J. F. Notman when he was assigned to 93 Squadron, 326 Wing, RAF in December 1944. Notman writes that he "at the time was flying bomber escort to U.S. Mitchells over Yugoslavia from Rimini in Spitfires. We were told at the time that if shot down the [locals] would first of all cut off our balls before asking questions – if we showed the card, they might not! Fortunately, I never put it to the test."[39] It is doubtful that the fate mentioned actually awaited airmen in the European theater despite the presence of a significant Muslim population in Serbia and Bosnia, then part of Yugoslavia. The strong oral traditions in RAF squadrons regarding the origins of Goolie Chits in India and Iraq is likely responsible for this notion among British air crews in the European theater. The text of the card is nearly identical to that on the large cloth chits, reading, "Please communicate my particulars to the Allied Military Mission in Moscow." The first word on the card is actually a vernacular of "I beg

Fig. 6.27: Ben L. Donahue, who piloted a B-24 with the 781st Bomb Squadron, 465th Bomb Group (Heavy), 15th Air Force, was issued this combination ID card and blood chit in January 1945. On the front of the card, below the American flag, is the title "U.S. Army Air Force", and the words "Identification Card" in 7 languages printed over the Great Seal of the United States, and the bearer's photo and signature. Below these is printed a message in English and Russian. On the back is a slightly different message in English and in languages for Poland, Czechoslovakia, Yugoslavia, Hungary, Austria and Germany. None of the messages promises a reward. (REB 3222) *International Blood Chit Museum, a gift of Ben L. Donahue.*

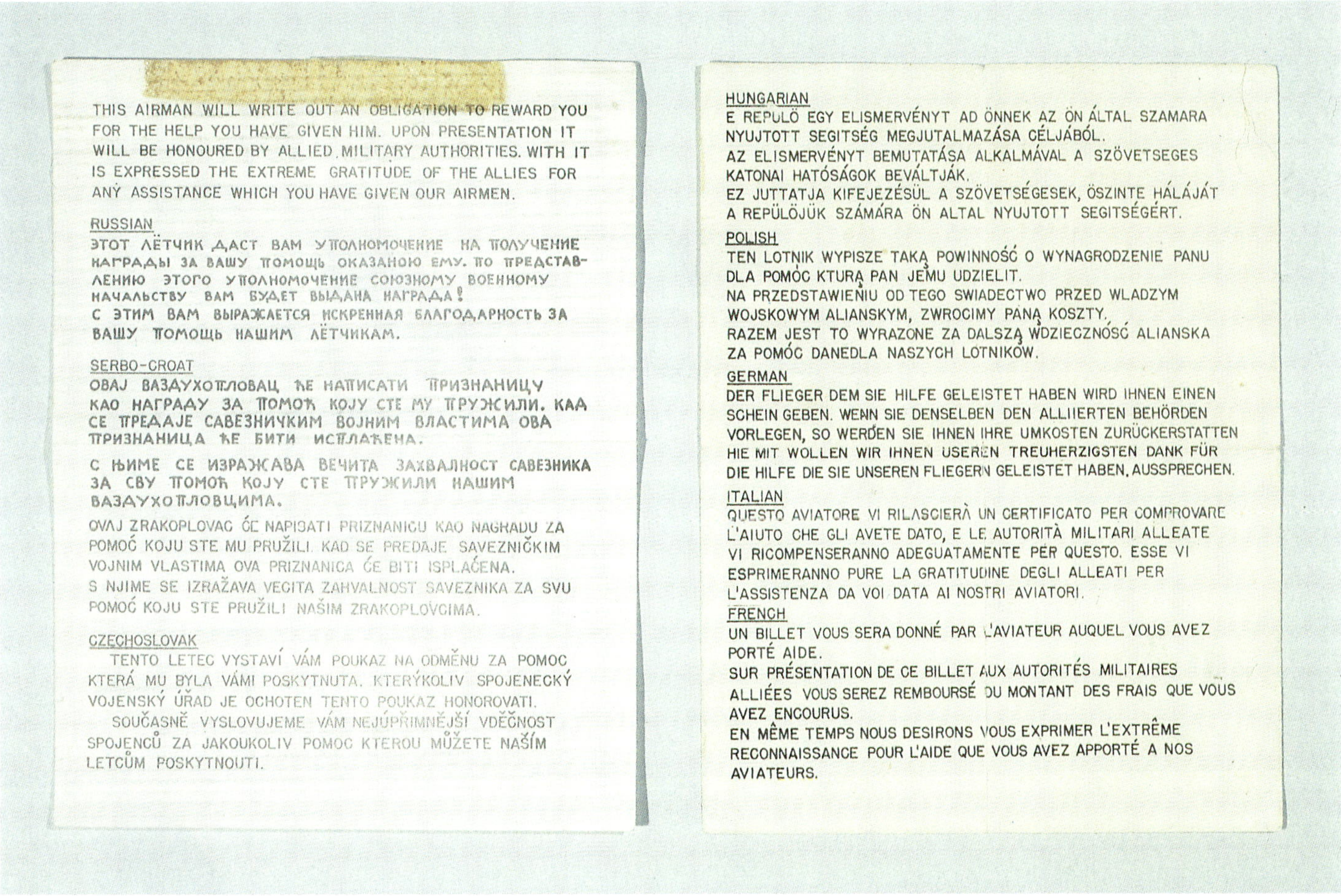

THIS AIRMAN WILL WRITE OUT AN OBLIGATION TO REWARD YOU FOR THE HELP YOU HAVE GIVEN HIM. UPON PRESENTATION IT WILL BE HONOURED BY ALLIED MILITARY AUTHORITIES. WITH IT IS EXPRESSED THE EXTREME GRATITUDE OF THE ALLIES FOR ANY ASSISTANCE WHICH YOU HAVE GIVEN OUR AIRMEN.

RUSSIAN

ЭТОТ ЛЁТЧИК ДАСТ ВАМ УПОЛНОМОЧЕНИЕ НА ПОЛУЧЕНИЕ НАГРАДЫ ЗА ВАШУ ПОМОЩЬ ОКАЗАНОЮ ЕМУ. ПО ПРЕДСТАВЛЕНИЮ ЭТОГО УПОЛНОМОЧЕНИЕ СОЮЗНОМУ ВОЕННОМУ НАЧАЛЬСТВУ ВАМ БУДЕТ ВЫДАНА НАГРАДА!
С ЭТИМ ВАМ ВЫРАЖАЕТСЯ ИСКРЕННАЯ БЛАГОДАРНОСТЬ ЗА ВАШУ ПОМОЩЬ НАШИМ ЛЁТЧИКАМ.

SERBO-CROAT

ОВАЈ ВАЗДУХОПЛОВАЦ ЋЕ НАПИСАТИ ПРИЗНАНИЦУ КАО НАГРАДУ ЗА ПОМОЋ КОЈУ СТЕ МУ ПРУЖИЛИ. КАД СЕ ПРЕДАЈЕ САВЕЗНИЧКИМ ВОЈНИМ ВЛАСТИМА ОВА ПРИЗНАНИЦА ЋЕ БИТИ ИСПЛАЋЕНА.

С ЊИМЕ СЕ ИЗРАЖАВА ВЕЧИТА ЗАХВАЛНОСТ САВЕЗНИКА ЗА СВУ ПОМОЋ КОЈУ СТЕ ПРУЖИЛИ НАШИМ ВАЗДУХОПЛОВЦИМА.

OVAJ ZRAKOPLOVAC ĆE NAPISATI PRIZNANICU KAO NAGRADU ZA POMOĆ KOJU STE MU PRUŽILI. KAD SE PREDAJE SAVEZNIČKIM VOJNIM VLASTIMA OVA PRIZNANICA ĆE BITI ISPLAĆENA.
S NJIME SE IZRAŽAVA VECITA ZAHVALNOST SAVEZNIKA ZA SVU POMOĆ KOJU STE PRUŽILI NAŠIM ZRAKOPLOVCIMA.

CZECHOSLOVAK

TENTO LETEC VYSTAVÍ VÁM POUKAZ NA ODMĚNU ZA POMOC KTERÁ MU BYLA VÁMI POSKYTNUTA. KTERÝKOLIV SPOJENECKÝ VOJENSKÝ ÚŘAD JE OCHOTEN TENTO POUKAZ HONOROVATI.
SOUČASNĚ VYSLOVUJEME VÁM NEJÚPŘIMNĚJŠÍ VDĚČNOST SPOJENCŮ ZA JAKOUKOLIV POMOC KTEROU MŮŽETE NAŠÍM LETCŮM POSKYTNOUTI.

HUNGARIAN

E REPULÖ EGY ELISMERVÉNYT AD ÖNNEK AZ ÖN ÁLTAL SZAMARA NYUJTOTT SEGITSÉG MEGJUTALMAZÁSA CÉLJÁBÓL.
AZ ELISMERVÉNYT BEMUTATÁSA ALKALMÁVAL A SZÖVETSEGES KATONAI HATÓSÁGOK BEVÁLTJÁK.
EZ JUTTATJA KIFEJEZÉSÜL A SZÖVETSÉGESEK, ÖSZINTE HÁLÁJÁT A REPÜLÖJÜK SZÁMÁRA ÖN ALTAL NYUJTOTT SEGITSÉGÉRT.

POLISH

TEN LOTNIK WYPISZE TAKĄ POWINNOŚĆ O WYNAGRODZENIE PANU DLA POMÓC KTURĄ PAN JEMU UDZIELIT.
NA PRZEDSTAWIEŃIU OD TEGO SWIADECTWO PRZED WLADZYM WOJSKOWYM ALIANSKYM, ZWROCIMY PANĄ KOSZTY.
RAZEM JEST TO WYRAZONE ZA DALSZĄ WDZIECZNÓSĆ ALIANSKA ZA POMÓC DANEDLA NASZYCH LOTNIKÓW.

GERMAN

DER FLIEGER DEM SIE HILFE GELEISTET HABEN WIRD IHNEN EINEN SCHEIN GEBEN. WENN SIE DENSELBEN DEN ALLIIERTEN BEHÖRDEN VORLEGEN, SO WERDEN SIE IHNEN IHRE UMKOSTEN ZURÜCKERSTATTEN HIE MIT WOLLEN WIR IHNEN USEREN TREUHERZIGSTEN DANK FÜR DIE HILFE DIE SIE UNSEREN FLIEGERN GELEISTET HABEN, AUSSPRECHEN.

ITALIAN

QUESTO AVIATORE VI RILASCIERÀ UN CERTIFICATO PER COMPROVARE L'AIUTO CHE GLI AVETE DATO, E LE AUTORITÀ MILITARI ALLEATE VI RICOMPENSERANNO ADEGUATAMENTE PER QUESTO. ESSE VI ESPRIMERANNO PURE LA GRATITUDINE DEGLI ALLEATI PER L'ASSISTENZA DA VOI DATA AI NOSTRI AVIATORI.

FRENCH

UN BILLET VOUS SERA DONNÉ PAR L'AVIATEUR AUQUEL VOUS AVEZ PORTÉ AIDE.
SUR PRÉSENTATION DE CE BILLET AUX AUTORITÉS MILITAIRES ALLIÉES VOUS SEREZ REMBOURSÉ DU MONTANT DES FRAIS QUE VOUS AVEZ ENCOURUS.
EN MÊME TEMPS NOUS DESIRONS VOUS EXPRIMER L'EXTRÊME RECONNAISSANCE POUR L'AIDE QUE VOUS AVEZ APPORTÉ A NOS AVIATEURS.

Fig. 6.28: The chit on the left in this illustration was originally attached to Ben Donahue's ID card-chit (Fig. 6.27) with tape, but has become detached. This chit adds the valuable promise of a reward to the ID Card-Blood Chit, and additional messages in English, Russian, Serb-Croat (two versions, presumably Serbian and Croatian), Czechoslovakian, Hungarian, Polish, German, Italian and French. The spelling of "honoured" confirms this item was produced in Britain, as were most other pre-capture aids used by crews in the European and Mediterranean Theaters of Operation. The stripes on Donahue's chit appear to be the result of it having been kept in a photo album. (REB 3224) *International Blood Chit Museum, a gift of Ben L. Donahue.*

your pardon" or "I am sorry", as in "excuse me", while the first word on the cloth chits is "please".[40]

Some air crews received a combination ID card and Blood Chit. The example shown in Fig. 6.27 was issued to then Lt. Ben L. Donahue, a B-24 pilot with the 15th Air Force's 781st Bomb Squadron, 465th Bomb Group, based in Italy. Donahue received this chit in January 1945, but reports that they were issued prior to that time and continued to be issued through June 1945. The card does not offer a reward for assistance. The reverse of a similar card is shown to the right of Donahue's. The request for assistance is given in English and Russian on the front of the card, and in languages for Poland, Czechoslovakia, Yugoslavia, Hungary, Austria and Germany on the back of the card.

To augment the ID Card-Blood Chit described above, another two-sided chit was issued which provided the valuable offer of a reward for assistance. The example shown on the left in Fig. 6.28 was originally attached to Donahue's ID card/chit with tape, but has become detached. The two chits were issued to Donahue together, but this type of chit was also issued individually in some areas. On the right is the back of an identical chit. The languages on this chit are English, Russian, Serb-Croat (two versions, presumably Serbian and Croatian), Czechoslovakian, Hungarian, Polish, German, Italian and French. The text alludes to the system of writing Promissory Notes, also called "chits", and giving them to persons who provided. Blank pieces of paper (not illustrated) identical in size to this chit were issued to crews for this purpose.

A British version of the ID card/chit, shown in Fig. 6.29, was also issued. It is identical to the American version except that it displays the Union Jack and the Royal Arms, and refers to the airman as "British." This card was part of a group of items from an airman whose service records date from just before VE-Day to the early years of the Cold War, prompting speculation that this sort of card may have been used by air crews through the late 1940s.

The myriad political relationships, along with competing Allied interests and alliances, often made eastward evasion difficult and complex. The experiences of an RAF air crew member, Flight Sergeant William "Scottie" Young, illustrate the fact that even when equipped with the proper evasion language and identification aids, evasion in eastern Europe and the USSR was a precarious business

ROYAL AIR FORCE

IDENTIFICATION CARD
УДОСТОВЕРЕНИЕ ЛИЧНОСТИ
DOWOD OSOBISTY
LEGITIMACE
LEGITIMACIJA
SZEMÉLYAZONOSSAGI IGAZOLVANY
LEGITIMATION

SIGNATURE

I AM A BRITISH AIRMAN
PLEASE TAKE ME TO YOUR COMMANDING OFFICER AND NOTIFY NEAREST BRITISH OR AMERICAN MILITARY MISSION IN BELGRADE, BUCHAREST, POLTAVA OR OTHER NEARBY PLACE. ALSO, PLEASE ARRANGE FOR TRANSPORTATION.
THANK YOU

SHOW THIS TO RUSSIANS

Я БРИТАНСКИЙ ЛЕТЧИК
ПОЖАЛУЙСТА ПРЕДСТАВТЕ МЕНЯ ВАШЕМУ КОМАНДИРУ И УВЕДОМИТЕ БЛИЖАЙШУЮ БРИТАНСКУЮ ИЛИ АМЕРИКАНСКУЮ ВОЕННУЮ МИССИЮ В БЕЛГРАДЕ, БУХАРЕСТЕ, ПОЛТАВЕ ИЛИ В ДРУГОМ БЛИЖАЙШЕМ МЕСТЕ. ТАКЖЕ РАСПОРЯДИТЕСЬ О ПЕРЕДВИЖЕНИИ.
БОЛЬШОЕ СПАСИБО!

Fig. 6.29: The British version of the ID card-chit is identical to the U.S. version except for the flag, coat of arms, and fact that the bearer is identified as "British." There is some speculation that this type of card/chit may continued to have been issued during the cold war in the late 1940s. (REB 1244) *Courtesy of Bob Gill*

at best. Scottie approached Russian truck drivers with his Blood Chit held in his hands over his head, shouting "Ya Anglichahnin" as instructed. The Russians, armed with rifles, would not accept his chit as a bona-fide identification, telling him that the Germans had thousands of forged chits.[41] No evidence has been uncovered in the research material for this book to indicate that the Germans had forgeries of these chits, but the incident underscores the deep distrust that the Soviets held for their Allies.

While language aids did not guarantee a friendly reception, by and large they did seem to produce positive results. Ben Donahue, whose chits are shown in Fig. 6.27 and 6.28, recalls that "several crews of our unit did use their chit and it did receive the expected response."[42]

Marvin C. Bigelow, a former 8th Air Force fighter pilot, recalls, "I was stationed at Duxford Air Base with the 83rd Fighter Squadron, 78th Fighter Group, 8th Army Air Force . . . A number of the missions I flew took me to Berlin escorting B-17s . . . we sometimes came within sight of the Russian-German front. As a result, we were given the option of carrying the . . . chit if we wished to do so. If we were hit and going down, the Russian front was much closer than France and several of the squadron took that route home."[43]

Malcolm E. Corum, who served with 391st Squadron, 34th Bomb Group and carried the American chit of the type shown in Figs. 6.25 and 6.26, writes that he is "aware of one crew from the 34th Bomb Group (18th Squadron) that crash-landed in eastern Germany or western Poland in 1945. They were returned to Britain some 30 days later . . . I am also sure that other crews from the 8th Air Force were RAMP in the same manner as a result of wearing Blood Chits on missions later in the war."[44] The term "RAMP" is the acronym for Recovered Allied Military Personnel, and is often used as a proper noun or verb.

While Corum refers to "wearing" the chits, they were not worn sewn to jackets or flight suits as were some chits in the Far East early in the war. These Russian language chits were normally car-

Fig. 6.30: Shown here are two of the chits printed on paper-like fabric, one American and one British, enclosed in their plastic holders. One has a black rubber tongue to which the cotton cord is attached, and the other has a green canvas tongue that can be attached to the button of a pocket per instructions on the chit. *International Blood Chit Museum*

ried in a plastic case that was often worn around the neck on a cotton cord, despite the printed instructions that told the wearer to "carry this folder and its contents in left breast pocket." Illustrated in Fig. 6.30 are an American and a British chit enclosed in plastic holders.

Language aids continued to be produced and issued to air crews until the end of the war in Europe. Just two months before the end of the war in Europe, P/W&X issued their projected requirements for the next quarter.[45] This projection, recognizing that operations over enemy territory involved flights toward advancing Soviet armies, included 34,000 Russian language Blood Chits:

> Requirements for 1st, 2d and 3d Air Divisions for next three month period are listed below. Figures listed do not include equipment on which indents have been submitted.

Division	Map Wallets CDEF	Aids Boxes (curved)	Ration Boxes	Universal Phrase Cards	R. Blood Chits
1	3500	3500	1000	2000	15,000
2d	3000	2000	- - -	1300	11,000
3d	1000	1000	1200	1000	8,000

One final evasion item was produced for use in the Soviet Union and territory under its control. It was completed too late for its introduction into the European theater during WWII, as illustrated by an excerpt from the MIS-X War Diary for August and September 1945 that reads: "Pointie Talkie No. 6. 25,000 copies of the 100,000 have been received, which were on procurement. The remainder has been cancelled."[46]

Pointie Talkie No. 6 is a 26 page booklet that was, as noted inside its cover, "Prepared for Air Combat Personnel by The Military Intelligence Service, Washington, D.C." The cover of this booklet, shown in Fig. 6.31, features the flags of the United States and the Soviet Union, printed in color. The booklet illustrated is in the Special Collections Library of the U.S. Air Force Academy. The back cover features provisions for the owner to enter his name, serial number and commanding officer. It is unusual that the commanding officer's name would be written on a document carried into combat, as this information might help the enemy if the airman were captured. Apparently it was felt that the danger to airmen downed in Soviet territory was sufficiently great to warrant its inclusion as an aid to confirming the airman's identity. At the top of the back cover is the Great Seal of the United States, and at the bottom is the notation "U.S. Government Printing Office - 1945".

The Pointie talkie consists of three parts: Part I is a standard format Pointie talkie "questions and answers" section, Part II consists of useful vocabulary in English, Russian and phonetic Russian, and Part III is a table of figures. Part I is broken down into sections such as "Upon Landing in U.S.S.R.", "Finding an Interpreter", "Medical Care", and so on.

Additional language materials, such as Language Guides and Phrase Books, were available to American servicemen. Language

Fig. 6.31: The cover of Pointie Talkie No. 6 features the American and Soviet flags. It contains three parts: a standard "questions and answers" section, a list of useful vocabulary in English, Russian and phonetic Russian, and a table of figures. The Pointie Talkie was prepared by MIS-X for use in Soviet controlled territory, but was completed too late in the war for distribution to air crews. (REB 3922) *U.S. Air Force Academy Photo, courtesy of the Special Collections Branch, U.S. Air Force Academy Library*

guides were used in conjunction with phonograph records to teach servicemen the languages spoken in areas to which they would be posted. As instructional material, they were not intended to be taken on operations.

Phrase Books contained useful phrases in foreign languages for a variety of situations, although not specifically for evasion in enemy occupied territory. The four Phrase Books shown in Fig. 6.32 are typical of these items. They are the comprehensive 228 page TM 30-243 *French Phrase Book*, produced in January 1943, and three less comprehensive phrase books containing 127 pages each that were produced at a later date. These are the TM 30-602 *French Phrase Book*, TM 30-606 *German Phrase Book*, and TM 30-644 *Russian Phrase Book*. A useful feature of TM 30-243, omit-

Fig. 6.32: Four Phrase Books, typical of the types available to American servicemen, are shown here. The earliest of these, the cloth-covered *TM 30-243 French Phrase Book*, was produced in January 1943. It contains instructions that indicate the book was intended to be carried by combat personnel, advising the soldier to "acquaint yourself with those sections you may have need of before you go on a particular duty. If you are going on patrol in a captured town or on reconnaissance, read those sections carefully, and then put the book in your pocket open at the proper place — you may need a foreign phrase in a hurry." It also contains a special section entitled "These Words May Save A Life", detailed in Fig. 6.33, which would have been useful in first encounters with local peoples. The other paperback Phrase Books, printed at a later date, do not contain special sections. The typical layout of these later books is illustrated in Fig. 6.34. *International Blood Chit Museum*

ted from the later Phrase Books, is the section appearing just inside the front cover, illustrated in Fig. 6.33, printed in red ink and entitled "These Words May Save A Life". This section would have been useful in first encounters with local peoples, and the layout of this section is typical of the remainder of the book. For comparison, the layout of TM 30-606 is illustrated in Fig. 6.34. Compare both of these formats to that of the World War I booklet, *The Soldiers' French Phrase Book*, illustrated in Chapter (Fig. 2.6).

Other types of Phrase Books, such as that shown in Fig. 6.35, were also made. The book illustrated was prepared by the ETOUSA G-2 Section (Intelligence) in March 1943, and contains two parts. Part I is English to German, French and Spanish. Part II is English to German only and is primarily for use with captured prisoners.

As the enemy was driven from foreign countries, local entrepreneurs produced language material for sale to Allied servicemen. The 70-page *Greek Spoken*, shown in Fig. 6.36, is such a book. Subtitled *A Complete Guide Especially Written for the English and American Troops*, it is divided into three sections: Greek Grammar, Conversations, and an English-Greek Dictionary. This booklet belonged to a P. Webb, RNVR.

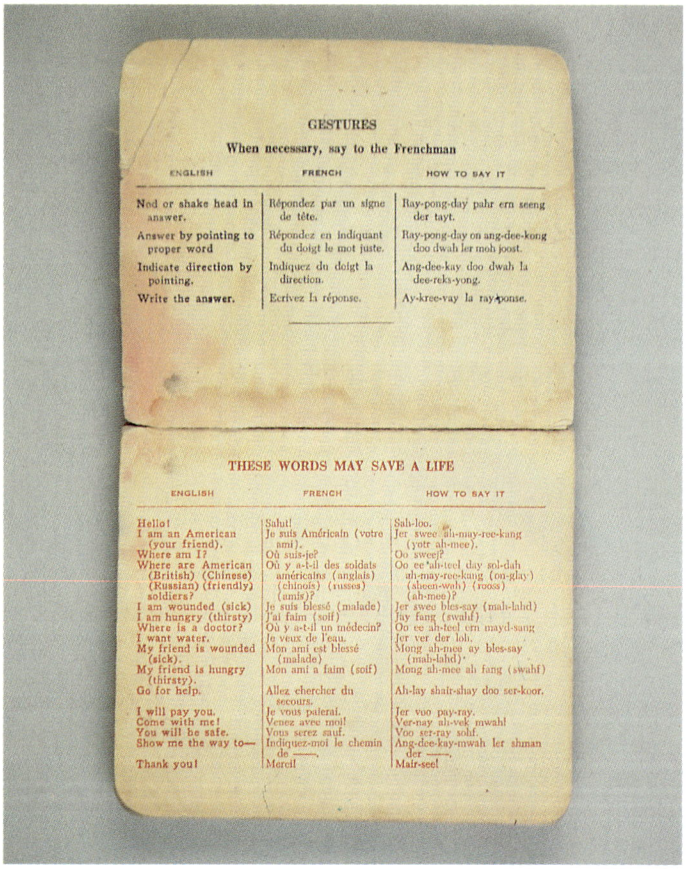

GESTURES

When necessary, say to the Frenchman

ENGLISH	FRENCH	HOW TO SAY IT
Nod or shake head in answer.	Répondez par un signe de tête.	Ray-pong-day pahr ern seeng der tayt.
Answer by pointing to proper word	Répondez en indiquant du doigt le mot juste.	Ray-pong-day on ang-dee-kong doo dwah ler moh joost.
Indicate direction by pointing.	Indiquez du doigt la direction.	Ang-dee-kay doo dwah la dee-reks-yong.
Write the answer.	Ecrivez la réponse.	Ay-kree-vay la ray-ponse.

THESE WORDS MAY SAVE A LIFE

ENGLISH	FRENCH	HOW TO SAY IT
Hello!	Salut!	Sah-loo.
I am an American (your friend).	Je suis Américain (votre ami).	Jer swee ah-may-ree-kang (yotr ah-mee).
Where am I?	Où suis-je?	Oo sweej?
Where are American (British) (Chinese) (Russian) (friendly) soldiers?	Où y a-t-il des soldats américains (anglais) (chinois) (russes) (amis)?	Oo ee ah-teel day sol-dah ah-may-ree-kang (on-glay) (sheen-wah) (rooss) (ah-mee)?
I am wounded (sick)	Je suis blessé (malade)	Jer swee bles-say (mah-lahd)
I am hungry (thirsty)	J'ai faim (soif)	Jay fang (swahf)
Where is a doctor?	Où y a-t-il un médecin?	Oo ee ah-teel ern mayd-sang
I want water.	Je veux de l'eau.	Jer ver der loh.
My friend is wounded (sick).	Mon ami est blessé (malade)	Mong ah-mee ay bles-say (mah-lahd)
My friend is hungry (thirsty).	Mon ami a faim (soif)	Mong ah-mee ah fang (swahf)
Go for help.	Allez chercher du secours.	Ah-lay shair-shay doo ser-koor.
I will pay you.	Je vous paierai.	Jer voo pay-ray.
Come with me!	Venez avec moi!	Ver-nay ah-vek mwah!
You will be safe.	Vous serez sauf.	Voo ser-ray sohf.
Show me the way to—	Indiquez-moi le chemin de ——.	Ang-dee-kay-mwah ler shman der ——.
Thank you!	Merci!	Mair-see!

Fig. 6.33: The special section of the *TM 30-243 French Phrase Book* entitled "These Words May Save A Life" is detailed here. It appears just inside the front cover and is printed in red ink. The layout is typical of the remainder of the book, which is printed in black ink. This 228-page Phrase Book is very comprehensive when compared to the other paperback Phrase Books illustrated in Fig. 6.32. *International Blood Chit Museum*

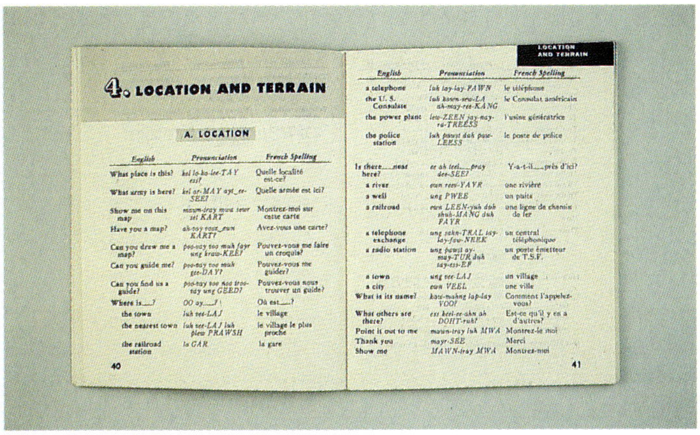

Fig. 6.34: These pages of the *TM 30-602 French Phrase Book* illustrate the typical layout of World War II paperback Phrase Books. Compare this layout to those of *TM 30-243* (Fig. 6.33) and of the World War I *The Soldiers' French Phrase Book* (Fig. 2.6). *International Blood Chit Museum*

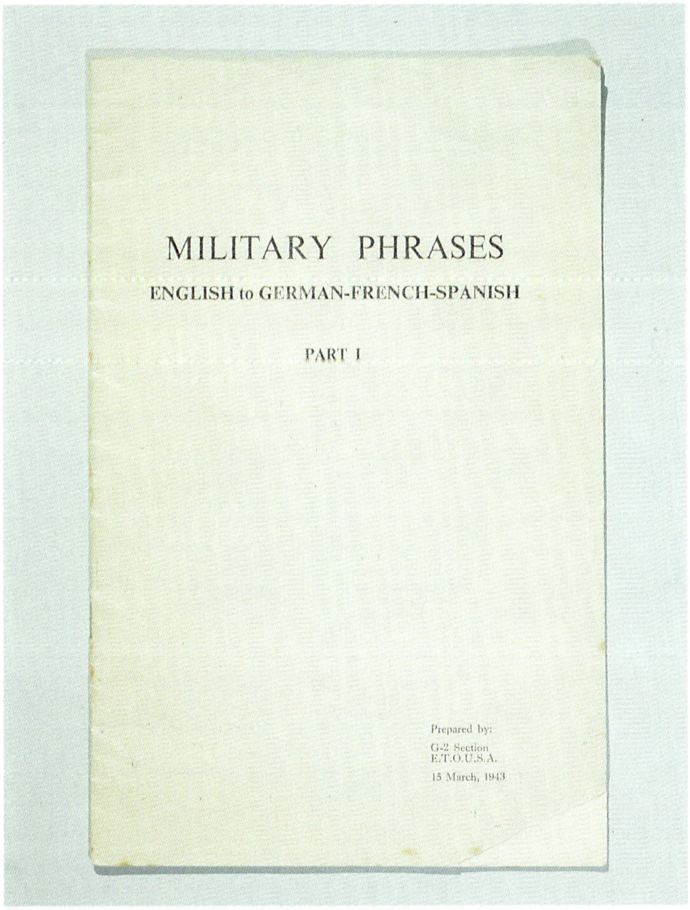

Fig. 6.35: G-2 Section (Intelligence) of ETOUSA produced this paper two-part Phrase Book in March 1943. Part I, formatted in English to German, French, and Spanish, covers a variety of situations. Although not made specifically for evasion purposes, it includes phrases such as "I had to make a forced landing", "I want a guide who knows the country well", and "Can you give me an interpreter?" Part II, for use in communicating with captured prisoners, is formatted in English to German. (REB 3910) *International Blood Chit Museum*

Fig. 6.36: As countries were liberated, local entrepreneurs produced books such as *Greek Spoken* for sale to Allied troops. It contains sections on grammar, conversation, and a dictionary in English, Greek, and phonetic Greek. Items such as this supplemented officially issued language material. *International Blood Chit Museum*

One final aid containing phrases is illustrated in Fig. 6.37. It is printed on a cotton handkerchief and provides both military and everyday phrases in English, French, and German. This souvenir handkerchief was found in the pocket of an RAF Battle Dress tunic and may have been produced in Britain for sale to servicemen following the invasion of France.

There is evidence that the Germans gave at least some of their airmen survival and evasion training and aids. After the Allied invasion, a detachment of the Luftwaffe's secretive Kampfgeschwader 200 (KG200) began plans to attack power stations supplying electricity to Soviet armament plants around Moscow. The attack was to use a force of Mistel (Misteltoe) "piggy-back" aircraft, most likely the Mistel I, a Bf 109 mounted on a converted Ju 88. Because the target was beyond the extreme range of the aircraft, plans called for flying the 1000 mile mission and returning to bases in German-held Latvia. The crews received advice from a botanist on how to survive on various tree barks, lichens, mosses and berries. They also received information about what regional wildlife was edible, though it is not clear if they were to carry weapons for hunting or material for traps. Language experts taught them some essential Russian words and phrases, and they were provided with a German-Russian dictionary. Each pilot also received a packet of rubles,

Fig. 6.37: An unusual language-aid, obviously not intended for evasion, is a collection of military and civilian phrases in English, French and German printed on a cotton handkerchief. Dating from late in World War II and probably of British origin, this item was found in the pocket of an RAF Battle Dress tunic, and may have been produced for sale to servicemen following the Normandy invasion. It can best be described as a souvenir item. (REB 9010) *International Blood Chit Museum*

survival rations, and supplies including 25 cigars. During the wait for favorable weather, an American bomber formation destroyed 18 Mistels, the unit's main force, at the detachment's airfield at Rechlin-Larz. The raid was postponed and then finally cancelled.

Those fortunate Allied crew members who flew their missions and returned to base may have thought of their evasion kits as interesting curiosities, but the less fortunate air crews who used their Blood Chits, Phrase Cards, cloth maps, compasses, and other aids to evade capture knew their true value.

The Allies' development and fielding of evasion aids played a vital role in the safe return of thousands of air crew members in the European theater. In some cases, these items were virtual life-savers, allowing Allied airmen to communicate with a local population who provided them with medical care, food, water, and eventual return to friendly military authorities.

While Eighth, Ninth and Fifteenth Air Force crews and their Bomber Command equivalents were climbing into heavy sheepskin flying clothes to endure the cold European skies as they struck at targets on the Continent, other airmen – Americans, Australians, New Zealanders, Dutch, and Chinese – were fighting a very different war. Fighting with what they had, they often flew obsolescent aircraft against an unexpectedly aggressive and successful enemy over tropical islands, vast oceans and the Asian mainland.

With the war in Europe at a close, the Allies would concentrate their full attention on the war against Japan. The Allies' evasion and escape organizations, and the aids and assistance they provided to air crews in Asia and the Pacific, would be a vital part of that effort. Nowhere in their colorful history would Blood Chits flourish as they did in Asia during World War II.

CHAPTER 7

A Debt Repaid World War II in Asia

In Asia, as in Europe, the British evasion and escape agency MI9 took the initial steps, establishing an advance base in New Delhi known as GS-I(e) in October 1941, under the command of Col. Robin Ridgeway. Two years later, in the Fall of 1943, GS-I(e) was renamed E-Group. Its Main Headquarters remained in New Delhi, and a No. 2 Headquarters was established in Calcutta. Advance Headquarters were established in Calcutta and in Columbo, Ceylon. E-Group also operated Jungle Schools in Colombo and in Assam, India.[1] From their Calcutta bases E-Group procured and issued evasion kits, cloth maps and Blood Chits, many of which were manufactured locally. Operationally, E-Group also organized evasion nets among the mountain tribes of Burma, including the Karen, the Naga, and the Shan.[2]

The British Army Aid Group (BAAG) was organized by a senior British officer, Lt. Col. (Sir) Lindsay Ride, who had escaped from a Japanese prisoner of war camp. With headquarters in Kwelin and numerous forward posts around Canton and Hong Kong, BAAG rescued Allied airmen in those areas, issued weather reports used by allied air forces operating in China, and did much to win the confidence of the Chinese people by providing medical care and food to local Chinese civilians. BAAG was taken under the umbrella of MI9 in May 1942.[3]

That the United States was ill-prepared early in the war to provide their air crews with necessary language aids is illustrated by the situation of the Doolittle Raiders, who were instructed to land their B-25 *Mitchell* medium bombers in Free China following their bombing raid on Tokyo in April 1942. Gen. "Brick" Holstrom, pilot of the number two aircraft, reported during a talk to the Northern California Chapter of the American Aviation Historical Society that he and the other Raiders were not provided any written lanuguage material to assist them in communicating with the local population in China. He noted that they were given limited verbal language instruction aboard *USS Hornet*, but that the phrases they learned were in a dialect that was not spoken in the part of China where they landed.

American E&E efforts in Asia got a late start due to the Allied agreement that placed emphasis on the war in Europe. In 1943 Brigadier Crockatt of MI9 suggested that MIS-X establish E&E operations in China. Following a MIS-X fact-finding mission to BAAG headquarters in Kwelin, led by 1st Lt. Barclay P. Schoyer who was fluent in Mandarin,[4] the Air Ground Aid Section (AGAS) was formed as the section of MIS-X responsible for directing E&E operations for the U.S. in China. Headquartered in Kunming, AGAS operations covered a great expanse of territory including "all areas of China, French Indo-China, and Thailand."[5] Additional headquarters were established in Luichow, Shangjao, Mintsing, Yungan, Sian and Yenan. Due to its operational effectiveness, BAAG continued to operate in the Canton and Hong Kong areas.

MIS-X Technical Section India was also established, with the "responsibility of maintaining liaison with E-Group in India and briefing and supplying American units stationed in India or Burma."[6]

The responsibilities and duties of AGAS personnel were nearly as vast and varied as the territory in which they operated. These included "equipping all combat units based in China with Escape and Evasion aids", "briefing all units stationed in China on E and E", "collection, evaluation, and coordination of all information concerning downed aircraft in China, French Indo-China, and Eastern Tibet", "search and rescue activities in China", "collection and dissemination of all E and E information", and "investigation and the payment of rewards."[7]

Col. A. R. Wichtrich was transferred from MIS-X Headquarters to command AGAS, and Lt. Schoyer was promoted to Major and returned to China where he supervised some of the AGAS forward posts.[8] More than 40 AGAS posts were located in very remote areas and were manned by two people - an officer and an enlisted radio operator. When an aircraft was reported down, AGAS HQ would radio its position to the nearest AGAS post, and the two-man team would then attempt to find the aircraft and bring the crew back to safety. Living conditions in these posts were difficult, transportation was often only by foot or on horseback, and it was nearly a full-time occupation for AGAS agents just to maintain their health.[9]

Other secret organizations assisted in the rescue of American air crews in Asia, including SACO, the Sino American Cooperative Organization, a joint Chinese and American force created by an alliance of the United States Navy and the Chinese military. The two individuals responsible for SACO were Vice Admiral Milton E. Miles of Naval Group China and Major General Tai Li, the director of intelligence services for the Nationalist government. Both the U.S. and China had veto powers over SACO's activities, and the group stressed cooperation. As the only U.S. Navy presence in

China, SACO's responsibilities included tracking Japanese ship movements and reporting these by radio to U.S. submarines; weather reporting, which included establishing a station in the Gobi Desert; and the training of Chinese guerrillas. SACO worked closely with the American 14th Air Force for whom it provided some communication and photo-interpretation services, and it also assisted AGAS in the rescue of downed air crews.[10]

A primary reason for the success of AGAS can be attributed to the careful selection of its officers, who were chosen on the basis of their personal character rather than for any previous experience in China.[11] The attitudes of many Westerners toward the Chinese people had been forged in the 1930s and earlier when foreigners and Chinese persons had not been allowed to mix freely. These attitudes rendered many "Old China Hands" unable to function effectively in organizations such as BAAG, AGAS, and SACO where American and British operatives lived with and depended upon their Chinese colleagues.[12]

The American OSS also assisted in rescuing downed airmen. Although it was not their primary responsibility, it has been reported that nearly 400 downed Allied airmen were rescued by OSS operatives in the China and Burma-India Theaters.[13]

Kuomintang (KMT) and guerilla forces also aided in the rescue of American airmen. Less known is the aid given by the Chinese Communist Party (CCP) in northern China, particularly in the Yangtze Valley where their New Fourth Army operated.[14] Although the cooperation of the Communists was known to their most bitter enemy, Chiang Kai-shek, he did not interfere with the arrangements of AGAS, the Twentieth Bomber Command (XX BC), and others, for the rescue of Allied airmen by the CCP.[15]

These arrangements were made through the U.S. Army's Dixie Mission, which arrived at CCP headquarters in Yenan, Shensi Province, on 22 July 1944. As detailed in the monograph *Dixie Mission: The United States Army Observer Group in Yenan, 1944*, by Col, David D. Barrett, the Mission's Commanding officer, AGAS was represented by 1st Lt. Henry C. Whittlesey, who effectively organized a program with the Communists for assisting downed American flyers. In early February 1945 Whittlesey, then a Captain, was killed by the Japanese while inspecting a village near the front lines. The Communists lost nearly an entire batallion in an unsuccessful attempt to recover his body.

As the war spread throughout the Asian mainland, Allied pilots, air crews, and the E&E organizations were faced with great differences in culture, geography, climate, and political history. The war in Asia involved not only China but also India, Burma, Thailand, French Indo-China, Malaya, and Korea. Each of these countries presented both shared and unique evasion issues.

China, a vast nation larger than Europe and half-again as large as the continental United States, presented many problems to evaders, including geography and climate. China contains two the world's greatest rivers, the Yellow and the Yangtze, some of the highest mountain ranges on earth to the west and south toward India, and a lengthy coast line. The climate varies from cold deserts in the north to damp tropical areas in the south.

Politically, wartime China was a confusing jigsaw puzzle of areas controlled by the Kuomintang, Communists, the Japanese, local warlords, and Chinese guerrilla organizations. Regarding the delicate political situation in China, Americans were cautioned: "You are not in China to get into politics or discuss politics . . . In fact, by order of the Commanding General, U.S. Forces, China Theater, you are NOT to discuss Chinese politics, or make any promises or give aid to any particular group or party, except on Army orders."[16] Adding to the confusion, bandits sometimes effectively controlled localities within these larger areas, and downed airmen were at risk of being killed for their valuables or simply because they were strangers.

Air crews were advised that the evader "will run into many difficulties in making his way through a strange country" and that "language difficulties will probably bother him", but that "his only real danger is that of disease. Even when in Chinese hands, he must watch his health, and strictly observe the precautionary points on food and water."[17] That evasion was possible in China in spite of the political, language, health, and other difficulties, even in Japanese controlled areas, was stressed to crews who were lectured that "organized Chinese guerrillas control most of the 'occupied territory', a few miles from railroads and waterworks", and that "the flyer who fails to get back to Free China will be exceedingly unlucky."[18]

India, by virtue of its location, its industrial, natural, and human resources, and its own military forces, played an important role in the Allied effort against the Japanese. With land routes blocked by the Japanese until the construction of the Ledo Road, the difficult and costly airlift to supply China over "The Hump" – the Himalayas – originated in India. From these bases a stream of aircraft flew supplies over "some of the wildest and least explored country on the globe".[19] American crews were briefed that in India and Southeast Asia "if airmen should come down in the forest and mountains, they may have to walk for a couple of days. But nearly every where else a village may be expected within reasonable distance. Having seen one and knowing where it lies, he can head for it. It is probable that, with the gift of a watch, torch [flashlight] or other small trinket, the villagers will help evaders all they can. Traditions of hospitality in the East are very strong."[20]

Burma was described as "an extremely rugged mountain country and its ranges spreading from border to border are cut through by some of the largest rivers on earth, one draining from the snows of the Himalayas . . . much of the back country is quite inaccessible . . . and the distances seem vast." As to the reception a downed American airman might expect in Burma, the official view was that "the chances are excellent that he will make contact with the natives immediately on landing. The additional chances that these natives will be friendly are also excellent . . . under normal circumstances Americans can count on their friendship."[21]

When World War II began, Thailand was an independent state, an exception to the colonial history of Southeast Asia. The Thai government had cooperated with the Japanese since 1939, and had declared war on the United States. In spite of this, evaders in Thailand were told to "head for Shan States in Northwest Burma, or through French Indo-China to Kunming . . . About 10% of the population . . . is Chinese, who bitterly hate the Japanese. Get in touch with them. The route will depend on such contacts. Remember, however . . . wealthy Chinese merchants may not prove to be as inclined to run the risk of giving aid to the evader as the poor classes."[22]

With the consent of the Vichy French government, the Japanese occupied French Indo-China in the summer of 1941. The situation an evader faced in Indo-China was complicated by nationalists who opposed both the French and the Japanese, and by an armed independence movement led by Communists. Regarding assistance from the population of Indo-China, air crews were advised that "while some of the natives in the larger cities have accepted the Japanese doctrines, the country people may be depended on to help . . . the probabilities of evasion depend on the success in the early stages of avoiding the Japanese and contacting the friendly natives. The natives will help if it is possible for them to do so . . . The land routes are due north into the Yunnan province of China. Obtain a guide if possible, and pay him for his services."[23]

Malaya, a British colony rich in natural resources and in a strategic location, was seized by the Japanese early in the war. Because of its location and geography, airmen downed over Malaya faced long journeys by land or sea to return to friendly control. Again, in spite of difficulties, evasion was possible and crews were lectured that "by land, the only route is north through northern Thailand or southern Burma into China. The undergrowth is extremely dense and a guide will be necessary since the only paths through the jungle are known to the local inhabitants . . . It may be possible to escape by sea from the western side of the peninsula to Sumatra and thence to India or Australia depending on the winds."[24]

Korea had been Japanese territory since as early as 1894 when a defeated China was forced to cede the peninsula after controlling it since the 7th Century. The Japanese gained additional territory as a concession from the defeated Russians in 1905, and the Korean peninsula was annexed by Japan in 1910. Japan ruled Korea harshly, as it did all conquered territory, and the MIS-X Manual advised air crew members that assistance from Koreans "should be forthcoming because of their hatred of the Japs"[25] but that "Korea is firmly controlled and overrun by Japanese soldiers" and that an evader in Korea needed to move "westward toward the Yellow Sea" where "a boat will be necessary to reach China and contact made there with guerrilla bands for further aid." Evaders were also advised to avoid urban areas and not to head north toward Russia because of strong Japanese troop concentrations in Manchukuo.[26]

The risks for those operating in the skies over Asia created an urgent demand for expanded survival and evasion training and equipment. The responsibility for meeting these needs fell on the established British and American evasion and escape organizations in Asia.

E-Group, MIS-X, and AGAS realized that in China, as elsewhere, Allied airmen had little chance of avoiding capture and returning to friendly control without the help of the local populace. This included many peoples in addition to the Han population which, in spite of their many political divisions, shares a common ethnic heritage. One of the many ethnic minorities was the Lolo Tribe, who inhabited a mountainous region in southeastern China. AGAS sent two missions to the Lolos, both headed by Captain Frank "Moon" Mullen.[27] It was later reported that "an agreement made by this agency with the Lolo people of western China provided that the tribal chieftain whose people rendered aid was to receive 'six yards of colored cloth, some wine and two catties of salt' for each American returned to safety and each member of the party which escorted the Americans was to received CN $300 per day."[28] A cattie is a unit of weight equivalent to about 20 U.S. ounces, and CN denotes the currency of Free China, the Chinese National yuan.

Adding to the difficulties created by the variety of peoples in China was the fact that Westerners could not readily distinguish the differences between Chinese and Japanese persons. Furthermore, not all Japanese in China and other parts of Asia were in the military. Civilians followed the army, particularly in urban areas. There were also Japanese agents who posed as Chinese, paid Chinese traitors, and Chinese "puppet troops" who worked for the Japanese. Puppet troops were often used in areas that the Japanese themselves could not spare their own personnel to control. Although fed and paid by the Japanese, they could be made to cooperate with guerrilla forces. The Japanese would punish puppet troops for any outbreaks in their territories, so puppet troops would sometimes appease guerilla forces rather than risk punishment by the Japanese.[29]

In spite of the varied peoples and customs that air crews might encounter, airmen were provided with some basic rules for dealing with all Chinese. These included "Be easy and friendly . . . prod them on, but gently. Once certain of their good intentions, let them run their own show. Don't try to interfere – a carefully worked-out plan may be ruined . . . If an evader finds it difficult to get their help, he should appeal to their self-respect, their honor and patriotism – also their pocketbooks. Simply be an American, in the best sense. They will not think less of an evader if he breaks a rule or two, if they are convinced he wishes to respect them and to be friendly with them."[30]

A U.S. Navy pilot who walked out of China offered further advice about contact with the Chinese: "Keep your faith in them

Fig. 7.1: These 25, 50, and 100 Chinese CN notes issued by the Bank of China were the official currency of the Kuomintang government. The notes shown were printed by the American Bank Note Company for the Chinese government in 1940. Note the "Chungking" overstamp just below the "100" on the back of the bottom bill, denoting that the currency was backed by the Nationalist government which had moved to Chungking in October 1938, after the fall of Wuhan. *International Blood Chit Museum*

Fig. 7.2: The CRB currency shown here was called "puppet money" as it was issued by the Central Reserve Bank of the puppet government established by the Japanese in Nanking. Essentially occupation money, it took about 40 CRB to equal the value of one CN, so large denomination bills such as the 500 and 1000 CRB notes shown were commonly used. Both CN and CRB notes, along with bank notes of other banks such as the Central Bank of China and the Communications Bank, were all in circulation at the same time. *International Blood Chit Museum*

and never cease to trust them completely as they apparently will never betray you as long as you let them suggest your course and your activities even though it does try your patience to the limit and you feel you could do better by yourself. They know China and you don't."[31]

The KMT government in China was also active in promoting air crew rescue, and late in the war Chiang Kai-shek issued an order, Yu No. 8437, detailing rewards for the rescue of downed flyers. The full text of the order is given in Appendix I.

The British and American E&E agencies eventually developed an extremely sophisticated E&E system that included techniques, training, and equipment to allow a downed airman to survive, to evade capture, and to communicate with local inhabitants. The thousands of Blood Chits and Pointie Talkies issued to Allied air crew members, together with countless leaflets instructing native populations how to assist downed airmen that were dropped over Asian countries, were all a part of that system.

The Americans and British shared much of the responsibility for furnishing evasion and survival items to air crews in Asia. In the early years of American involvement, the British provided much of the E&E material issued to American forces: "Equipment includes the Jungle Kit for operations over India, Assam, The Himalayas, Burma and Southeast Asia. The British are furnishing the standard evasion kits to both the 10th and 14th Air Forces and the A.T.C. Blood chits are also used."[32]

One universally valuable evasion item in all theaters was Evasion Currency. Official literature and lectures encouraged evaders in Asia to pay helpers for their services, and to use money to entice hesitant or reluctant persons to provide assistance. The training proved its worth for evaders in the field. One downed airman who returned to his unit reported, "At one point we stopped for an interminable time while the three Chinese engaged in what sounded like an argument. Obviously I was the subject of it, but I had no idea what the argument was about. Possibly they were arguing over which one was to get the credit for rescuing me? I reached into my pocket and gave each of them a fifty-rupee note. That seemed to satisfy them. On we went."[33]

In China, American currency was worthless. No Chinese would accept American money because he might pay for it with his life if it was found on him by the Japanese. The value of the differing Chinese currencies varied greatly: one Chinese National yuan, called "CN", was worth about forty yuan of "puppet money", called "CRB" for the Central Reserve Bank that issued it.[34] Inflation also played havoc with currency values in China, and between April and June of 1945 the exchange rate for U.S. dollars to CN doubled, soaring from 1:600 to 1:1200.[35] Nonetheless, CN was a vital evasion aid. Some of the CN notes of Free, or unoccupied, China are shown in Fig. 7.1, while some of the CRB notes of the Japanese-installed puppet government are shown in Fig. 7.2. These, and notes issued by other banks in China, were all in circulation at the same time.

In late summer 1944, AGAS-China recommended to CPM Branch of G-2 specific minimum amounts of money to be included in evasion material. AGAS recommended that each air crew member "should have at least $5000 CN in $100 CN notes". Further, AGAS recommended that crewmen operating over French Indo-China "should have at least 500 Piastres, obtainable only over here in 5, 10, 20, and 50 Piastre denomination notes."[36]

Paper currency was not universally valued, and in remote areas coins proved far more useful. One evader who returned from Burma related, "I showed them a 20 rupee note and they just dropped it on the floor. The silver 1/2 rupee pieces they understood."[37]

An item found to be of far more value than either coins or currency to native peoples in some areas was opium. Small balls of the gummy resin were included in evasion kits beginning in early 1944. The value of the drug among some of the indigenous tribes, and a recommendation for its use, appears in an official evasion report of 17 January 1944. This report stated that "there will be made available to the pilots another commodity which is much more highly prized by the natives than money. Further details on this will be given the pilots in the near future."[38]

When General Curtis LeMay of the Twentieth Bomber Command found that some of the more remote tribes would not help return downed airmen unless paid with opium, he ordered his people to obtain it. The officer in charge of purchasing, knowing that he could not enter the drug on his manifest, wrote that the payment was for the purchase of fertilizer.[39]

That opium was a valuable addition to evasion material was confirmed as early as May 1944, in a walkout report that related, "The chemical again proved its worth. On the fifth day out Lt. Robbins ran upon a small lean-to and shortly upon a native Kachin whom he persuaded to lead him through the jungle. After a bit the Kachin was anxious to be on his own way. The chemical proved a powerful persuader and inducement for the boy to continue with him. The boy took him to Kawlangyang where other Kachins went on with him to Arang."[40]

In addition to the actual payment of money or opium, the mere promise of a large reward could and did entice reluctant natives to

Fig. 7.3: Shown here is very unusual a post-AVG Rescue Patch. Its most striking features are that the chop is centered and it has no "may guo" characters. This is virtually unheard of on Rescue Patches issued to USAAF aircrews in China. Its serial number has been changed by hand from 0418 to 2418, prompting speculation that it may be an unissued or reissued AVG chit. This chit belonged to George Brouillette, who served with an unspecified unit of the 14th Air Force. Its text is identical 14th VBS and AVG Rescue Patches without the "may guo" characters: "This foreign person has come to China to help in the war effort. Soldiers and civilians, one and all, should rescue, protect, and provide him medical care." (REB 2220) *International Blood Chit Museum*

provide needed assistance. One successful evader related, "Two men looked at my pointie-talkie and when they came to the phrase 'if you get me through behind the Chinese lines, you will be rewarded' they seemed very much interested, and made signs as if they planned to take charge."[41]

As they did in Europe, the British issued Purses to airmen, and this item was added to the inventory of evasion aids issued to American crews as well. In the Far East, Blood Chits were sometimes included in the Purses:

> BLOOD CHITS. These are also contained in Money Purses, and permit an Allied air crew member to identify himself. At the top of each piece of silk is either a Union Jack or a Stars and Stripes according to the nationality of the holder. This flag may also be a valuable subsidiary aid for an evader to identify himself to our own troops; held up or across the chest it is visible for some distance.[42]

The British also followed their practice of writing notes promising a reward to helpers. These notes were, as in the European theater, sometimes referred to as "chits", a practice that has caused confusion among some present day researchers unaware of the differences between Promissory Notes and Blood Chits. Promissory Notes were also contained in Purses issued to airmen:

> PROMISSORY NOTES. Blank sheets of paper with the Coat of Arms printed at the top are also enclosed in this Money Purse for use as Promissory Notes. An evader should write brief details of the help he has received from anyone whom he desires to reward and should append his signature, his own name, rank and number only. No amount is to be specified on these sheets, but the statement should promise a suitable reward at the first available opportunity.[43]

Detailed instructions regarding the use of evasion aids, including Blood Chits, were given to all Allied air crew members. Royal Air Force and Fleet Air Arm crews were lectured, "If you are an aircrew member you will have what is called a 'blood chit', indicating your nationality . . . The blood chits have a statement in 17 languages to this effect printed on them. Use them, therefore, if there is anyone likely to understand."[44] American aviators also received briefings about evasion aids, including Blood Chits, and how they were to be used in China and throughout Asia. These instructions were specific:

> An evader should identify his nationality and objective.
>
> Fliers should be in possession of a chit promising reward for safe delivery to unoccupied territory. This chit should be printed on good paper and made to look as impressive as possible i.e. should be "chopped" with seals of high Chinese and Allied authorities.
>
> This chit should be produced as soon as possible after landing and proffered with the words "PAO CHANG" (Headman) and "CHING KAH" (reward).
>
> Have the highest regard for the bloodchit and any other identification equipment furnished. The Chinese in Occupied China will be suspicious: - they may even suspect an evader of being a spy.
>
> As soon as natives (they will probably be farmers) appear on the scene their attention should be drawn to the Chinese or

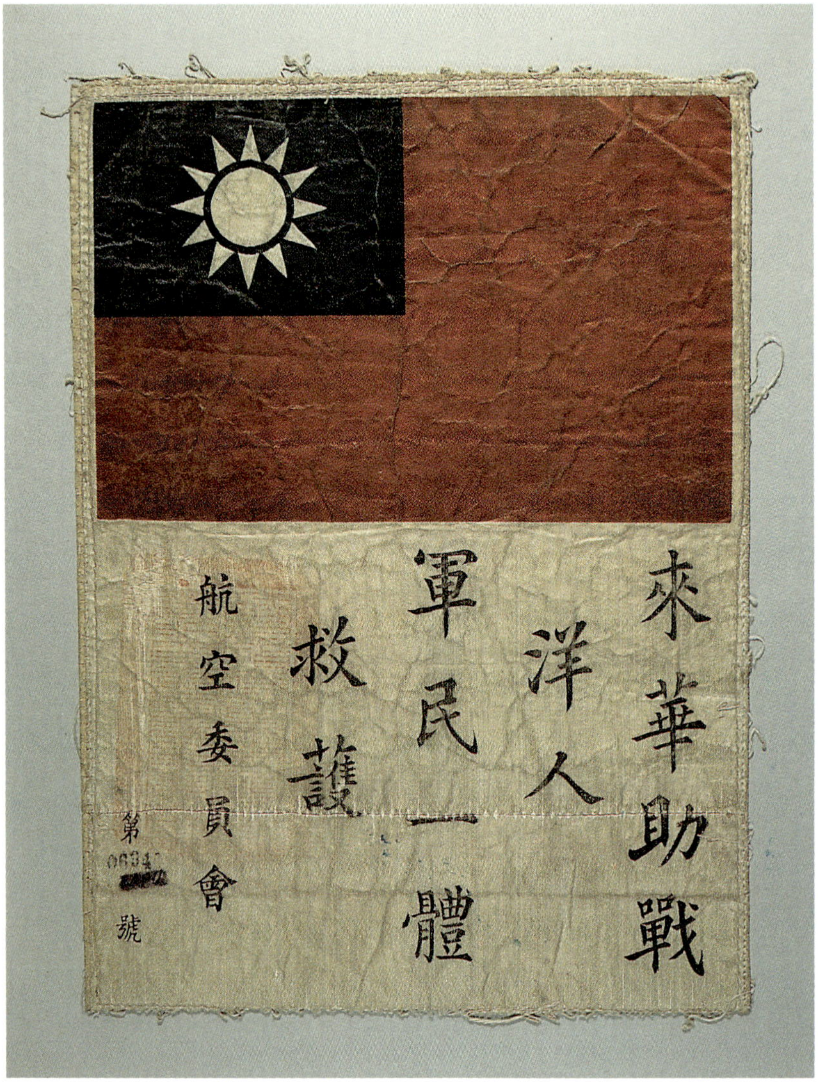

Fig. 7.4: This unusual Rescue Patch has many features nearly identical to the Rescue Patches of both 14th VBS and AVG. Only two characters accompany the serial number, as on 14th VBS patches and some of the AVG patches. It has the same dimensions of the AVG Rescue Patches, which are larger than the later Aero Commission chits. The text is written in a delicate hand, as is the text on William Labussière's 14th VBS chit and on some AVG chits as well. It was re-numbered from 1427 (crossed off in purple ink) to 0694. Obviously a very early example, it may have been produced with those for the AVG but never issued, then re-numbered and stamped with the chop following the disbandment of the AVG. The line across the chit two-thirds down in the text area is stitching where it was sewn inside a flight jacket. This example illustrates the heavy wear a chit sewn to a garment is subject to. The "may guo" stamp has all but disappeared due to wear. (REB 2222) *International Blood Chit Museum*

> Allied flags which, together with an explanation written in Chinese characters, are stitched to the flying jacket.
>
> Repeat several times the words "MAY GWO" (American) or "YING GWO" (English).[45]

The Commission for Aeronautical Affairs under Mme. Chiang Kai-shek's leadership continued to issue Blood Chits, originally called Rescue Patches, to American flyers following the disbandment of the AVG. The basic design of these chits remained unchanged from the earlier items illustrated in Chapter 4, with the Nationalist Chinese flag at the top and the standard text below it.

Shown in Fig. 7.3 is a highly unusual post-AVG Rescue Patch. It features a centered chop, it lacks the two "may guo" characters in vertical parentheses, and the serial number has been changed by hand from 0418 to 2418. The above suggests that this chit, issued to 14th AF flyer George Brouillette, may be an unissued AVG Rescue Patch that was re-numbered and issued after the AVG was disbanded. Its text is identical to that of the earlier 14th VBS and AVG Rescue Patches without the "may guo" characters, examined in detail in Chapter 4:

Fig. 7.5: This example of a World War II Aero Commission Rescue Patch is notable because the serial number is parallel to the side rather than to the bottom of the patch, and the "may guo" characters are stamped in black ink instead of the usual purple ink. It belonged to Richard Griggs, who wrote his name on it. The text reads, "This foreign person (an American) has come to China to help in the war effort. Soldiers and civilians, one and all, should rescue, protect, and provide him medical care." This Rescue Patch also illustrates the fact that the ink used for the text was not waterproof. (REB 2224.1) *International Blood Chit Museum*

Fig. 7.6: Many variations exist in the Chinese Aero Commission Rescue Patches. On this patch, numbered 3287, the chop is centered on the left side of the text rather than being at the top left. Also notable is the fact that the "may guo" stamp is a darker purple color than most. (REB 2224.2) *International Blood Chit Museum*

> This foreign person has come to China to help in the war effort. Soldiers and civilians, one and all, should rescue, protect, and provide him medical care.[46]

Several minor changes were made to the chits issued after the disbandment of the AVG, such as stamping the "may guo" characters onto the chit instead of writing them by hand, and moving the chop from the center of the text to the left side.

Unlike the mass-produced Blood Chits of the U.S. and Britain, the Chinese Aero Commission chits showed many small variations, and a model by which to judge examples is virtually impossible. The examples illustrated in this work are but a few of the many variations to be found.

The Rescue Patch shown in Fig. 7.4 is unusual in that it has many features identical to the Rescue Patches of both the 14th VBS and AVG, including its size and the very fine hand in which the text is written. This chit may have been produced for but not issued to the AVG, and then later renumbered from 1427 to 0694, and chopped in the appropriate location for issue to Army Air Force personnel following the disbandment of the AVG. The line across the text area of the chit is where it was sewn into the lining of a flight jacket.

Another chit, shown in Fig. 7.5, is notable because the serial number, 3038, is stamped parallel to the side of the patch, and the "may guo" stamp is in black rather than the usual purple ink. This example also shows that the ink used for the text is not waterproof and will bleed when the chit becomes damp. The name of the person to whom it was issued, Richard Griggs, is written beneath the chop.

In yet another variation, the "may guo" stamp on the Rescue Patch shown in Fig. 7.6 is a darker purple than normal. The chop is centered on the left side of the text rather than being at the top left as was the norm. Minor variations like these make it impossible to come up with an exact model for a "typical" early World War II Chinese Aero Commission Rescue Patch.

If there were a "pattern" early style Aero Commission chit, the example shown in Fig. 7.7 might serve as such. The "may guo" stamp is done in the purple ink typically used for it, and the chop is at the upper left corner of the text area. The character meaning "help", third down in the first (right-hand) column, identifies this as an early chit.

Fig. 7.7: This Rescue Patch might be called the "pattern" for the early style World War II Aero Commission chits having a text identical to the earlier AVG chits. The chop is at the upper left corner of the text area, the serial number, 04853, is centered between the characters as intended, and the "may guo" stamp is done in purple ink. The third character down in the first (right-hand) column of text, meaning "help", identifies this as an early chit with a message identical to that on the AVG Rescue Patches having the "may guo" characters. (REB 2224) *International Blood Chit Museum*

Fig. 7.8: This unissued chit belonged to John Cella, E&E Officer of the Twentieth Bomber Command. It is typical of the later style Chinese Aero Commission Rescue Patches, whereon the third character down in the first (right-hand) column of text was changed from "help" to "participate." The revised text reads, "This foreign person (an American) has come to China to *participate* in the war effort. Soldiers and civilians, one and all, should rescue, protect, and provide him medical care." As is typical of Rescue Patches with the revised text, the "may guo" stamp in purple ink does not fit into the space at the bottom of the second (from right) column and is stamped partially over the second character down. (REB 2226) *International Blood Chit Museum, a gift of John L. Cella*

The Aero Commission chit shown in Fig. 7.8 belonged to John L. Cella, Escape and Evasion Officer and Search and Rescue Officer for Twentieth Bomber Command. Never issued, it is in the condition in whichhe received it from AGAS for issue to XX BC personnel. The chit features a revised text. The third character in the first (right-hand) column has been changed from "help" to "participate". The text change was made late in the war, but before the end of 1944, and reads:

> This foreign person (an American) has come to China to *participate* in the war effort. Soldiers and civilians, one and all, should rescue, protect, and provide him medical care.[47]

While the Aero Commission was issuing its chits to American flyers in China, the British were producing a variety of evasion aids for issue to their crews. Shown in Fig. 7.9 are some of these items which will be discussed in detail later in this chapter. Blood Chits were produced locally on paper backed with linen and on silk. They were also printed on rayon in Britain. Pointie Talkies, like that in Malay, were printed in-theater, and others, like the one in Chinese, were manufactured in Britain. The instruction booklet for the Mark III Jungle Kit was printed in India. A silk evasion map printed in India and a rayon map printed in Britain are shown, along with an RAF evasion compass on a cord to be worn around the neck.

The instruction booklet for the British Mark III Jungle Kit contains five sections. The first describes the contents of the kit, which consisted of two tins, one containing evasion aids and the other containing first aid items, and the contents of the Mk. III flying overalls with pockets containing various survival and evasion items. The second section describes the uses of the medicines included in the kit. The third section gives remedies for health and medical problems that might be encountered. The fourth section contains hints on jungle living. Three fold-out pages of phrases in English

Fig. 7.9: Some of the evasion aids issued to British crews operating in the skies over Asia are shown here. These include several styles of Blood Chits made of silk, rayon, and paper backed with linen. Also shown are two Pointie Talkie booklets, the instruction booklet from a Mark III Jungle Kit, and a compass enclosed in a waterproof plastic pouch to be worn around the neck. Two cloth evasion maps, one printed in black and white on silk in India, and one printed on rayon in Britain, are also shown. *International Blood Chit Museum*

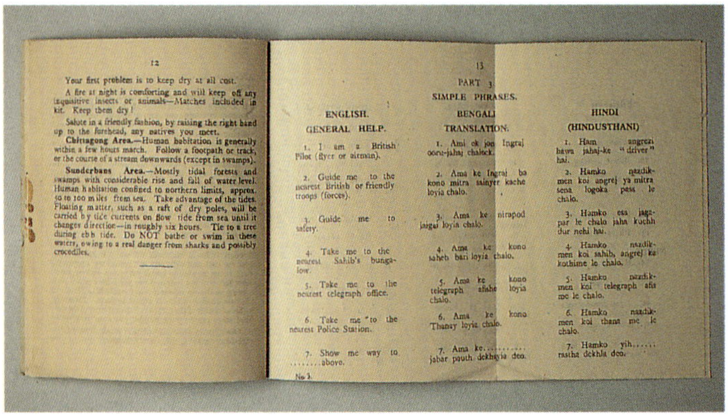

Fig. 7.10: The instruction booklet included in the British Mark III Jungle Kit contains three fold-out pages of phrases in English and in phonetic Bengali and Hindustani (Urdu). One of these pages is shown here. Other sections of the booklet describe the kit's contents, the uses for the medicines included in the kit, "Likely Troubles and their Remedies", and "Do's and Dont's." The booklet was printed in India. *International Blood Chit Museum*

and in phonetic Bengali and Hindustani, one of which is shown in Fig. 7.10, make up the fifth section. Phrases are included for seeking a guide, food and water, and medical assistance.

In Asia, as in Africa, the British produced Goolie Chits on linen-backed paper at a very early date. Four of these paper and linen chits are illustrated. That shown in Fig. 7.11 gives its message in Bengali above the Royal Arms and in English below it. C. H. Anderson, a hurricane pilot flying against the Japanese in the Burma Campaign during 1943-44, was issued this type of chit and always carried it "just in case".[48]

The Goolie Chit shown in Fig. 7.12 was intended for use in Burma. It is printed on paper and glued to a linen backing. The four languages above the Royal Arms are Burmese (upper left), W. Shan (upper right), Sgaw Karen (lower left) and S. Shan (lower right). Jimmy Bell, a member of the CBI Hump Pilots Association, was issued one of these Burma chits along with a Bengali chit (Fig. 7.7) while serving with 31 and 62 Squadrons RAF in India and Burma from 1942 to 1945. He writes, "When you were flying on operations there were two things you always made sure that you flew with, and that was your parachute and 'Goolie Chits'."[49] D. W. Walker also reports that these chits were issued to No. 62 Squadron RAF.[50]

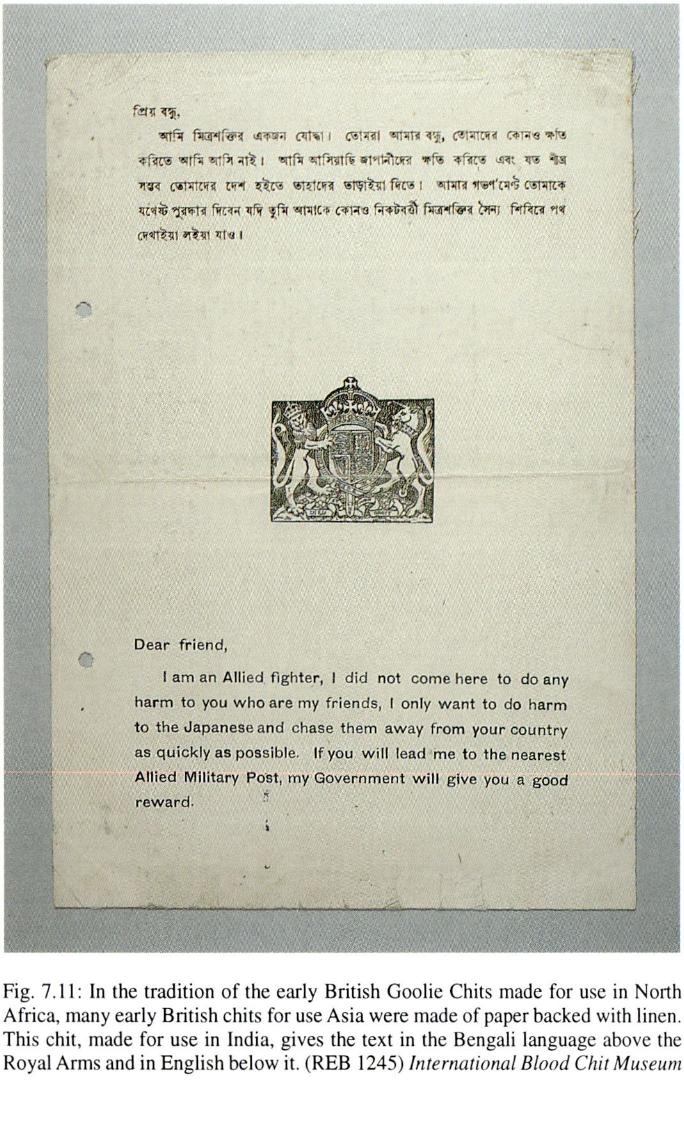

প্রিয় বন্ধু,

আমি মিত্রশক্তির একজন যোদ্ধা। তোমরা আমার বন্ধু, তোমাদের কোনও ক্ষতি করিতে আমি আসি নাই। আমি আসিয়াছি জাপানীদের ক্ষতি করিতে এবং যত শীঘ্র সম্ভব তোমাদের দেশ হইতে তাহাদের তাড়াইয়া দিতে। আমার গভর্ণমেন্ট তোমাকে যথেষ্ট পুরস্কার দিবেন যদি তুমি আমাকে কোনও নিকটবর্তী মিত্রশক্তির সৈন্য শিবিরে পথ দেখাইয়া লইয়া যাও।

Dear friend,

I am an Allied fighter, I did not come here to do any harm to you who are my friends, I only want to do harm to the Japanese and chase them away from your country as quickly as possible. If you will lead me to the nearest Allied Military Post, my Government will give you a good reward.

Fig. 7.11: In the tradition of the early British Goolie Chits made for use in North Africa, many early British chits for use Asia were made of paper backed with linen. This chit, made for use in India, gives the text in the Bengali language above the Royal Arms and in English below it. (REB 1245) *International Blood Chit Museum*

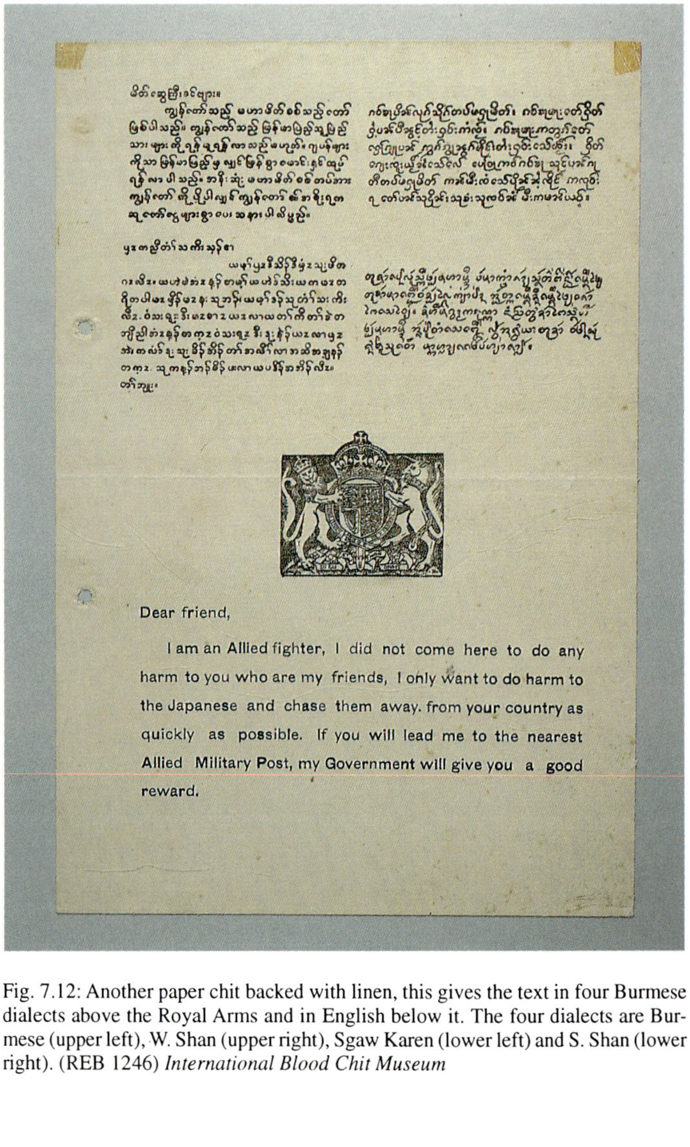

Dear friend,

I am an Allied fighter, I did not come here to do any harm to you who are my friends, I only want to do harm to the Japanese and chase them away. from your country as quickly as possible. If you will lead me to the nearest Allied Military Post, my Government will give you a good reward.

Fig. 7.12: Another paper chit backed with linen, this gives the text in four Burmese dialects above the Royal Arms and in English below it. The four dialects are Burmese (upper left), W. Shan (upper right), Sgaw Karen (lower left) and S. Shan (lower right). (REB 1246) *International Blood Chit Museum*

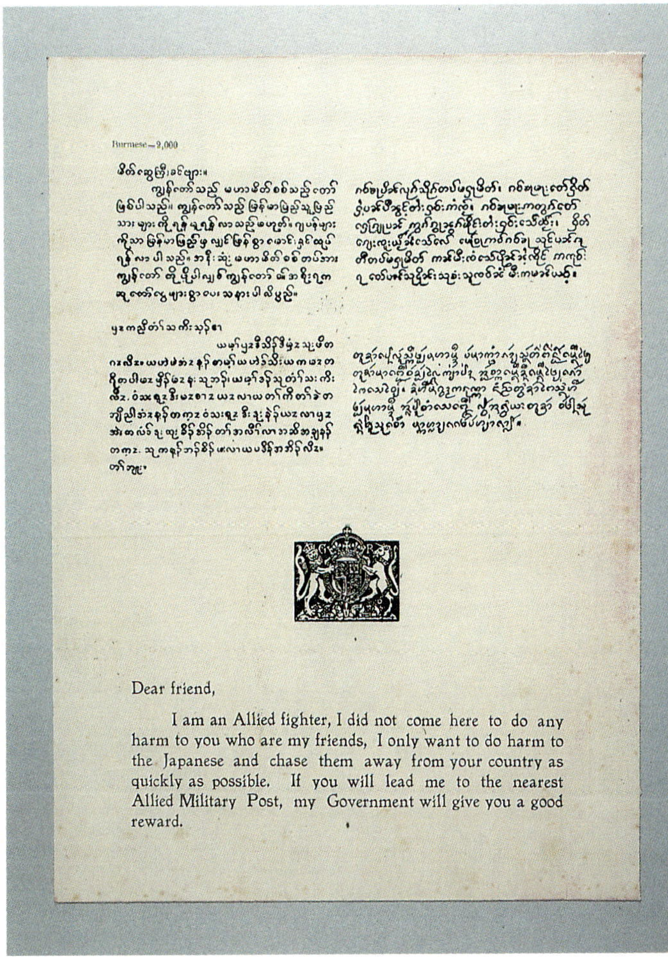
Burmese—2,000

Dear friend,

I am an Allied fighter, I did not come here to do any harm to you who are my friends, I only want to do harm to the Japanese and chase them away from your country as quickly as possible. If you will lead me to the nearest Allied Military Post, my Government will give you a good reward.

Fig. 7.13: An unissued variation of the British paper and linen chit for use in Burma is shown here. The size of the Royal Arms is smaller and the text style is distinctively different than that on the chit shown in Fig. 7.12, and there is a notation at the top that 2000 were produced. This chit belonged to John L. Cella, E&E Officer of the XX BC, who received it from E-Group in Calcutta for issue to U.S. crews. (REB 1246.1) *International Blood Chit Museum, a gift of John L. Cella*

Fig. 7.13 shows another paper and linen Goolie Chit made for use in Burma. This chit belonged to John L. Cella, the Twentieth Bomber Command's Escape and Evasion Officer and Search and Rescue Officer. Never issued, it is exactly as received by him from E-Group in Calcutta, which provided British chits for issue to American as well as British and Commonwealth aircrews. Similar in layout to the Goolie Chit shown in Fig. 7.12, it was made at a later date has some distinct differences. The size of the Royal Arms is smaller, the type faces are different, and there is a notation at the top that 2000 were produced. The English and Burmese texts are identical to those on the chit shown in Fig. 7.12.

Another in the series of linen-backed paper Goolie Chits is shown in Fig. 7.14. As its message relates, this example was made for use in Thailand. In addition to the standard request for assistance and promise of reward, this chit contains a list of useful words and phrases at the bottom, a common feature on some of the British North African chits.

The earliest efforts of MIS-X Technical Section India were focussed on producing aids in languages found along the Hump route. The first of these aids was a set of three canvas-backed cards in Urdu (Hindustani), Chinese, and Burmese, designated DEB-1. Language aids were given DEB designations for use by MIS-X China, MIS-X India, and MIS India, while evasion maps, compasses, hacksaws, and wire cutters were given, respectively, DEM, DEC, DES, and DEW designations.[51] The meaning of the letters "D" and "E" was not discovered during research for this work, but they may represent something like "Device, Evasion". The last letter of each represents the first letter in "blood chit", "map", "compass", "saw", and "wire cutter".

The DEB-1 chits were at least in the planning stage, if not in production and distribution, by late summer 1943 as indicated by a memo from HQ India Task Force:[52]

Blood Chits.	
Tri-language Cards (Chinese, Burmese, Hind.)	DEB-1
Loose Leaf Type (Burmese)	DEB-2
Kachin Phrases	DEB-3

This will be used in all correspondence.

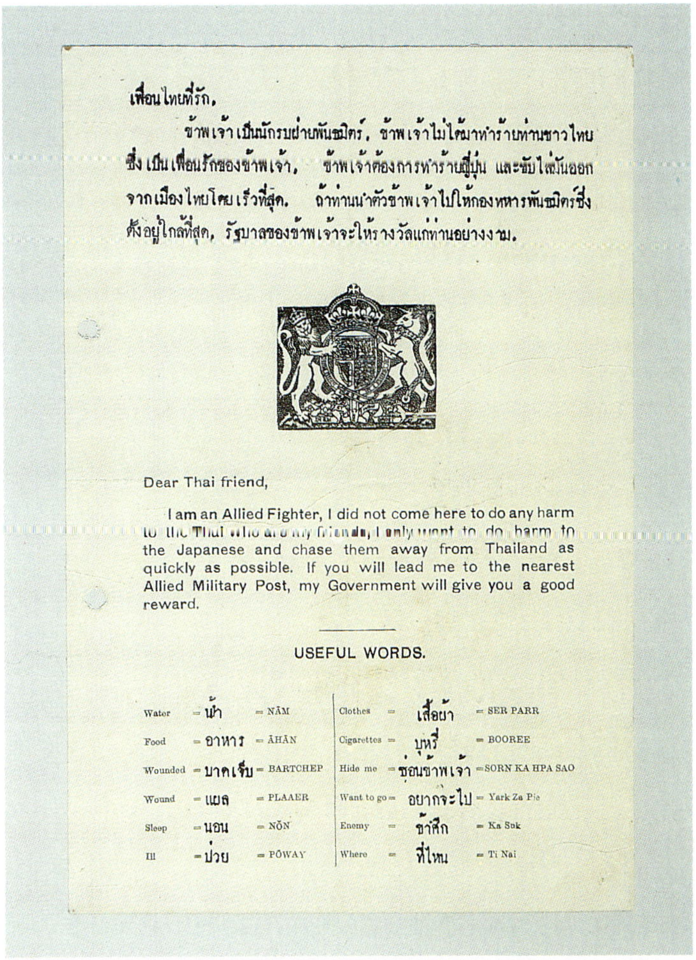
เพื่อนไทยที่รัก,

ข้าพเจ้าเป็นนักรบฝ่ายพันธมิตร, ข้าพเจ้าไม่ได้มาทำร้ายท่านชาวไทย ซึ่งเป็นเพื่อนรักของข้าพเจ้า, ข้าพเจ้าต้องการทำร้ายญี่ปุ่น และขับไล่มันออกจากเมืองไทยโดยเร็วที่สุด. ถ้าท่านนำตัวข้าพเจ้าไปให้กองทหารพันธมิตรซึ่งตั้งอยู่ใกล้ที่สุด, รัฐบาลของข้าพเจ้าจะให้รางวัลแก่ท่านอย่างงาม.

Dear Thai friend,

I am an Allied Fighter, I did not come here to do any harm to the Thai who are my friends, I only want to do harm to the Japanese and chase them away from Thailand as quickly as possible. If you will lead me to the nearest Allied Military Post, my Government will give you a good reward.

USEFUL WORDS.

Water	= น้ำ	= NĀM	Clothes =	เสื้อผ้า	= SER PARR
Food	= อาหาร	= ĀHĀN	Cigarettes =	บุหรี่	= BOOREE
Wounded	= บาดเจ็บ	= BARTCHEP	Hide me =	ซ่อนข้าพเจ้า	= SORN KA HPA SAO
Wound	= แผล	= PLAAER	Want to go =	อยากจะไป	= Yark Za Pie
Sleep	= นอน	= NŌN	Enemy =	ข้าศึก	= Ka Suk
Ill	= ป่วย	= PŌWAY	Where =	ที่ไหน	= Ti Nai

Fig. 7.14: This British paper and linen chit was made for use in Siam (Thailand). It features, in the tradition of several paper and linen chits made for use in North Africa, a list of useful words and phrases at the bottom in addition to the standard chit message. (REB 1247) *International Blood Chit Museum*

Fig. 7.15: Two sets of DEB-1 canvas-backed card chits made by MIS-X Technical Section India are illustrated here. The cards on the left are full-sized and are shown with their original plastic holder, while those on the right were trimmed to fit into the theater-made holder, also shown. The bottom card on the right shows the canvas backing, which was generally green or brown in color. The languages are Urdu, Chinese and Shan. These cards were used by crews flying over "The Hump" between India and China. (REB 3224) *International Blood Chit Museum*

Chinese, Shan, and Urdu languages. Although produced at a later date, these are obviously a direct descendant of the DEB-1 cards. These are not backed with canvas.

Another set of four cards patterned after DEB-1, shown in Fig. 7.17, was also made at a later date. These are smaller in size and are printed on a harder paper than the cards shown in Fig. 7.16. They were issued in a green cloth pouch with the AAF insignia and the words "Army Air Force" printed on it in white. The corners of the cards were trimmed by their owner to facilitate insertion into and removal from the pouch.

Unfortunately, no example of DEB-2, a set of cards in various Burmese dialects, was available for illustration. The cards are thought to be of a similar construction as the DEB-1 cards, but without a canvas backing.

In China, AGAS not only distributed Rescue Patches provided by the Aero Commission, but it also cooperated with PO Box 1142 in the U.S. and with E-Group and MIS-X Technical Section in India in the design of other language materials which were produced in the U.S. and in India. One of the projects undertaken in early 1944 involved combining contents of the standard paper navigation chart for the Hump route with a Blood Chit so that two of the most important evasion documents – cloth map and Blood Chit – could be issued as one item. By the end of February 1944, the design of this map was well under way, as disclosed by a memo from PO Box 1142 to Major Wichtrich of AGAS:

> The K'un-ming, Yunan, China to Chabua, Assam, India No. 133 Strip Map to be reproduced on silk cloth with attached blood chits has been submitted, through channels, for production. The finished cloth will be 23" x 25" in size and will cover approximately 33% more area in its North and South dimension than the old 133 Strip Map, being based on the January 1944 edition of said No. 133 Strip Map. In addition to the blood chits submitted with Major E. B. Wilcox's letter of 6 January 1944, an additional "chit" in the Bengali language will be printed on the new cloth map.[53]

Shown in Fig. 7.15 are two sets of DEB-1 card chits. On the left are full-sized cards with the two-ended plastic holder in which they came, and on the right is a set of cards trimmed-down slightly to fit into their theater-made holder. The bottom card on the right is upside-down to show the canvas backing, which came in both tan and green colors. As per the above memo, the languages on the cards are, from top to bottom, Urdu (Hindustani), Chinese, and Shan (a Burmese dialect).

Later, a set of four card chits tied together with string, shown in Fig. 7.16, was issued in a green canvas pouch with the AAF winged star printed on it in black. The back of the first card (in English) also has the winged star printed on it. The cards are in the

Major Wilcox, referred to in the above quotation, served with E-Group in Calcutta. The final product of this effort was printed on rayon, not silk. AAF Cloth Chart No. 133 covered the Hump route from Kunming, China to Chabua, India, and was the only cloth map to incorporate a Blood Chit. The right border of the eastern portion of the map, shown in Fig. 7.18, is a Blood Chit in seven languages: Bengali, Hindustani (Urdu), Kachin, Lisu, Burmese, Chinese, and English. This combined evasion map and chit was logical as the supply route over the Hump was heavily travelled and crossed territory where many languages were spoken.

Russell A. "Rusty" Bales is a member of ATC-Hump Pilots, Ltd., a pilots-only group. His aircraft was shot down over northern

Fig. 7.16: This set of four un-backed card chits tied together with string was a later development, obviously patterned after the DEB-1 chits (Fig. 7.15). They are shown with their canvas holder bearing the winged star of the USAAF. The winged star is also printed on the back of the card in English. (REB 3226) *International Blood Chit Museum*

Fig. 7.18: AAF Cloth Chart No. 133 shows terrain covered by the Hump route from Chabua, India, to Kunming, China. The chit printed in the border is in seven languages used along the Hump route: Bengali, Hindustani (Urdu), Kachin, Lisu, Burmese, Chinese, and English. This was the only item produced by the United States during World War II that combined two of the most important evasion aids: cloth map and Blood Chit. (REB 3230) *International Blood Chit Museum*

Fig. 7.17: Another set of four card chits, similar to that in Fig. 7.16, was printed on a hard, white cardstock, and issued with a canvas pouch. The cards were trimmed by the user so they could be more easily inserted into the pouch. (REB 3228) *Courtesy of Bob Gill*

Burma on 23 October 1943, and Bales was instrumental in the design of a new Blood Chit by virtue of several hand-written chits obtained by him during the course of his walkout. He writes:

> I had been shot down in the approximate center of the Hump. I and my crew were fortunate enough to be able to bail out, landing on the western slope of the high ridge to the west of the Mekong. That would put us between the Mekong River and the Salween River, in northern Burma.
>
> During the course of our dilemma we came across some friendly natives . . . and they assisted us into an English outpost, which to the best of my ability, I was able to identify as Ridam, Burma. That outpost was being evacuated due to the incursion of Japanese troops into that area. We were shot down on October 13, 1943, and we arrived at Ridam on October 23.
>
> There were two English Intelligence Majors at that base. I had two crewmen who were injured so they very kindly al-

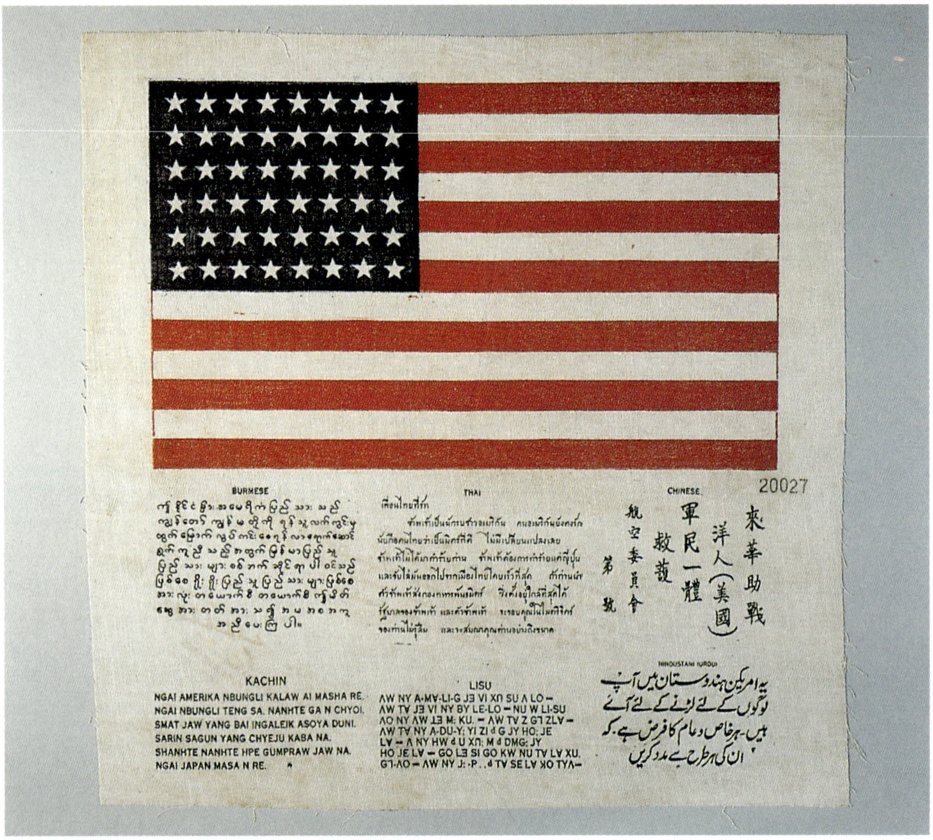

Fig. 7.19: This is the first Blood Chit printed in the U.S. It was produced by MIS-X in about March 1944 and is printed on cotton. Three of the messages on it are identical to those on the DEB-1 card chits produced earlier in India. The other three languages on the chit may be a result of Russel A. "Rusty" Bales' walkout in late 1943, during which a British intelligence officer in Burma gave him a set of six hand-written chits that he turned in upon his return to base. (REB 3232) *International Blood Chit Museum*

lowed those two to go ahead on their horses, on the 24th. We evacuated the base on the 25th, on foot. We arrived at the Fot Hertz Valley on the 11th of November.

During that march out I showed the two Englishmen the Nationalist Flag and message to assist the person with the flag-blood chit. The flag had only the Chinese message, and I asked them if they were familiar with the Burmese language. I was very fortunate, as between them they were able to give me six renditions of the Chinese message, covering about every language that a Hump Pilot and crewman would be likely to encounter.

I am so sorry that I did not get the identity of those two officers, also that I did not keep the original writings. As it happened, when I arrived at the Eastern Sector Headquarters of the Air Transport Command, I gave those originals to a Lt. Col John R. Kilgore.

Although I do not have a set of those cards, they were printed on cards in a group of six and handed out to ATC Hump Crews. Later a Blood Chit was prepared, showing an American Flag and those messages.[54]

The card chits referred to were evidently not those of DEB-1 or DEB-2, because these existed prior to the time his aircraft went down. In March of 1944, however, MIS-X Technical Section in the United States did produce a Blood Chit, illustrated in Fig. 7.19, with an American flag and six languages for use by ATC crews. As this was the only chit produced in six languages, it is thought to be the one referred to by Bales. Three of the messages are identical to those on the DEB-1 cards, so perhaps three of the six hand-written messages carried back by Bales were combined with the messages of DEB- 1 to provide the remaining text for this Blood Chit.

This six-language chit "was printed in the United States about March 1944, and is the first Blood Chit put out by the MIS-X Section, C.P.M. Branch, G-2 Branch, for Airmen in China."[55] This American-made chit, printed on cotton, was significant because it signaled a trend that saw more and more chits made in the U.S. It is also important to note the late date of this very first chit made in the U.S., because some of the later U.S. chits have been mis-dated to as early as 1943.

T/Sgt. Everett E. Swanson, who was top gunner on Major Weschler's crew in *Princess Patsy* and *Superstitious Aloysious* in the 40th BG, 20th AF, was issued one of these cotton chits around May of 1944 in India. He recalls that these chits were issued through early 1945.[56] Horace O. Bedsole, who was in China from September 1944 to July 1945 as a bombardier with the 373rd and 425th BS, 308th BG, was also issued one of these chits.[57]

The cotton 6-language chit produced in the U.S. did not have a "DEB" number, although it was sometimes mistakenly referred to as "DEB-10", which it somewhat resembled. DEB-10 was actually a 5-language chit designed and produced in India for use by Hump crews:

DEB-10 - American Flag 5 Language Blood Chit. This chit was designed by Major A. R. Wichtrich of A.G.A.S., China and the production of 5,000 of them was undertaken by MIS-X, India during the early part of 1944. This blood-chit was printed on silk cloth with a cotton cloth backing. All materials, printing and tailoring were procured locally thru Reverse Lend-Lease channels.[58]

Under the terms of Reverse Lend-Lease, first defined in September 1942, the British agreed to supply items and services for the Allied war effort not only in the Britain, but also in its Colonies. This included India, and Blood Chits required by American air crews there were procured from the British under this system.

An example of the DEB-10 chit is shown in Fig. 7.20. Although made locally in India, it is an officially issued chit. The five languages are, top to bottom beginning on the left, Burmese, Sgaw Karen, Chinese, W. Shan and S. Shan / N. Thai. This chit bears serial number 01390, although not all chits of this type were numbered. William A. Rooney was a forward area Intelligence Officer with the 40th Bomb Group in China for about 5 months, and with XX BC in India and China for all of the 19 months they were in that theater. He retained his three chits, one of which is this type.[59] A variation of the 5 language DEB-10 India-made chit, shown in Fig. 7.21, identifies the Burmese dialects with labels.

Fig. 7.21: A variation of the 5 language DEB-10 India-made chit is shown here. Unlike that shown in Fig. 7.20, the Burmese dialects on this version are labeled. (REB 3234.1) *International Blood Chit Museum*

Another silk chit, the American version of which was designated DEB-11, was also procured through the British via Reverse Lend-Lease. Like the DEB-10, it was produced in India with local labor:

DEB-11 - American Flag 17 Language Blood Chit. This chit was designed by Major Wilcox and S/Ldr. Vint of "E" Group. It is printed on silk cloth and is not "backed" so it is not adaptable to be worn on a garment. To date 10,000 of these chits have been produced in the same manner as were the DEB-10. 15,000 more of the DEB-11 are now on order.[60]

As indicated by the preceding quote, DEB-11 was designed by E-Group, the British E&E organization headquartered in Calcutta. Not surprisingly, the first versions of this chit were produced for issue to British and Commonwealth forces and featured the Union Jack.

Two versions of the British chit, illustrated in Figs. 7.22 and 7.23, were produced.

Fig. 7.20: This chit, known to MIS-X as DEB-10, was made in India. It was procured from the British through Reverse Lend-Lease, and although locally made, it was issued to American air crews flying the Hump. It is made of silk and has a cotton backing. (REB 3234) *International Blood Chit Museum*

Fig. 7.22: This 17-language British chit was designed by Major Wilcox and S/Ldr. Vint of E-Group. It was printed on silk in India and issued to British air crews operating in the Far East. This chit, featuring blue text, is one of two versions. (REB 1252) *International Blood Chit Museum*

The text portion of the chit in Fig. 7.22 is printed in blue ink. The languages in the left column are French, Tamil, Sumatra, Thai, E. Shan and N. Thai, and W. Shan; those in the center column are English, Jawi (Urdu), Chinese, and Haka; those in the right column are Kachin, Malay, Lazio (Chin), Bengali, Annamite (Vietnamese), Burmese and Karen. Messages written in the Roman alphabet such as that for the Haka, the "boat people" of southern China, are not actual written languages but must instead be read aloud by the user. D. W. Walker carried one of these chits while posted to No. 62 Squadron, RAF.[61]

The text on the version shown in Fig. 7.23 was printed in black ink, which is easier to read than the blue ink. The languages in the each column of this chit are identical to those on the chit with the blue text, except that the locations of the Burmese and Karen messages have been reversed.

The 17-language chits printed with the American flag at the top for use by U.S. forces were, as noted above, designated as DEB-11 by MIS-X. Like the DEB-10 chit, these were officially issued items produced in India by vendors to the British. The example shown in Fig. 7.24 has a blue text identical to that on the British version with blue text. A similar blood chit containing 16 languages has also been reported.

A two-sided American version of DEB-11, shown in Figs. 7.25 and 7.26, clearly shows the change from one color of text to the other. Fig. 7.25 shows the side of the chit with the blue text and flag, with one corner folded over to reveal the black text on the reverse. The juxtaposition of the Burmese and Karen sections are evident on this chit. Fig. 7.26 shows the entire back of the chit with black text.

What seems to be the final version of the DEB-11 chit, printed on rayon, is illustrated in Fig. 7.27. Although the text is printed in blue, the languages are arranged in the same order as those on the silk chits with black text. This chit belonged to Maj. Ralph L. Stevenson, who received it when he was posted to HQ India-China Division, Air Transport Command (ICD-ATC) in 1945. The fact that he obtained his chit during the last year of the war seems to indicate that this rayon chit is the final version of the series. The Burmese text was changed from that on the earlier silk chits, and this change caused a delay in the printing as detailed in this entry from the War Diary of MIS-X India: "Due to the fact that it was decided to change the Burmese Section of the 17 language blood chit, the printer did not get started on the printing of the 15,000 ordered by MIS-X from AG Publications during October [1944], until the last week of December. It is expected that delivery of these Chits will start early in January [1945]."[62]

The photo in Fig. 7.28 was taken by Major Stevenson. It shows an ATC airman inspecting a survival and evasion kit. Note the booklet "Survival" near his left hand, and to his left a folded rayon DEB-11 Blood Chit (Fig. 7.27), confirming that these chits were included in the kits issued to air crews of the India-China Division, Air Transport Command, in 1945.

The British produced a second style of the 17-language chit. On this, the Union Jack appears in the upper left corner rather than across the entire top of the chit. The example shown in Fig. 7.29 is an early version, printed on silk in India. The edges are sewn, and at the top there is a remnant of the glue that held the silk to a paper backing during the printing process. This glue also commonly appears on the edges of silk Government of India Survey maps. This chit is faded, but is still a fine example of the rare early version on silk.

The majority of these second style 17-language British chits were, like that shown in Fig. 7.30, printed on rayon. These were

Fig. 7.23: The second variation of the British 17-language chit features black text, and the locations of the Burmese and Karen messages have been reversed from their positions on the chit with the blue text. (REB 1253) *International Blood Chit Museum*

identical to the earlier silk version, but being rayon they were expected to hold up better under difficult field conditions. They were produced in large numbers, probably in Britain.

Bob Gilroy was issued one of these chits, and writes, "I was a pilot in RAF Transport Command towards the latter part of the war flying the route from England to Singapore via the Middle East, India and Ceylon in Avro Yorks. We were issued . . . 3 types of 'goolie chits' in the hope they would help to dissuade the native women along the route from practicing their notorious brand of amateur testicular surgery! Luckily, thanks to Rolls Royce reliability and total absence of enemy activity, I never had the need to use them."[63] It is likely that the reference to the male organs is a result of the oral traditions in the RAF that kept alive the original reasons for the creation of Goolie Chits, rather than any serious danger of this faced in the Far East during World War II.

Merv M. McGrane, an RAAF wireless operator on B-24 Liberators with 358 Sqdn RAF flying missions over Burma and Indo-China, writes, "All crew members carried escape kits – I still have mine – which included a 'blood flag' . . . with the Union Jack on the top left hand corner . . . in 17 different languages (including English)."[64]

One of the most beautiful chits to come out of the India-Burma Theater of Operations was printed on silk and backed with cotton. Shown in Fig. 7.31, it features the "Peacock Flag" of British Burma, below which is a message in Burmese. These chits were issued to British crews by E-Group, which also provided them to American E&E sections for issue to American crews that operated over Burma. One of these chits is sewn to the inside back lining of the A-2 jacket of Gen. Earl S. Hoag on display at the U.S. Air Force Museum (March 1995). Gen. Hoag commanded ATC's Africa-Middle East, India-China, and European Wings. As noted before, exact word-for-word translations are not always possible. The three slightly different translations obtained for the text of this chit are best summarized as follows:

> This foreigner, an American, has come to help free us from our enemy. We therefore urge that every citizen of Burma, whether civilian or military personnel, attend to our friend's needs with all their efforts.[65]

Fig. 7.24: The American version of the 17-language E-Group chit was called DEB-11 by MIS-X. Although locally made in India, it was issued to American airmen. This chit has a blue text identical to that on the corresponding British chit shown in Fig. 7.22. (REB 3236) *International Blood Chit Museum*

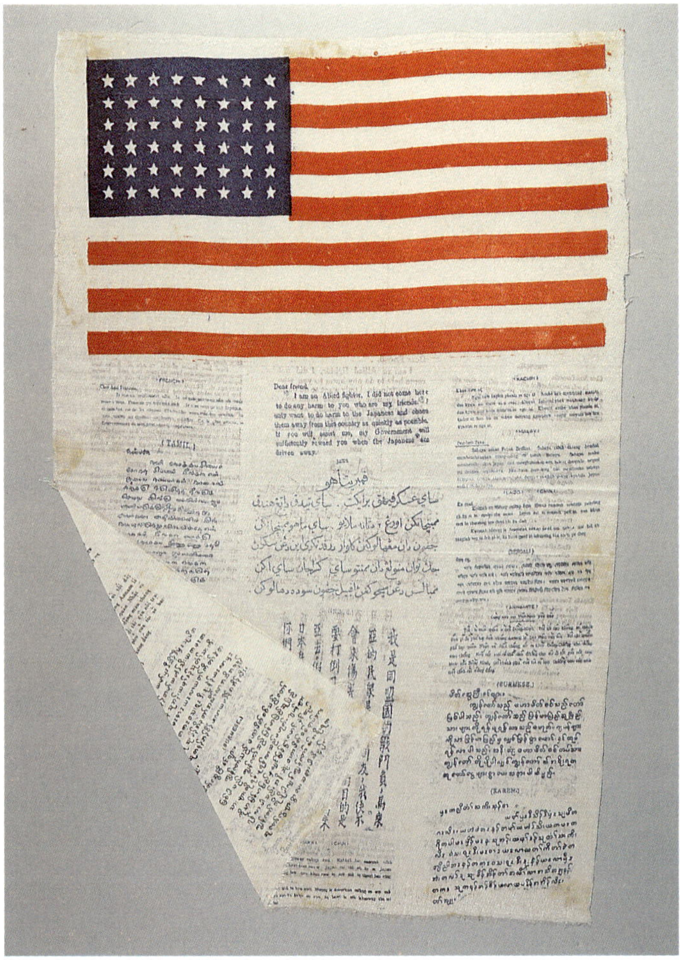

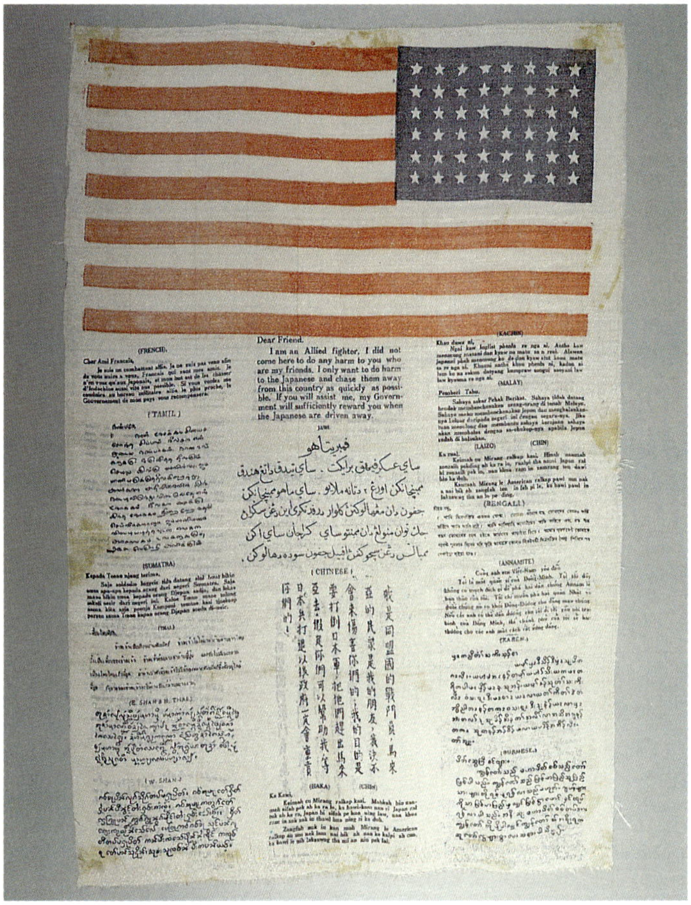

Fig. 7.25: The change from blue text to black text is evident on this two-sided DEB-11 chit. The front of the chit, shown here, has a blue text identical to the blue text on the British version. The text in black ink has been printed on the back of the chit, illustrated in Fig. 7.26. (REB 3238.1) *International Blood Chit Museum*

An accounting of the quantities of chits issued by MIS-X India as of February 1945 is given in the War Diary for MIS-X, HQ IBT.[66] Several language aids were issued in such limited quantities that they seem to be experimental pieces, and unfortunately no record as to their nature has been found:

LIST OF MATERIALS ISSUED BY MIS-X, INDIA, TO DATE

Item	Quantity
American Flag Blood Chits, 6 languages	2061
DEB-2	2351
DEB-3	1000
DEB-6	3
DEB-7	3
DEB-8	3
DEB-9	75
DEB-10	6726
DEB-11	12787
DEB-12	1

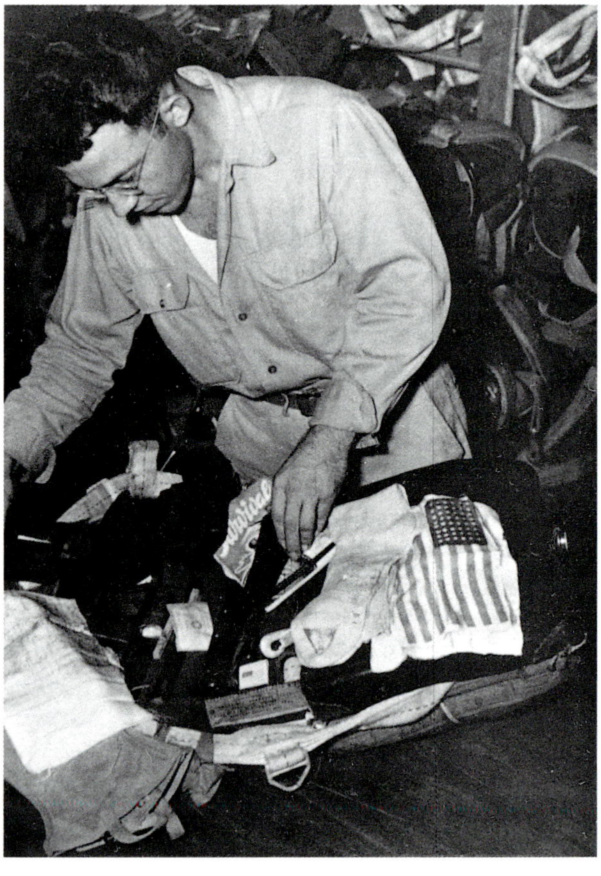

Fig. 7.28: A rayon DEB-11 Blood Chit was included in the survival and evasion kits prepared for ICD-ATC air crews, as shown in this photo of an airman inspecting a survival kit taken by Maj. Ralph L. Stevenson in 1945. Note the folded blood chit. *Courtesy of Ralph L. Stevenson Jr., son of Maj. Stevenson*

OPPOSITE: Fig. 7.26: The back of the two-sided DEB-11 chit features a black text that is identical to the black text on the corresponding British chit, including the positioning of the Burmese and Karen messages. The black text may have been adopted because it is much clearer and easier to read than the blue text. *International Blood Chit Museum*

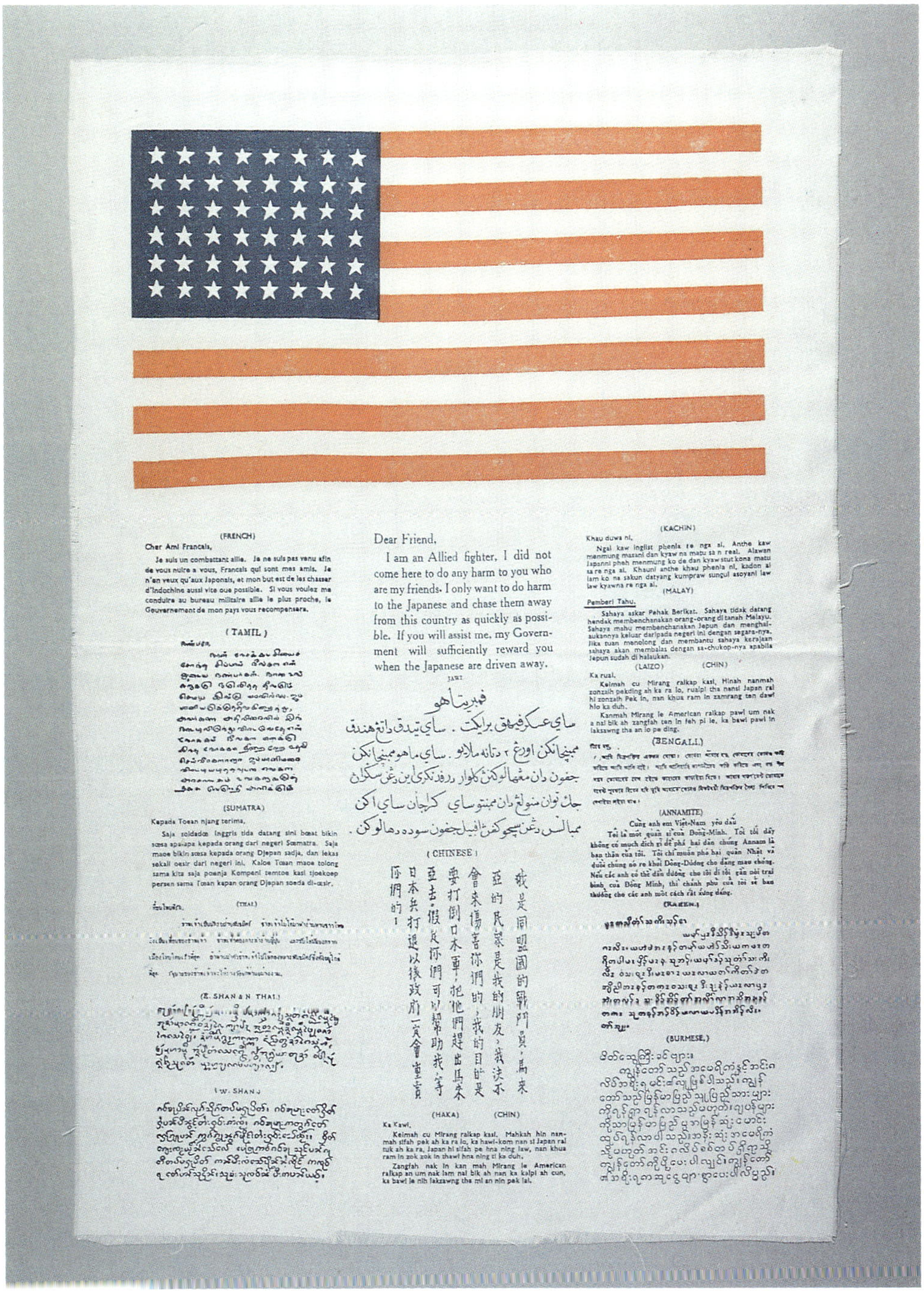

Fig. 7.27: The final version of the DEB-11 chit was printed on rayon, and is substantially larger than the silk versions. Although the text is printed in blue ink, it is nearly identical to the black text on the silk DEB-11 chits. The Burmese text has been revised from that on the earlier silk versions. This chit belonged to Maj. Ralph L. Stevenson, who received it in India in 1945 where he served at Headquarters of the India-China Division, Air Transport Command (ICD-ATC). (REB 3240) *International Blood Chit Museum*

As detailed elsewhere, DEB-2 was a set of card chits in Burmese, DEB-3 consisted of Kachin phrases, DEB-10 was a 5-language silk chit backed with cotton, and DEB-11 was a 17-language unbacked silk chit.

As the war progressed and evasion aids were required in ever-greater quantities, MIS-X Technical Section at Fort Hunt, Virginia, took a more active role in their production. The sheer number of chits urgently needed by the growing number of Americans involved in the conflict required that they be mass-produced in the U.S. Although the first chit produced in the U.S. was printed on cotton, virtually all of those to follow were on the acetate-rayon fabric especially developed for evasion maps and Blood Chits. The manufacture of Blood Chits and cloth charts on this fabric in United States during World War II is discussed in detail in Appendix II. The first of these new rayon Blood Chits was an updated version of the India-made DEB-10 chit, urgently needed for Hump crews:

> Inclosed are three samples of the American Flag Blood Chit made up to duplicate the sample forwarded some weeks ago. About 10,000 will be available this month. The serial numbers will range from 2,001 upward.
>
> A change was made in the Chinese script because several experts here, including Dr. Rock, stated that the wording on the originals was somewhat unfavorable to the Chinese. The longer statement on the samples inclosed is practically identical to the other script.[67]

The MIS-X rayon chit produced in the U.S., shown in Fig. 7.32, followed the silk DEB-10 chits produced in India and can be referred to as the "Rayon DEB-10". The Chinese text has been reworked from the Indian chits to fit into a a smaller space, and revised per the above quote. As is typical of the rayon DEB-10s, the serial number is stamped in the purple ink that was so popular in Asia during World War II. Capt. Russell A. "Rusty" Bales carried one of these chits, numbered 5551, in his jacket pocket.[68]

While the rayon DEB-10 chits were being produced, plans were made to produce in the U.S. a copy of the Rescue Patch being is-

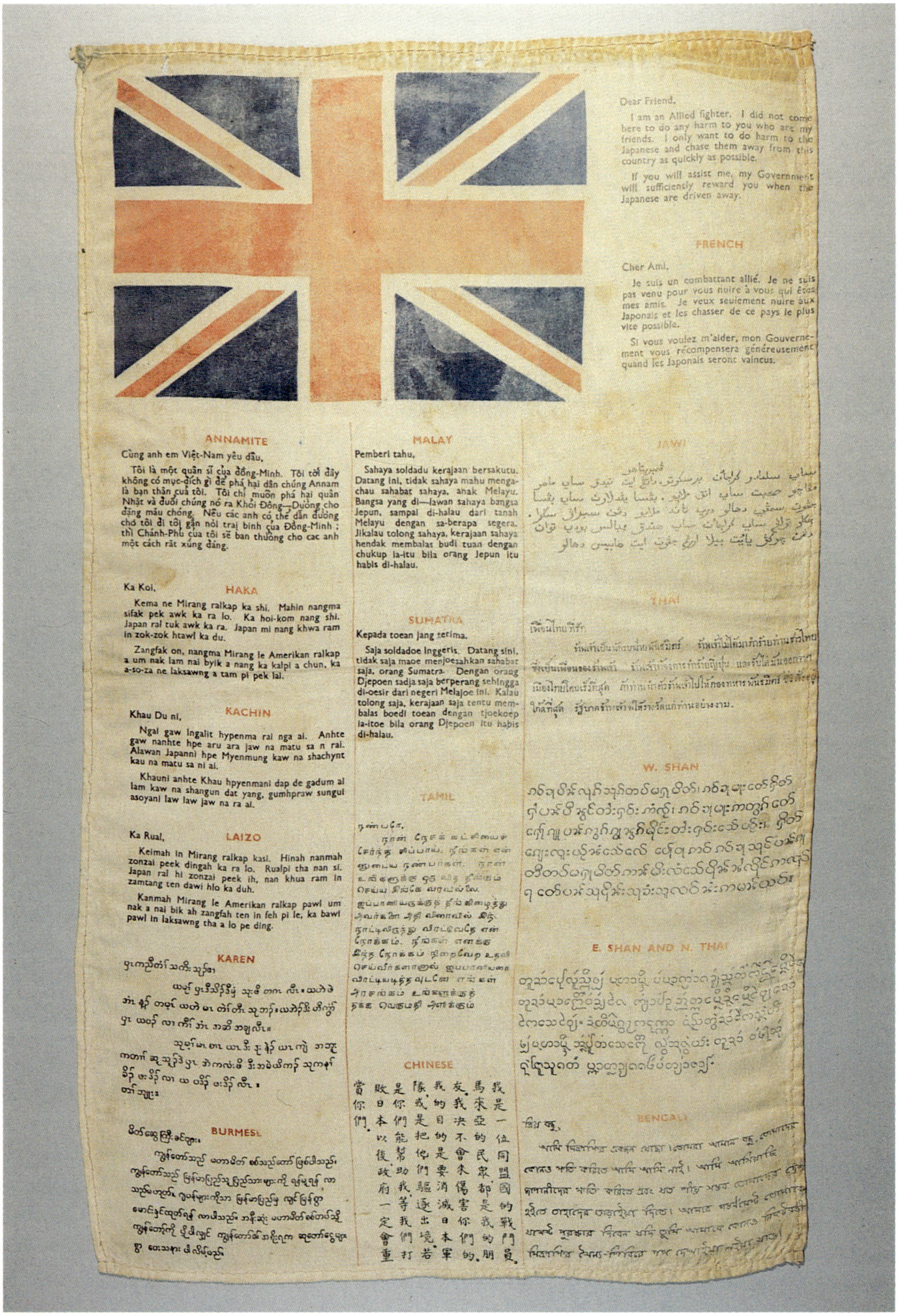

Fig. 7.29: A second style of the British 17-language chit featured a small Union Jack in the upper left corner. Although badly faded, this example is the rare early version of this chit, printed in India on silk. A remnant of the glue that was used to hold the silk to a paper backing during the printing process is evident at the top of the chit. (REB 1256) *International Blood Chit Museum*

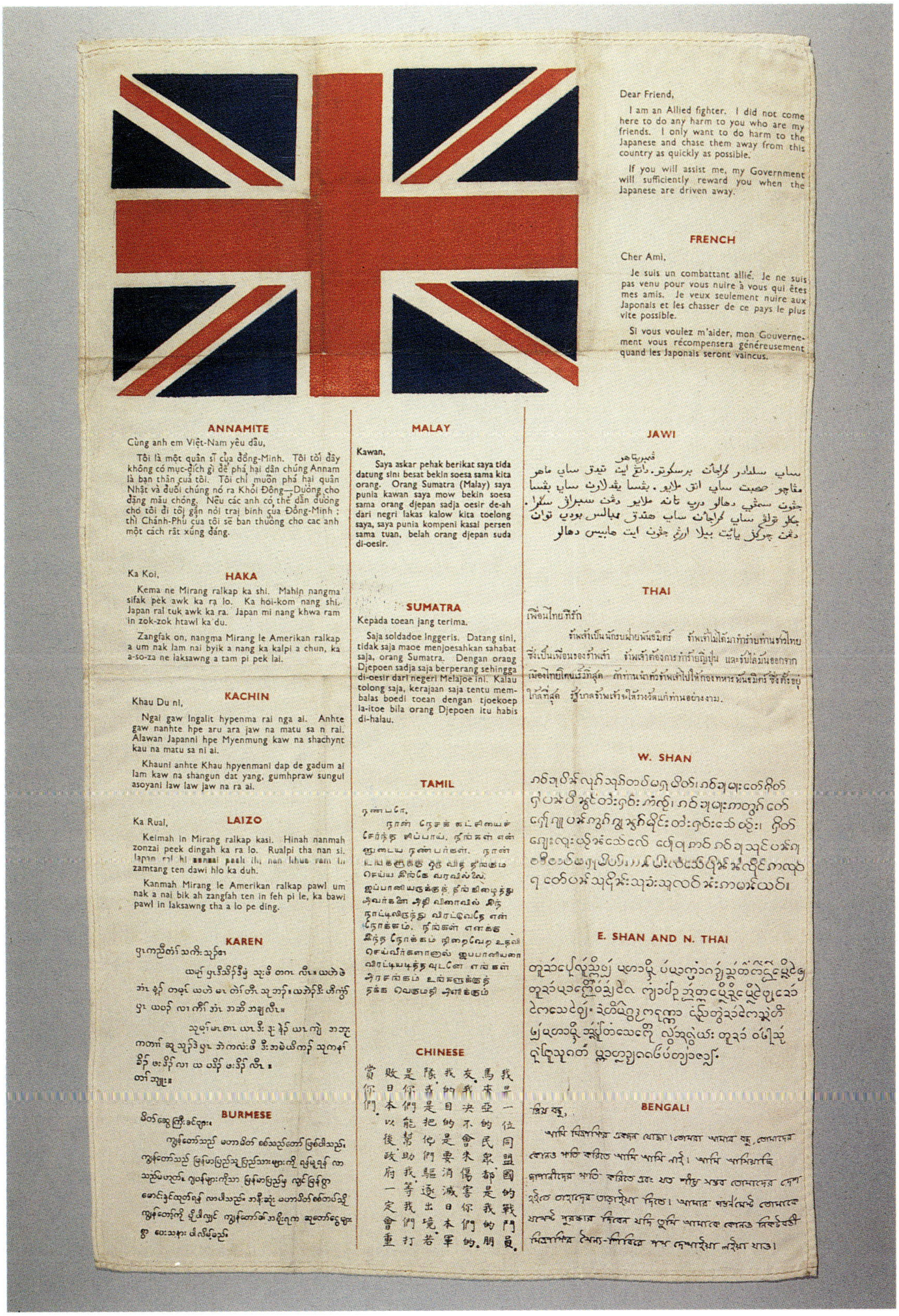

Fig. 7.30: The more common version of the second style 17-language British chit was printed on rayon. It is identical to the earlier silk version, but could withstand more rugged use than its predecessor. Large numbers of this rayon version were produced, probably in Britain. (REB 1256.1) *International Blood Chit Museum*

Fig. 7.31: This British silk chit for use in Burma is arguably the most beautiful of all the official-issue chits to be produced during the war. Featuring the British "Peacock Flag" for Burma, it was printed on silk and backed with cotton. E-Group issued it to British airmen, and provided copies to American E&E organizations for issue to American crews operating over Burma. (REB 1258) *International Blood Chit Museum*

sued by the Chinese Aero Commission. On 20 August 1944, AGAS sent a memo to MIS-X addressing the issues of serial numbers and reproducing the chop on the proposed MIS-X copies of Chinese Aero Commission chits. The memo refers to American and Chinese Identification Flags (AIF and CIF), other terms for Blood Chits:

> It is recommended by this Headquarters, that CPM, Washington, D.C., commence numbering all American Identification Flags with 4,000 and all Chinese Identification Flags with W1 (The "W" to precede all their serializations and identifying the number and the flag as produced in Washington, D.C.)
>
> In reproducing the CIF, the photographer and the duplicator should pay special attention to the Central Government chop on the lower, left hand portion of the CIF. The chop makes the CIF official and in the eyes of many Chinese who may be of help to downed-flyers is of prime importance.[69]

Two days later, on 22 August 1944, AGAS sent several of the Aero Commission chits with the revised text (Fig. 7.8) to MIS-X at Fort Hunt so production of copies of them in the U.S. could begin. They again addressed the issue of serial numbers:

> We are still waiting for our maps and American flags. I believe this has all been straightened out. One thing important is, to start numbering the American Flag from 4,000 on.
>
> We are sending you some Chinese flags for reproduction. Start numbering these from W-1 on.[70]

In early September 1944, PO Box 1142 was still awaiting a sample chit from AGAS, and notified them that the "sample China flag has not arrived. Until it does, nothing can be done on its production."[71] From information contained declassified correspondence, the sample Aero Commission chits appear to have arrived at

Fig. 7.32: This U.S. made rayon chit is patterned after the silk India-made DEB-10 chits (Figs. 7.20 and 7.21). The languages are, top to bottom beginning at the left, Burmese, S. Shan, Chinese, Sgaw Karen and W. Shan. The Chinese text has been changed from that on the previous chits, and the serial number was stamped with the purple ink that was so widely used in Asia. It is meticulously printed on rayon and sewn to a sturdy drill backing, and is one of the most beautiful of the U.S.-made chits. (REB 3242) *International Blood Chit Museum*

Fig. 7.33: This chit is the first type made by MIS-X in the U.S. for use in China, and is a copy of late style Chinese Aero Commission Rescue Patch (Fig. 7.8). No number appears in the lower right corner of these early chits, although the number "(1)" appeared on later printings. For this reason this style chit will be referred herein to as the MIS-X "type 1." Its message is identical to that on the late style Aero Commission chit: "This foreign person (an American) has come to China to participate in the war effort. Soldiers and civilians, one and all, should rescue, protect and provide him medical care." The chop is that of the Kuomintang government's Ambassador to the United States, and the serial number is preceded by the letter "W" to denote that it was issued in Washington, D.C. The two characters at the upper left of the text area translate as "borrowed", and according to declassified documents are included not as an offer of a reward, but because the chit is issued in the name of the Aero Commission but not by it. (REB 3244) *International Blood Chit Museum*

Fort Hunt by early October 1944. The resulting chit, an example of which is shown in Fig. 7.33, was the first chit made by MIS-X specifically for use in China. It was a virtual copy of the Rescue Patch with revised text issued by the Aero Commission in China, except that it was printed on rayon instead of silk. The message is the same as on the Aero Commission chit, reading:

> This foreign person (an American) has come to China to participate in the war effort. Soldiers and civilians, one and all, should rescue, protect, and provide him medical care.[72]

The five characters to the left of the message read, as on the Chinese version, "Aeronautical Commission". The serial number is in its correct position with respect to the three characters that accompany it. At the upper left of the text area are two characters that do not appear on the original Chinese version. These characters translate as "borrowed". It is specifically cautioned by Agnes Chan, the translator, that "borrowed" as used here does not imply that a reward would be paid for the return of the bearer, but is an indication that the chit was issued in the name of one agency by another.[73] This interpretation of the meaning of the term "borrowed" is confirmed by a memo from AGAS China which states that it was "recommended that chits have an additional black Chinese line in smaller type added, roughly reading 'Issued for Chinese Commission of Aeronautical Affairs.'"[74] The term, therefore, is used because the chit was issued *for* the Aero Commission but not *by* it. The chit was, in effect, "borrowed" from the Aero Commission and issued by the United States.

This first U.S. rayon chit for China will be referred to herein as the MIS-X "type 1" chit because the number "(1)" appeared in the lower right corner in at least one later printing, although the number does not appear on the first printing. The serial number of the chit shown is W 232, pursuant to the request by AGAS that numbers on these chits be preceded with the letter "W" to denote that

Fig. 7.34: A second style of rayon chit was Produced by MIS-X for use in China. The early version of this chit, shown here, lacks the number "(2)" in the lower right corner that appeared on later printings. This type of chit will be referred to herein as the MIS-X "type 2." The use of "W" serial numbers is continued on these chits. The text has been revised from that on the type 1 chit, and reads, "The American Air Force has come to China to help us in the war effort. Hope that our soldiers and civilians one and all will rescue, protect and provide them medical care. National People's Government Aeronautical Commission. Borrowed. The 'th' Number." (REB 3246) *International Blood Chit Museum*

they were issued in Washington, D.C. The chop is that of the Nationalist Chinese Ambassador to the United States, and was stamped onto the chit at the Chinese Embassy in Washington, D.C., by MIS-X personnel.

This chit, and those that followed it, were single pieces of fabric rather than two fabrics sewn together. This was done because, as outlined in a memo from PO Box 1142 referring to its experience in producing the rayon DEB-10, "considerable time and labor is involved in sewing on the cotton backing. If this can be eliminated . . . it will speed up production. The edges of the rayon could be hemmed or finished with a seal by solvent, the same as done with all cloth maps."[75]

The second type of MIS-X chit produced for use in China, referred to herein as the "type 2" MIS-X chit, featured a revised text. The example shown in Fig. 7.34 is typical of the first printings of this type, having no number "(2)" in the lower right corner, although the number appeared on later printings. The use of the "W" serial numbers continues, as it did on subsequent MIS-X chits bearing the Nationalist government's flag.In July 1945, the U.S. government provided the following information regarding the chit's message:[76]

Translation of Chinese Flag Blood Chit #2 is given below:
American military aviator
Come to China to help in war
Our military and civilians in one
Body are to rescue and protect
National People's Government
Aeronautical Commission
Number________________

This translation more closely resembles the text of the type 1 chit than that of the type 2. The type 2 MIS-X chit more accurately reads:

> The American Air Force has come to China to help us in the war effort. Hope that our soldiers and civilians one and all will rescue, protect, and provide them medical care. National People's Government Aeronautical Commission. Borrowed. The "th" Number.[77]

Note that the word "hope" as used in this message does not carry a mandate to perform the acts simply because it is a request of the government. In modern, or vernacular, Chinese this might be implied, but chits and other official documents were written in classical Chinese, so "hope" means only that and nothing more.[78] The two characters meaning "borrowed" appear in the upper left corner of the text area on this chit just as they did on the type 1. Again, this does not imply that a reward is offered, but that the chit was issued by the U.S. in the name of China's Aero Commission.

MIS-X type 1 chits from a later printing, an example of which is shown in Fig. 7.35, are identical to the first version (Fig. 7.33) except that the number "(1)" appears in the lower right corner. For an unknown reason the serial number, W 65010, is in the midst of a range of numbers used on the type 2 chits. Perhaps MIS-X decided to print more of the type 1 chits while in the midst of printing the type 2 chits, and simply used a convenient range of serial numbers for this purpose.

Later printings of the type 2 MIS-X chits, an example of which is shown in Fig. 7.36, have the number "(2)" in the lower right corner. This type of chit was produced in quantity, and serial numbers range well above W 200000. Note that the serial number on this chit is located in its correct position with respect to the the Chinese characters at the left side of the chit.

Fig. 7.35: On this type 1 chit the number "(1)" appears in the lower right corner. This was apparently printed much later than the first type 1 chits, as the serial number, W 65010, is well within the range of numbers used for type 2 chits. It is thought that MIS-X may have decided to print a belated run of the type 1 chits and simply chose a convenient range of serial numbers for them. (REB 3244.1) *International Blood Chit Museum*

Some of the type 2 chits were printed with their serial number at the bottom instead of in the correct position on the left side. One of these is illustrated in Fig. 7.37. The chop mark is extremely faint on many of the chits with the serial number at the bottom, as it is on this example. William A. Rooney, Intelligence Officer for the 40th BG, was issued one of these chits with the serial number, W 75640, at the bottom.[79]

The type 2 chit continued to be issued to aviators in China even after Japan signed its unconditional surrender aboard *USS Missouri* in Tokyo Bay. Marine Aviator Roger Perreault of the 1st Marine Air Wing writes, "After the war ended in August 1945 we were sent from the Philippines to Okinawa for a short time, and then to Tsingtao, China. The air Squadron's specific job was to fly recon missions into the interior of China each day and observe and plot the progress of the Japanese troops that were being marched to the seaports for transfer to Japan. On arriving in China in October 1945 we were all issued new blood chits . . . written in Chinese only. At this time China was in political turmoil with no less than nine factions trying to control Tsingtao, including the Russians, Chinese Communists and numerous warlords. We were being shot at on our recon flights, probably by the Chinese Communists, we never knew for sure." The chit Perreault was issued in China during October 1945 was a type 2 chit, number W 30130.[80]

Illustrated in Fig. 7.38 is the direct descendant of the two versions of the two India-made silk DEB-10 chits (Figs. 7.20 & 7.21) and the U.S.made rayon DEB-10 chit (Fig. 7.32). The last in the line of 5-language chits, this was the third type of rayon-acetate chit produced by MIS-X, and the number "(3)" appears in the lower right corner. These type 3 MIS-X chits were apparently produced sometime in the first or second quarter of 1945. The languages are in the same order as the previous MIS-X cotton-backed rayon chit, but the Chinese text has been modified to fit into a smaller space.

Fig. 7.36: The number "(2)" appears in the lower right corner of later printings of the type 2 MIS-X chits, unlike the early version. The serial number on this chit is in its correct position, between the Chinese characters on the left side, and the numbers eventually reached above 200,000. Like the type 1 chits, these were issued in the name of China's Aero Commission. (REB 3246.1) *International Blood Chit Museum*

Fig. 7.37: The serial number on this type 2 chit was printed at the bottom of the chit rather than in its correct position on the left side between the appropriate Chinese characters. The chop is faint on many of the chits with the numbers at the bottom, as it is on this example. (REB 3246.2) *International Blood Chit Museum*

This chit belonged to Maj. Ralph L. Stevenson who obtained it at HQ ICD-ATC in 1945. The serial number is printed at the bottom in red ink.

The fourth and final type of chit produced by MIS-X, bearing the number "(4)" in its lower right corner, is mentioned in a memo at the end of July 1945: "Inclosed herewith are ten American Flag Blood Chits (4) as requested by letter AGAS/181700/946, dated 18 Jul 45. At the present time these are available for shipment in quantities up to 40,000."[81]

This style of chit, referred to herein as the type 4 MIS-X chit, is illustrated in Fig. 7.39. The serial number, 30180, is printed in red at the bottom of the chit. The message is given in seven languages: French, Thai, Lao, Annamese, Chinese, Korean and Japanese. Annam was the northernmost of the three provinces of French Indo-China, and the language labeled as Annamese is what we now call Vietnamese.

This Blood Chit has been the object of ridicule among less knowledgeable persons who scoff at the inclusion of a message in Japanese, for it is rightly doubted that Japanese nationals would have provided assistance to downed air crew members. The inclusion of Japanese, however, was of immense benefit in places like Korea and the Marshall Islands, under Japanese rule since 1910 and 1914 respectively, where Japanese was the official language and the speaking of native tongues was vigorously discouraged. Moreover, organizations such as AGAS and the OSS did not operate in Korea at all, so downed air crews there were truly on their own and needed every possible advantage that could be provided.

Fig. 7.38: The third type MIS-X rayon chit, referred to herein as the MIS-X type 3, was last in the line of 5-language chits made both in India and the U.S. (7.20, 7.21, and 7.32). Serial numbers were printed in red at the bottom. The languages are in the same order as the previous MIS-X cotton-backed rayon chit (Fig. 7.32), but the Chinese text has been modified to fit into a smaller space. The chit illustrated belonged to Maj. Ralph L. Stevenson who obtained it when he served at HQ ICD-ATC in 1945. (REB 3248) *International Blood Chit Museum*

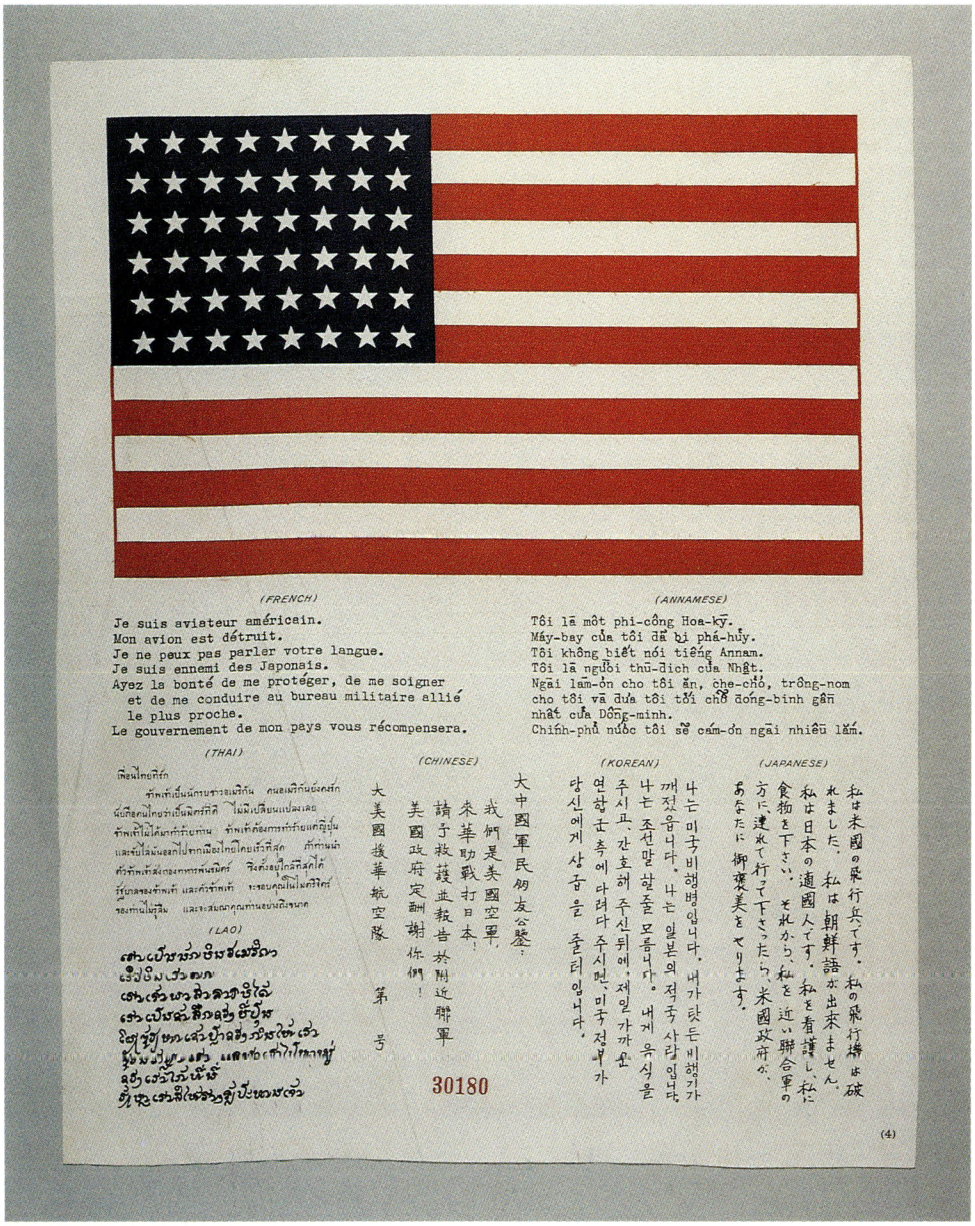

Fig. 7.39: The fourth and final MIS-X rayon-acetate chit was printed in seven languages. Annam was the northernmost province of French Indo-China, and the language labeled "Annamese" is what we now call Vietnamese. Serial numbers on the MIS-X type 4 chits are printed in red at the bottom, and early versions such as this one are identified by the fact that the notation "N$^{\underline{o}}$" does *not* precede the serial number. The inclusion of Japanese was important as it would have been of great value in Korea and the Marshall Islands, under Japanese control since 1910 and 1914. (REB 3250) *International Blood Chit Museum*

Even during the Korean War, more than five years after the Japanese were driven from Korea, Americans reported that they were able to communicate with Korean nationals using only the limited Japanese they had picked up while stationed in Japan. Far from being ridiculous or humorous, the inclusion of the Japanese language on the type 4 chit was logical and necessary.

A minor change in the type 4 MIS-X chit was made, wherein the notation "№" was placed in front of the serial number. An example is shown in Fig. 7.40. This appears to have been done at one point and the practice retained from that time onward. All higher numbered chits have this notation while the lower numbered chits do not. This chit was produced in extremely large quantities, as the serial number, № 210054, on the example illustrated, attests. One of these type 4 chits, № 121313, was issued to William A. Rooney, IO of the 40th BG.[82]

The U.S. government provided the following in July of 1945:

American Flag Blood Chit #4 has the following translation:

a. French, Annamese, Korean, Japanese and Lao messages: I am an American airman. My plane is destroyed. I cannot speak your language. I am an enemy of the Japanese. Please nourish and protect me and take me to the nearest Allied military post. My government will reward you.
b. Chinese message
To Chinese military and civilian friends Greetings: We are members of the American Air Force, and have come to China to help fight Japan. Please rescue and protect me, and notify the nearest Allied military (authorities). The American government will surely Reward and thank you.

c. Thai Message.
I am an American. Americans love the Thai very much. I did not come to do you any harm. I only want to do harm to the Japanese and want to get them out of Thailand as rapidly as possible. If you will send me to a friendly military detachment, and the nearest one, my government and I will thank you and will not forget. We will be forever indebted to you.[83]

The translation of the French message is adequate, although a more exact translation is:

I am an American aviator. My aircraft is destroyed. I cannot speak your language. I am an enemy of the Japanese. Have the goodness to protect me, take care of me, and guide me to the nearest Allied military post. The government of my country will pay you.[84]

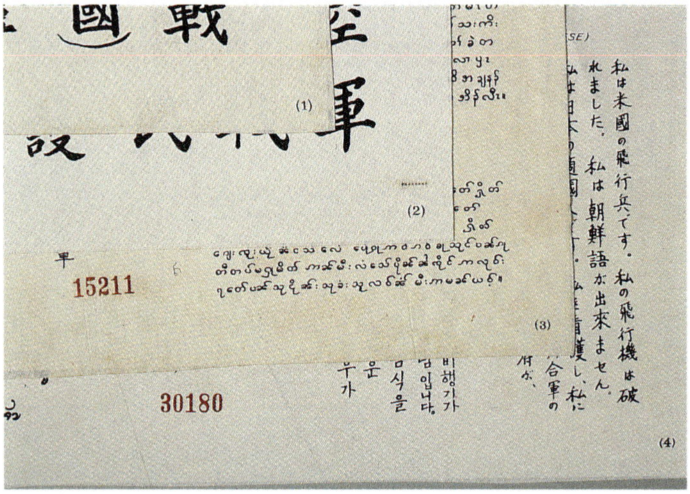

Fig. 7.41: The four numbers that appear in the lower right corners on MIS-X rayon chits are detailed here. Early versions of the type 1 and 2 chits did not feature these numbers. The numbers are referred to in MIS-X and AGAS correspondence, but their purpose other than identifying the different styles of chits, if any, is unknown. *International Blood Chit Museum*

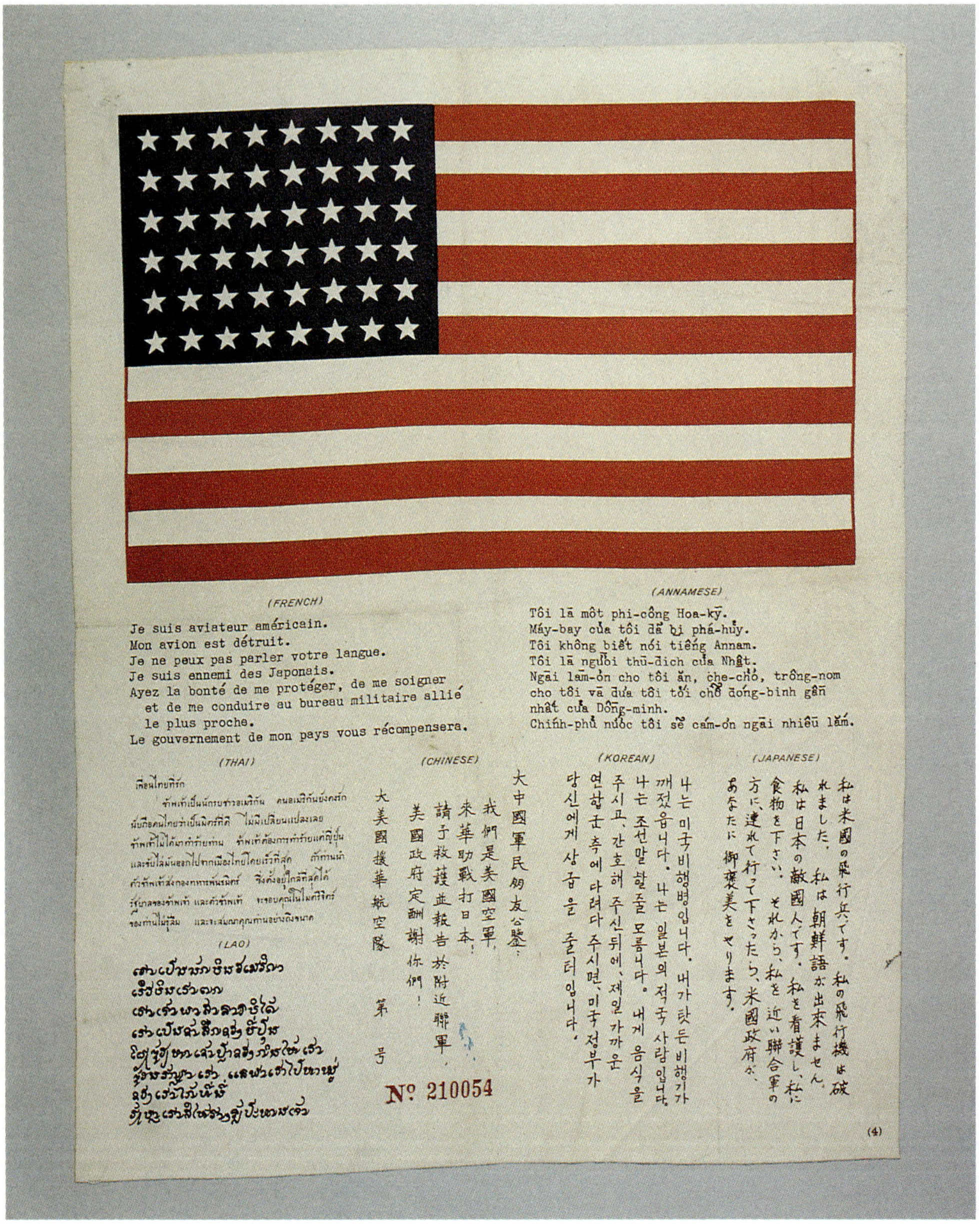

Fig. 7.40: Later versions of the MIS-X type 4 chit are identified by the notation "N$^{\underline{o}}$" which was placed in front of the serial number. The high serial number on this example indicates that large quantities of the type 4 chit were produced. (REB 3250.1) *International Blood Chit Museum*

Fig. 7.42: This flag, printed on silk and sewn to a cotton backing, was made in India. It was probably made around the same time as the DEB-10 chits (Figs. 7.20 & 7.21) as it resembles them in both materials and workmanship. It was another in the series of theater-made officially issued items. Although it had no message, the American flag often identified the airman before any word was read or spoken. (REB 3252) *International Blood Chit Museum*

A detail showing the numbers that appear in parentheses in the lower right corner on the four types of unbacked rayon MIS-X chits appears in Fig. 7.41.

The great variety of Blood Chits produced in China, India, Britain and the U.S. allowed crews to carry items somewhat tailored to their areas of operation. It is also well documented that crewmen often carried more than one type of chit. Less known is the fact that instructions were given *not* to use certain chits in certain areas. Specifically, in early January 1944, American air crews were instructed not to use Chinese Blood Chits in Burma:[85]

1. Many aircrews of this command have made it a practice to wear Chinese flags in flights over Burma.
2. As a result of interviews with men who have "walked out", it is concluded that the Chinese flags are of no value in Japanese-occupied Burma. The Chinese flag is a distinct disadvantage because friendly natives may mistake the writing for Japanese characters.
3. In view of this fact it is desired that members of this command flying missions over Burma do not wear the Chinese flag. It is felt that the American flag is now well

Fig. 7.43: Some of the India-made flags were converted to Blood Chits in China by the addition of text. The standard blood chit text was written in two horizontal rows instead of four vertical columns, without the "may guo" characters. The chit shown here was presented to comedian Joe E. Brown in China by the 449th Fighter Squadron when he was on tour in Asia entertaining American troops. The reverse, shown in Fig. 7.44 , was inscribed by Squadron members. (REB 3254) *International Blood Chit Museum*

enough known that it will serve as adequate identification in case of a forced landing in Northern Burma.

By Order of Colonel EGAN

While AVG personnel were originally instructed to attach their chits to their flight clothing, and this tradition continued among American air crews in Asia for a time, these instructions changed as greater numbers of evaders returned and told of their experiences. Of concern was the fact that a chit worn on the outside of flying clothes offered a brighter image to those people, friend and foe alike, trying to find or track the evader. By 1944 the instructions for the wearing of chits on garments had been formalized: "Specific instructions should be issued and the instructions rigidly enforced, that both the AIF and the CIF are to be sewn inside the flight jackets of the wearers. These flags, however important once on the ground, make too good a target. . ."[86] Horace O. Bedsole, bombardier assigned to the 308th BG, 373rd and 425th BS, writes, "My China chit is still sewn on my A-2 jacket back . . . but I honestly don't remember ever wearing it on any mission, it was more for looks."[87]

The crews of 20th Bomber Command in India and China received specific instructions on how to wear their Chinese Aero Commission Blood Chits:

> This should be sewn on the inside of the flight jacket. It consists of the Chinese flag and Official Request of the Chinese Government to return the pilot safely. It bears the seal of the government. It is issued only by Chungking and by serial number. The pilot may sew an American flag on the INSIDE of the jacket as well. Under no circumstances should the flags be on the outside of the jacket. They make too good a target and hinder concealment. Those seen worn on the outside are usually done so by non-operational jerks.[88]

Blood Chits issued to American air crew members were generally issued by serial number, and the specific number issued to each individual was recorded. Individual units turned in multiple copies of reports listing each person to whom a chit was issued along with its serial number. At a higher level, MIS-X Sections retained lists of chits by blocks of serial numbers and units to whom they had been sent. An example of a unit report accounting for 250 silk Chinese Aero Commission Rescue Patches issued to individual air crew members appears in Appendix IV, and an example of a MIS-X headquarters list of chits issued to various units appears in Appendix V.

This accounting of chits by serial number required mountains of paperwork at all levels, a fact perhaps not well appreciated in an age of computers and photocopiers. While complex under combat conditions and an administrative nightmare, this system provided accurate if cumbersome documentation for payment of rewards for aiding downed crews.

Former Twentieth Bomber Command officer John L. Cella, who served as the command's Escape & Evasion and Search & Rescue Officer at Kharagpur, India, and at Chengtu, China, recalled that he "distributed the chits, along with other E&E materials to all crew members. I recall our men were instructed to sew them to the inside lining of their flight jackets. I don't recall their ever being referred to as 'blood chits'. We just called them Chinese flags . . . my recollection is that they were initially obtained directly from the Chinese, probably through the AGAS in Kunming. I do remember a careful record was kept of to whom each chit was issued by the serial number"[89]

The use of serial numbers facilitated reward payments, which are discussed in detail in Appendix III, and allowed units to prevent the misuse of chits. The practice also assisted in the proper identification of the remains of deceased airmen since it was forbidden to take identification materials other than dog tags into combat. One memo reported that "the Chinese Identification Flags are issued to flying personnel by serial number. The serial numbers are then filed by the unit I.O. are also sent to HQ, AGAS-China, for filing. This is important since, as in several instances within the last two weeks, the CIF and AIF and its serial number has been the *only* identification of flying personnel killed in crash landings or impotent bail outs."[90] In one specific instance, a memo from the China Theater Search Detachment that reported that "the grave of an unidentified pilot has been located by the Changsha Search Team at Fya-Ch'ung . . . a village two miles east of Huang Hua-Shid which is 15 miles east of Changsha. A Chinese Flag Identification which was worn on this flier's jacket, Number 12331, has been found . . . It is requested that all available organizational and other records be checked for . . . identity of individual to whom Chinese Flag Identification Number 12331 was issued."[91]

It will be recalled that AGAS issued its own serial numbers on the 6-language cotton chits. Since these numbers conflicted with numbers already assigned by MIS-X, the following message detailing the assignment of serial numbers to U.S. flag Blood Chits was sent:

> Since April 1944 many American Blood Chits have been produced, all of which bore serial numbers, and the numbers mentioned above have already been duplicated. Numbers which have been assigned to American Flag Blood Chits in Washington are:
>
> 4001 thru 13550 - to 5-language chit
> 15001 thru 30000 - to 5-language chit
> 30001 thru 75000 - to 7-language chit
>
> Inasmuch as the above numbers have already been assigned and others will be assigned from time to time, it is suggested that MIS-X, Washington, be consulted before assigning serial numbers to Blood Chits.[92]

Examples of the 7-language type 4 chit with serial numbers well above 200,000 show that the range of numbers assigned to this chit was extended beyond that given in the above document.

The first production of the MIS-X type 1 through 4 chits can be dated to between September 1944 and April 1945 from declassified documents. The sample Aero Commission chit sent by AGAS had not arrived at PO Box 1142 as of 9 September 1944, so the production of the MIS-X version began after that date. The MIS-X chits are illustrated in the May 1945 issue of the *Air-Sea Rescue Bulletin*, which probably would have been produced in March or

April, so production of them had begun by then. Serial numbers were assigned by mid-February 1945, and the manner in which the information on serial numbers is written suggests that the chits may have been in production by then. These dates are important, because some MIS-X chits have been erroneously dated to as early as 1943. The dates also illustrate that MIS-X did a phenomenal job in a short time period, eventually shipping more than 300,000 chits for use by combat units through August 1945.

An extremely important point regarding the MIS-X chits is that although the type 3 chits went to Hump personnel in India, by far the vast majority of the type 1, 2, and 4 chits went not to India or China, but to Naval, Marine, and Army Air Force flyers in the Pacific. American units operating in China and India were, for the most part, furnished chits made in China and India, and American units operating in the Pacific late in the war were, for the most part, issued the U.S. rayon chits provided by MIS-X Technical Section at Fort Hunt.

According to the MIS-X War Diary for August and September 1945, of the total 142,230 "Chinese flag" chits shipped by them during the war, 122,611 went to Pacific Ocean Area (POA) Naval units and to Air Force units in the Pacific (AFPAC), and of the 173,505 "American flag" chits shipped by them during the war, 137,961 went to POA and AFPAC units.[93] Although the MIS-X rayon type 1, 2, and 4 chits were issued in far greater numbers to flyers operating in the Pacific, which is covered in Chapter 8, they are discussed here in Chapter 7 to show how they were developed directly from the earlier, theater-made chits.

In addition to Blood Chits, American flags without any accompanying text were used as aids to identification of American personnel. Figs. 7.42 through 7.47 show the development of officially issued flags.

Shown in Fig. 7.42 is a flag printed on silk and sewn to a muslin backing. It was made in India and procured through Reverse Lend-Lease for issue to U.S. aircrew. It was probably made around the same time as the DEB-10 chits, which it resembles in printing, fabric, and construction. These flags were used to identify the bearer as being an American. One of these flags is sewn into the right front inside lining of the A-2 jacket of Gen. Earl S. Hoag, who commanded the ATC's Africa-Middle East, India-China, and European Wings, on display at the U.S. Air Force Museum (March 1995). Some of these flags were converted to Blood Chits in China, but most were simply used as issued.

The theater-made Blood Chit shown in Fig. 7.43 is typical of the type made in China by modifying the above described flag. It

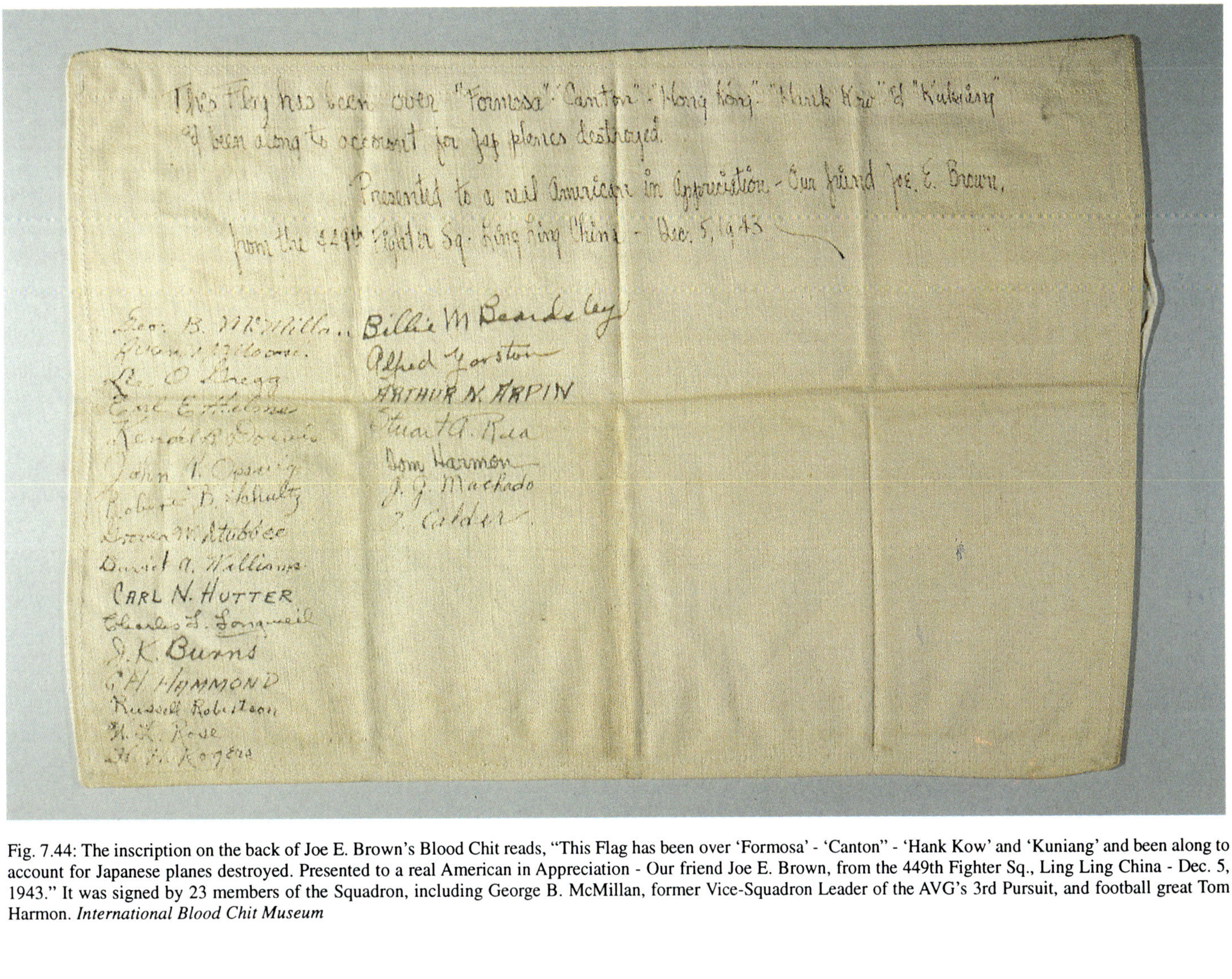

Fig. 7.44: The inscription on the back of Joe E. Brown's Blood Chit reads, "This Flag has been over 'Formosa' - 'Canton" - 'Hank Kow' and 'Kuniang' and been along to account for Japanese planes destroyed. Presented to a real American in Appreciation - Our friend Joe E. Brown, from the 449th Fighter Sq., Ling Ling China - Dec. 5, 1943." It was signed by 23 members of the Squadron, including George B. McMillan, former Vice-Squadron Leader of the AVG's 3rd Pursuit, and football great Tom Harmon. *International Blood Chit Museum*

Fig. 7.45: This flag, from an unknown source, has seen heavy use and is unraveling despite the glue placed along its edge. The fabric seems to be a silk similar to that used for some Asian-made items, but the quality of the printing suggests that it was made in the US. Regardless of its origin, it is another example of the flags that were used by American airmen as a means of identification. (REB 3256) *International Blood Chit Museum*

Fig. 7.47: The second style of MIS-X flag was printed on rayon-acetate fabric and featured a border along the ends of the stripes of the flag. This flag was issued to an unidentified airman in 1945 who, according to the note that accompanied it, traded it to another American serviceman for a Japanese war souvenir. (REB 3260) *International Blood Chit Museum*

was modified by the Chinese who wrote the standard chit message on it, without the "may guo" characters, in two horizontal rows instead of four vertical columns. The example illustrated was presented to comedian Joe E. Brown in China by the 449th Fighter Squadron. The back of this chit, shown in Fig. 7.44 , was signed by 23 members of the 449th, including George B. McMillan, former Deputy Squadron Leader of the AVG's 3rd Pursuit "Hell's Angles" Squadron, who was later killed in action, and by football great Tom Harmon.

Another flag, illustrated in Fig. 7.45, was printed on what seems to be rayon. This example has seen heavy use and is unravelling in spite of the glue placed along the edges. The origin of this flag is unknown – the printing quality suggests it was made in the United States.

The flag shown in Fig. 7.46 is printed on the same rayon-acetate fabric as the MIS-X Blood Chits. It was the first of two styles of MIS-X flags, and does not have a border along the ends of the stripes. An example of the second style MIS-X flag is illustrated in Fig. 7.47. It is also printed on rayon-acetate fabric, but there is a border at the ends of the flag's stripes.

As with MIS-X Blood Chits, the vast majority of MIS-X flags went to both sea and land based units in the Pacific. Of the total 24,117 flags shipped by MIS-X through the end of August 1945, 17,200 went to POA units, 6,916 went to India, and 1 went to an

Fig. 7.46: The first of two styles of officially issued MIS-X flags, this is printed on rayon-acetate fabric identical to that used for MIS-X Blood Chits. It has no border along the ends of the stripes. (REB 3258) *International Blood Chit Museum*

Fig. 7.48: This flag belonged to 14th A.F. member George Brouillette, who kept it folded and in a pocket. The previously illustrated flags are printed on one side only and were specifically intended for issue to aircrew members. This flag, however, is printed on two sides and was available for purchase by anyone stateside. Nonetheless, some of these flags were obtained and carried by air crews in Asia. (REB 3261) *International Blood Chit Museum*

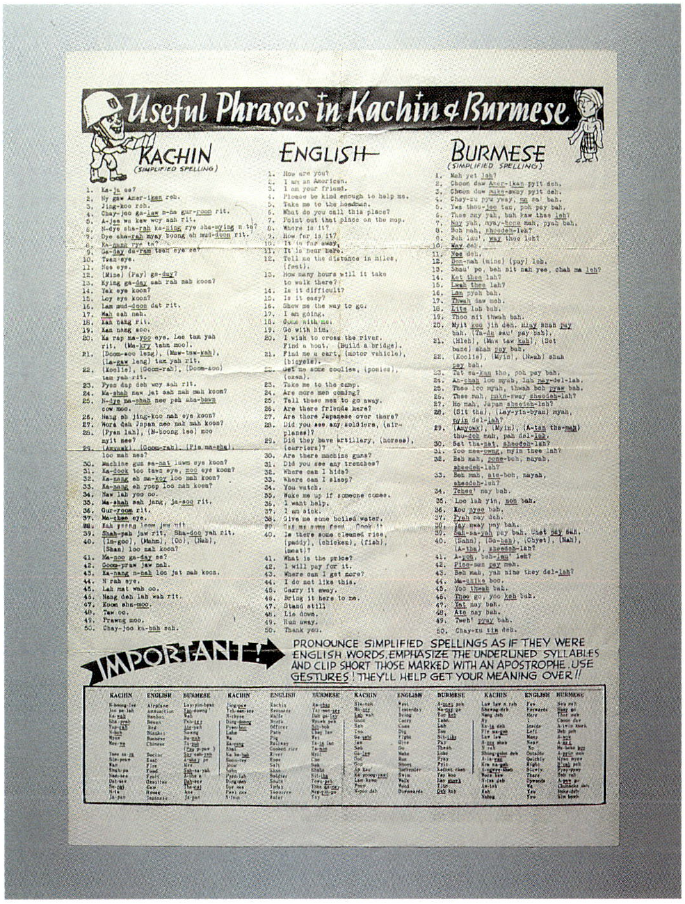

Fig. 7.49: This two-sided paper combined a phrase list with an introduction to Burmese cultures and customs. This paper was folded into eighths. Note that one of the small illustrations shows an airman with a U.S. flag on the back of his flight jacket. At least two versions of this were produced, one with the notation " (By U.S. Office of War Intelligence - XEM 4)" in the lower right corner beneath "With Useful Words & Phrases", and another without the notation. The styles of type font used on the reverse side are slightly different on the two versions. Perhaps because of its inclusion of Burmese cultural information, it was described in one evasion report as superior to other language aids carried. (With XEM notation: REB 3912, Without: REB 3912.1) *International Blood Chit Museum*

AFPAC unit.[94] The MIS-X flags are illustrated in this chapter to show their development from theater-made flags even though they were used in far greater numbers by airmen stationed in the Pacific.

Some flags carried by airmen were not specifically made for identification or evasion purposes. These were available for purchase by anyone in the U.S., although they seem to have also been procured and issued to some air crews. The example shown in Fig. 7.48 was carried by George Brouillette of the 14th A.F., whose Chinese Rescue Patch is shown in Fig. 7.3. The flag was folded and carried in a pocket. William Dreeland reports that this type of flag was issued, as part of a set that also included the type 2 and 4 MIS-X chits, to an enlisted member of a PBY squadron operating off the China coast.[95]

While a Blood Chit could introduce the evader to the local populace, explaining that he was an Allied flyer and that he needed assistance in returning to friendly control, more detailed communication required more complex language-related evasion aids. American B-29 crews were briefed, "It may be difficult to convey your desires to the Chinese by motions alone, so don't forget the pointie talkies."[96] Pointie Talkies are described along with Phrase Cards in an undated E-Group evasion bulletin:

PHRASE LISTS AND "POINTIE TALKIES". Phrase cards containing some fifty useful phrases are issued according to the area of operations. The easiest way of using these is to select the phrase required in English and then to point to the printing in dialect. There are other short publications such as "Malay in 400 Words" and "How to get Along with the Kachins, Shans and Burmans." "Pointie Talkies" are at present issued in Burmese, Siamese and Chinese, and Malay "Pointie Talkies" will be ready for distribution in January 1945. These are booklets some 60 pages in length and are an extension of the same idea of pointing to the text. As noted earlier, these booklets are normally contained in the Money Purse.[97]

The above referenced two-sided single sheet "How to Get Along With the Kachins, Shans and Burmans", illustrated in Fig. 7.49, features a list of Kachin Phrases on one side and detailed information regarding the various peoples of Burma on the other. Possibly because it included a wealth of information on the people of Burma in addition to language information in a phonetic format, this was considered by airmen to be one of the best language aids. The walk-out report of one successful evader reported, "The language card entitled 'How to get along with the Kachins, Shans and Burmans', was found very useful and superior to any of the other cards of this nature that were also carried."[98]

A great variety of Pointie Talkie (PT) booklets were produced for use in Asia by both British and American evasion and escape agencies. Shown in Fig. 7.50 are two British Pointie Talkie booklets, one for use with Chinese speaking people and one for people speaking Malay. Both languages were widely used in Asia and the Pacific. Chinese merchants were in virtually all countries of the area, and Malay was spoken in an area that stretched from the Britsh Malay States down through the Netherlands East Indies, including the western half of New Guinea. The Chinese booklet is made of rayon, and is held together with staples. It is written in English and Chinese, and is designed for use by downed airmen as illustrated by the sample pages shown in Fig. 7.51. Bob Gilroy, a pilot in RAF Transport Command, reports that he was issued one of these rayon Pointie Talkies.[99] The Pointie Talkie in Malay, on the right, has plain white paper covers and the pages are of newsprint. Several pages of this Pointie Talkie are shown in Fig. 7.52.

Illustrated in Fig. 7.53 are two Pointie Talkies produced for American air crews. On the left is what is considered to be the first

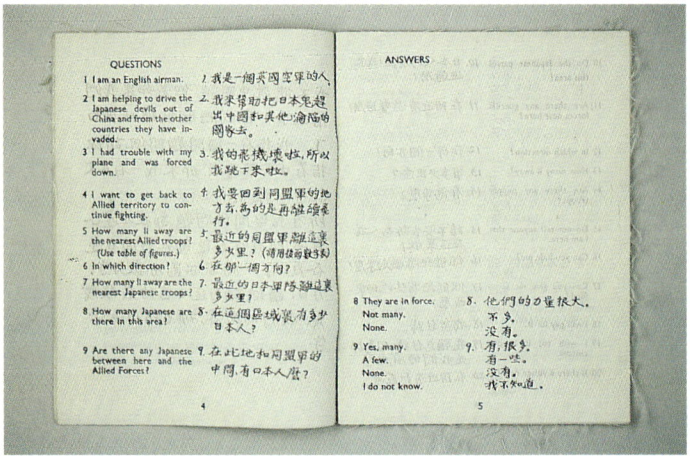

Fig. 7.51: The first pages of the British rayon *Pointie Talkie in Chinese*, shown on the left in Fig. 7.50, explain how to use the booklet, and the Question and Answer section begins with an explanation that the bearer is an English airman fighting the Japanese whose aircraft was forced down. There are several blank pages at the back labeled "Your Additions" for the airman to use for adding his own information. *International Blood Chit Museum*

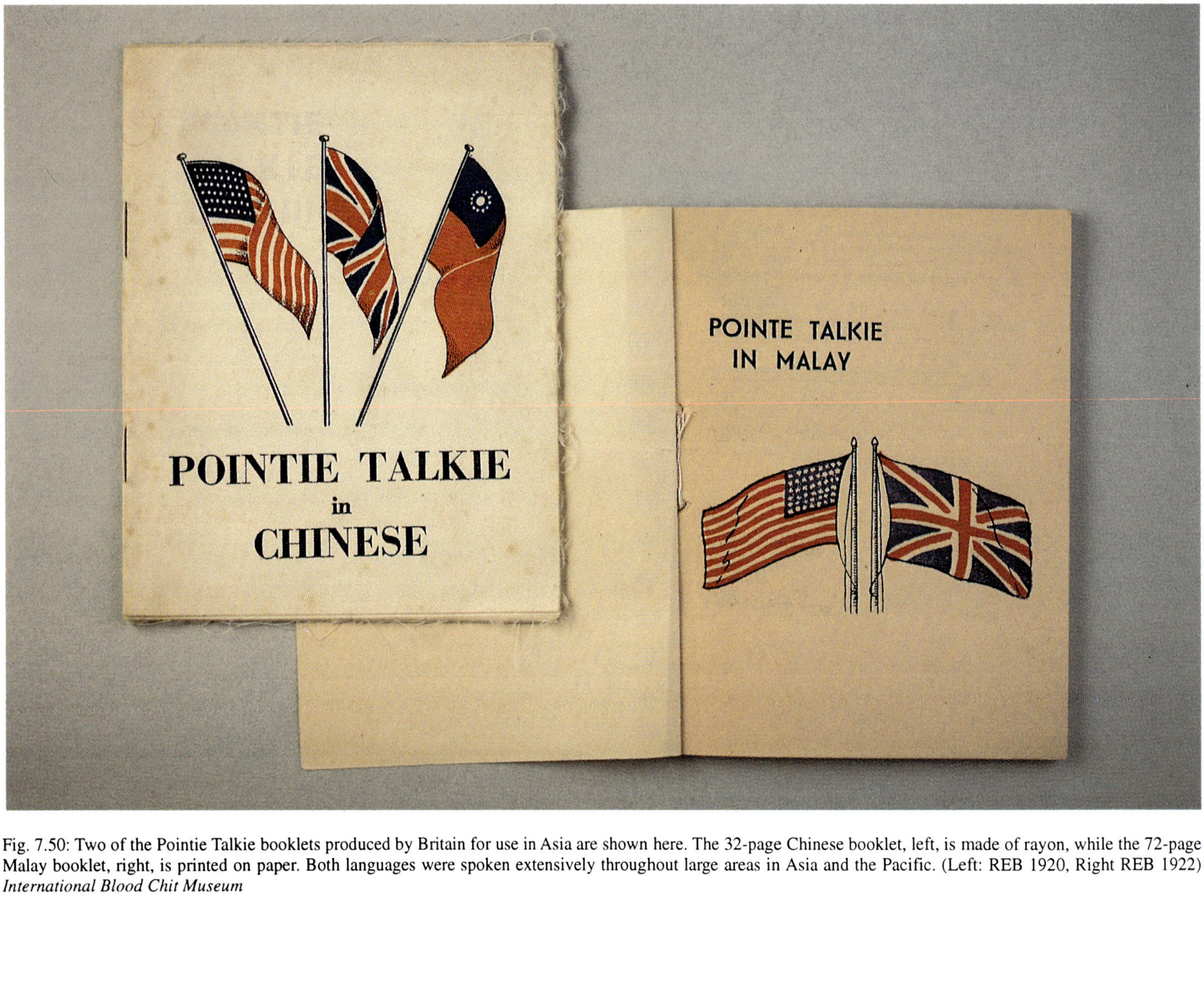

Fig. 7.50: Two of the Pointie Talkie booklets produced by Britain for use in Asia are shown here. The 32-page Chinese booklet, left, is made of rayon, while the 72-page Malay booklet, right, is printed on paper. Both languages were spoken extensively throughout large areas in Asia and the Pacific. (Left: REB 1920, Right REB 1922) *International Blood Chit Museum*

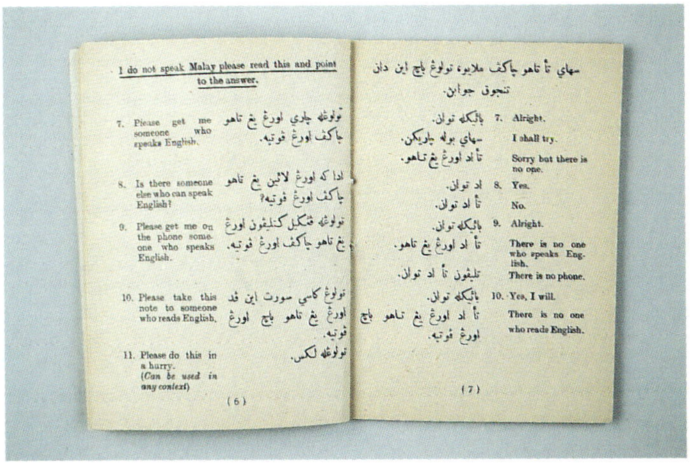

Fig. 7.52: After explaining how to use the booklet, the first pages of the British Pointie Talkie in Malay, shown on the right in Fig. 7.50, concentrate on finding someone who can speak English. The instruction, "I do not speak Malay please read this and point to the answer," is written at the top of each page. At the back is a list of place names, and a list of "Geographical and Useful Terms" such as Mountain Pass, Village, Horse, and Water. *International Blood Chit Museum*

Pointie Talkie booklet made for U.S. forces in China. It is made of paper with a purple colored cloth cover, although covers for this version are sometimes maroon or brown. It is organized to help the airman obtain assistance for evasion and other purposes such as servicing his aircraft, as illustrated by the pages shown in Fig. 7.54. A section entitled "Notes" acknowledges the problem of finding someone who can read while encouraging the airman to try his luck: "Remember that many Chinese are illiterate. Especially is this true in the countryside. But even in the countryside you can usually find someone in a crowd who can read these simple sentences."

On the right side of Fig. 7.53 is a U.S. made Pointie Talkie booklet that is virtually identical to the theater-made version to its left. The text of this booklet exactly copies the theater-made version page for page. The only differences between them are that the cover on this booklet is made of brown paper, the title page is in black and white, and on the last page is printed "U.S. Government Printing Office: 1944". Fig. 7.55 shows identical pages from both versions.

The Pointie Talkie booklet shown on the left of Fig. 7.56 was produced in China by the Office of War Intelligence (OWI) and the 21st Photo Reconnaissance Squadron. It has a blue cloth cover and

Fig. 7.53: These two Pointie Talkies are virtually identical, although that on the left was printed January 1943, probably in China, and that on the right was printed in the U.S. in 1944. The booklet on the left is considered to be an example of the first Pointie Talkie produced for American air crews in China, and sample pages from it are shown in Fig. 7.54. Identical pages from both are illustrated in Fig. 7.55. (Left: REB 3914, Right: REB 3915) *International Blood Chit Museum*

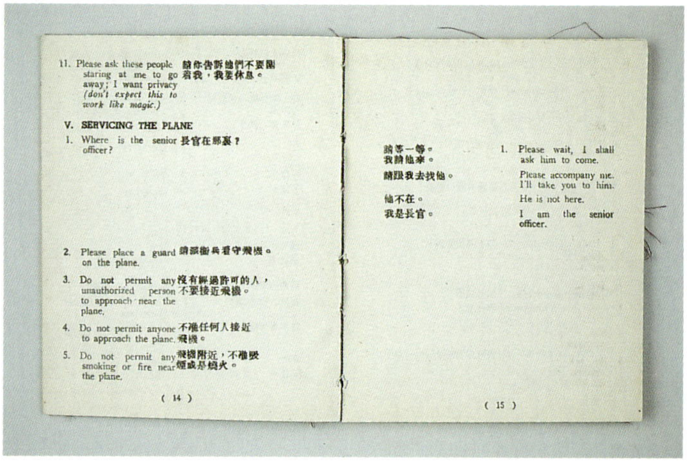

Fig. 7.54: The first Pointie Talkie booklet produced by U.S. forces in China includes, in addition to sections directly related to evasion, sections on non-evasion activities such as servicing aircraft. There is a considerable listing of place names near the back of the booklet, and several blank pages at the back are for personal additions. Onto one of these pages is taped a list of phonetic phrases, evidently provided by the airman's unit, including "Please take me to the New Fourth Army - Ching nee soong waw down Shin Sze Ju-un chu" and "Where is the 8th Route Army - Bah loo Ju-un dzai nah-lee", which aim at obtaining help from Chinese Communist forces. These phonetic phrases are of no use for pointing and talking, but they serve the purpose of seeking aid, and illustrate why the blank pages were included in the booklet.On the last page is printed "January 1943 Printing." There was at least one other printing of this booklet, done in August 1943. (REB 3914) *International Blood Chit Museum*

contains 87 pages for use in China. Some of the contents are identical to those in the PT shown in Fig. 7.53, and to this has been added material in Burmese, French, Annamese, Thai, Shan, Lolo, and Lao.

Both Pointie Talkies in Fig. 7.56 contain a Blood Chit written in Lolo, shown in Fig. 7.57. It is a copy of an order issued by the KMT government that incorporates some elements of a Blood Chit, and is accompanied by a phonetic version which could be read aloud to Lolos who could not read. The following English translation is also given:

> Translation from an order issued by Geralissimo Chiang Kai-shek's Field Quarters at Sichang for flyers operating in Lolo country to help to get rescue.
>
> This is an American flyer. He and his folks fly thousands of airplanes to China to help us to fight the Japanese. Already the Americans have killed many Japanese. Therefore, he is one of our best friends.
>
> In case this good friend of ours has to bail out because his machine fails we must help him. People must guard him and immediately escort him and his things to the nearest Magistracy to take care of or to take him directly to General Chang in Sichang who will in turn give the Lolos silver, salt, cloth and flag of honor in the way of rewards.
>
> If the natives neither take the men to report to the Magistrate nor to Gen. Chang, the natives will be bombed by airplanes and punished by strong troops, of the National Army. Hurting or insulting the flyers will be punished in the same way.
>
> CHIANG KAI-SHEK,
> (Original Chopped)
> Director of the National
> military council

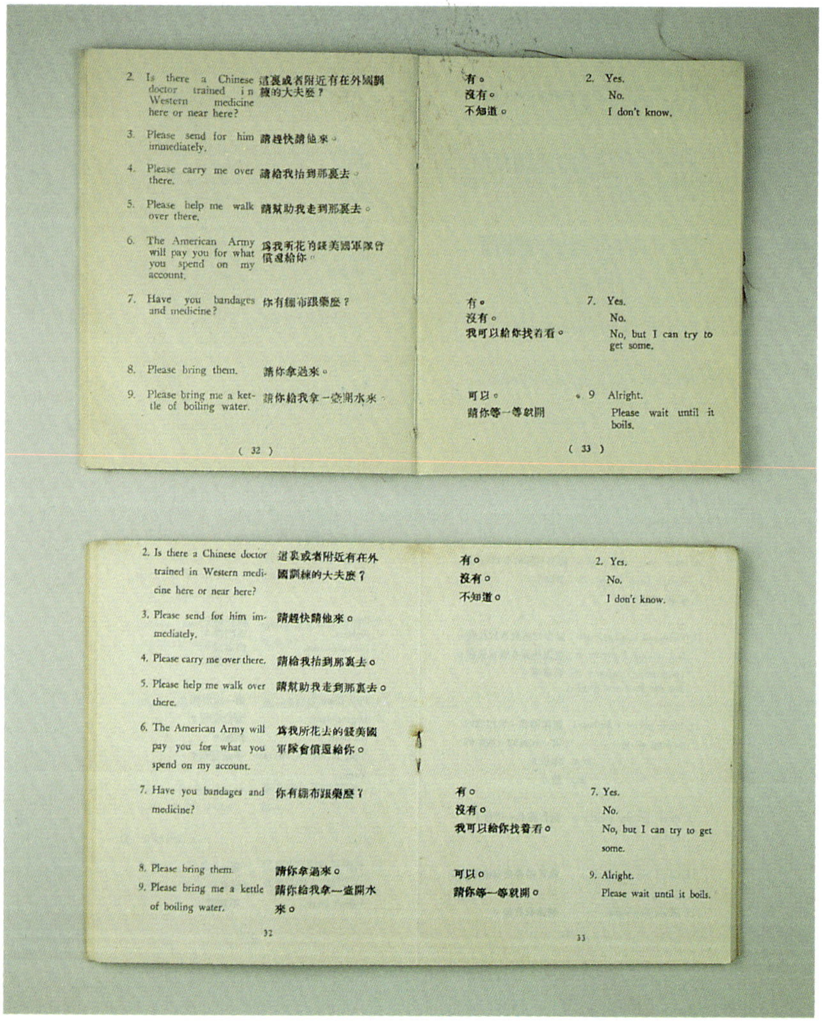

Much of the material in the Pointie Talkie on the right side of Fig. 7.56 is identical to the material in the PT shown to its left. In addition, a special section en-

Fig. 7.55: In 1944 the U.S. Government Printing Office produced an exact copy of the theater-made Pointie Talkie for China. The upper booklet is the theater-made version, and the lower one is the corresponding U.S. edition. All pages of both booklets virtually identical. *International Blood Chit Museum*

Fig. 7.56: Two Pointie Talkie booklets produced in China are shown here. That on the left is a product of a joint effort of the Office of War Intelligence (OWI) and the 21st Photo Reconnaissance Squadron, and that on the right is a product of a cooperative effort of the Air Ground Aid Section (AGAS) and OWI. Both contained material not found in earlier Pointie Talkie booklets. The Blood Chit for use among the Lolo tribe is shown in Fig. 7.57. The introductory page for a section for use in North China, from the right-hand PT, is illustrated in Fig. 7.58. (Left: REB 3916, Right REB 3918) *International Blood Chit Museum*

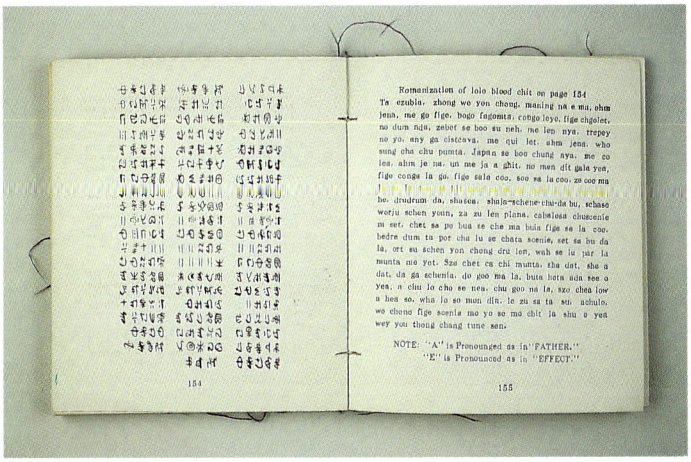

Fig. 7.57: The Lolo section of the OWI-21st Photo Recon Squadron and of the AGAS-OWI Pointie Talkie consists of a chit to be read by the Lolo people, and another that the airman could read aloud to people who were unable to read. The Lolo lived in a remote mountain region of southeastern China. *International Blood Chit Museum*

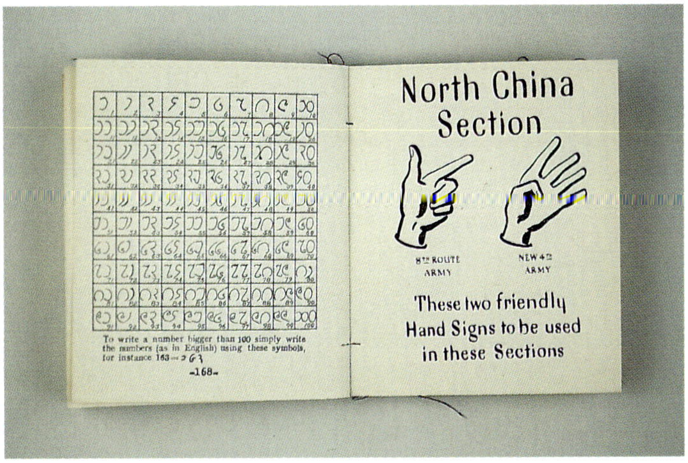

Fig. 7.58: This page from the 196-page AGAS-OWI Pointie Talkie introduces the section for use in North China, which was controlled by the Chinese Communist Party (CCP), in particular their New Fourth Army and Eighth Route Army. This 28 page section of the booklet is designed to obtain help from them. The airman could use the "friendly hand signs" to identify himself as a friend before any evasion aids were utilized. *International Blood Chit Museum*

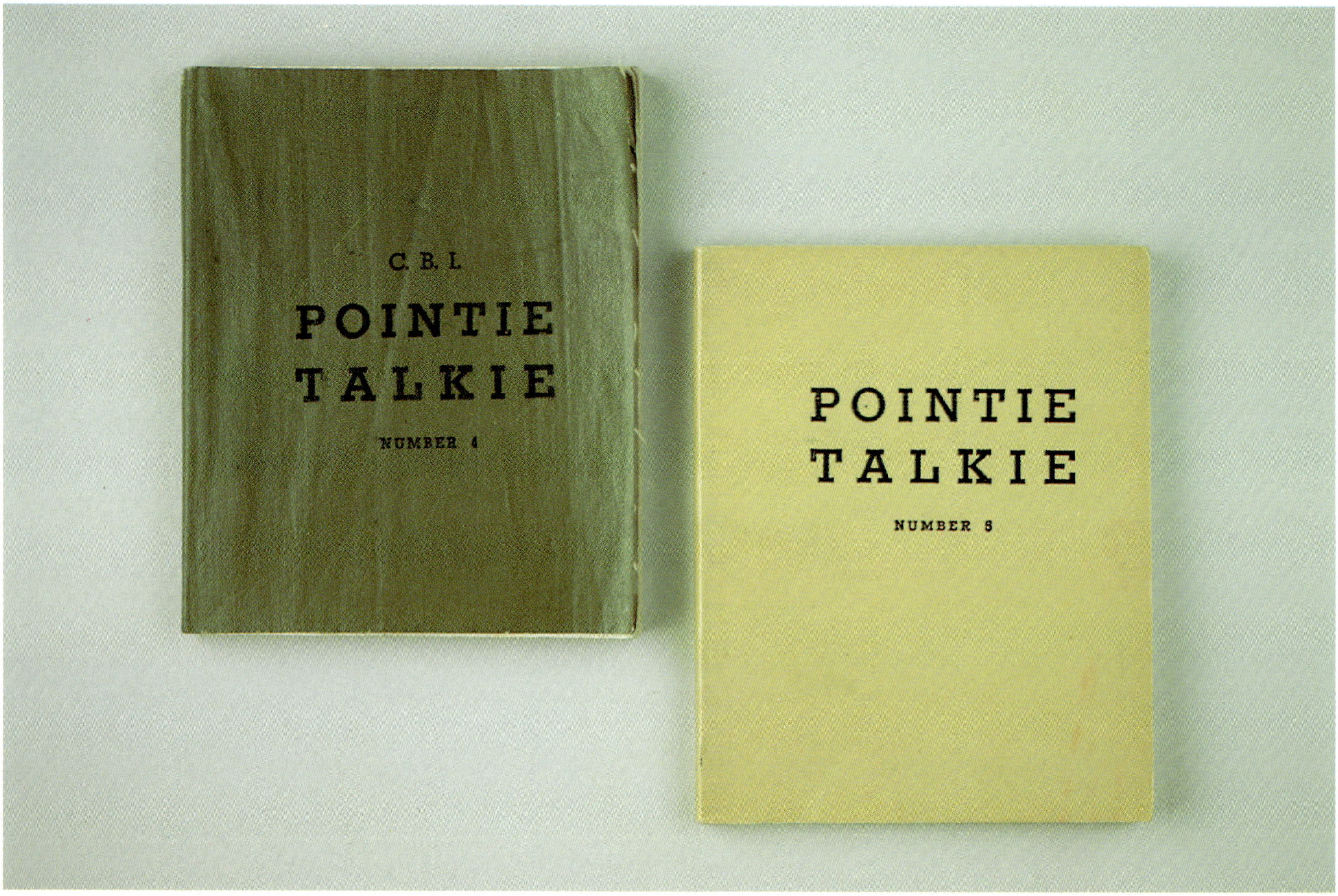

Fig. 7.59: The final two Pointie Talkie booklets made in the U.S. for use in Asia, *Pointie Talkie No. 4* and *Pointie Talkie No. 5*, are shown here. The first contains many features of the two theater-made Pointie Talkies shown in Fig. 7.56, while the latter contains these plus a special "Illustrations Section." Versions of *Pointie Talkie No. 4* were produced with the green cloth cover, illustrated, and a cream-colored paper cover like that on *Pointie Talkie No. 5*. (Left: REB 3920, Right: 3921, Not shown: *PT No. 4* with cream colored paper cover: REB 3920.1) *International Blood Chit Museum*

titled "North China Section" has been added at the end of the booklet. The introductory page of this section, shown in Fig. 7.58, features hand signs for use in this area.

The AGAS Journal for 1944 has several entries outlining the printing of Pointie Talkie booklets by AGAS, OWI and the 21st Photo Recon Squadron during April and May of that year. It may be that the OWI/21st Pointie Talkie shown in Fig. 7.56 is that referred to in the AGAS Journal:

2/17/44 Major Ogden BM called. Pointie Talkie being printed.
2/29/44 Pointie Talkie returned for final revision.
3/18/44 Printer came to check on P-T, and said he was starting to print 4000 today.
4/1/44 P-T was turned over to Dave Edwards, OWI, a proceed [sic] on plating and to do printing Monday.
4/3/44 On Lt. Taylor's visit to BMOI, he picked up Burmese translation of P-T.
4/4/44 Major Makepeace came in to enquire of Major Wichtrich re P-T: number being printed by OWI, distribution, etc. Lt. Taylor saw him and learned that he wants 400-500 for Y force personnel in the field.
4/4/44 A Lt. left prices for cutting and binding of the Pointie Talkies.
4/11/44 P-T booklet being photographed by 21st Photo Lab.
4/12/44 Picked up P-T from 21st Photo Sq.
5/9/44 Phone call from Air Freight saying there are 30 more pkgs paper. Lt. Taylor and Andy secured a Weapons Carrier to transport it to OWI.
5/19/44 1250 Chinese flags and 2000 P-T were sent.[100]

There is another entry in the Journal, from August, which refers to reproducing a Pointie Talkie booklet:

8/4/44 This radio from WARX (73947) relayed from Chungking: Re P-T, referred to in ARW ltr 5 July. Arr. for reprint has been made; finished sooner if you send hand-written orig mss. by air of that particular one, SHAN & BURMESE from which OWI printed. Sample not good for reprod., and hard to find type here. Send LOLO & THAI HILL changes. Meanwhile, Chinese sec. going ahead."[101]

On 22 August 1944 AGAS informed MIS-X "we are printing a Lolo section in our Pointie Talkie and printing Lolo Blood Chits to be dropped by plane over the Lolo Country."[102]

Around this time *CBI Pointie Talkie No. 4*, shown on the left in Fig. 7.59, was being compiled in the U.S. Like theater-made PTs, it contains sections in Chinese, Burmese, French, Annamese, Thai, Shan, Lolo and Lao plus a special section for North China. This Pointie Talkie was apparently compiled from a copies of several of the booklets produced in-theater by AGAS, OWI, and the 21st Photo Recon Squadron. Of interest is the fact that the phonetic Lolo material is identical to that in both the OWI-21st Recon Squadron and the AGAS-OWI booklets, even to the extent that it copies the sentence "Romanization of Lolo blood chit on page 154" word for word, even though in *Pointie Talkie No. 4* the lolo chit is not on page 154, but on page 170. The following memo regarding *CBI Pointie Talkie No. 4* was sent to Maj. Wichtrich on 9 September 1944: "The copy and layout for the Pointie-Talkie arrived. Previous to this the printer had set up all page forms from the sample book sent some time ago. Colonel Winfrey does not understand the purpose of your pasting the separate pages on large sheets, but he will duplicate the original OWI printing as closely as possible, adding of course all new material. Printers here would not work from the type of layout furnished."[103]

Fig. 7.60: *Pointie Talkie No. 5* incorporated this pictorial "comic-book" style Blood Chit for use with people who could not read. It was hoped that it would communicate that the bearer was a downed flyer assisting China to fight the Japanese, that he should be cared for and returned to friendly forces, and that those who helped him would be rewarded. *International Blood Chit Museum*

One final PT booklet for Asia, *Pointie Talkie No. 5*, was produced in the U.S. Shown on the right in Fig. 7.59, it has a brown paper cover and contains sections in all the languages used in *Pointie Talkie No. 4*. It was printed late in the war, and a special "Illustrations Section", was added in response to a suggestion made in April of 1945:

> The suggestion is offered that, inasmuch as most Chinese coolies, fishermen, junk operators, and farmers (the men with whom downed fliers make their first contacts in 99% of the cases) cannot read Chinese characters in the "Pointie-Talkie", a comic strip type sheet of paper be inserted in the "P-T" with graphic illustrations which would convey to a Chinese rescuer the essential ideas of identification and desired evasion. These illustrations could identify the downed pilot as an Allied flier, could show that he desires assistance in returning to American or Allied hands, and that the rescuer would be generously rewarded for his aid.[104]

The color illustration section from *Pointie Talkie No. 5* begins with a CBI shield, which had been engrained in the minds of the Chinese as a symbol of the American airmen who had come to help them through numerous aerial leaflets to this effect. Following the shield is the pictorial Blood Chit, the pages for which are shown in Fig. 7.60. This is essentially a comic-book Blood Chit. The "Foreword" to this pictorial section, which is printed on the page before the first panel, explains the purpose of the section:

> The following illustrations are to be shown to the Chinese who cannot read the Pointie Talkie or Blood Chit. They attempt to convey the following general ideas:
>
> 1. The bearer is an American flyer whose plane has been shot down.
> 2. The bearer is a friend of the Chinese whom he is helping in the fight against the Japanese.
> 3. The Chinese should protect the bearer and hide him from the Japanese.

Fig. 7.61: Six booklets containing language material typically available to U.S. servicemen in Asia are shown here. All are official U.S. publications, except for the booklet in the lower right which was produced in India for sale to Allied personnel.

The two top booklets introduce servicemen to the language, country, people, customs, and culture of an area. TM 30-333 *Chinese Language Guide*, top left, has a section showing the Chinese characters used on signs for "men", "women", "danger", "no admittance", and others, plus a list of words and phrases listed in alphabetical order. In addition ot cultural material, the *Pocket Guide to China*, top right, includes a list of useful words and phrases in English and phonetic Mandarin, the spoken dialect of most of China.

The lower four booklets are Phrase Books that could be carried to facilitate communications. The 134-page paperback *TM 30-634 Cantonese Phrase Book*, left center, is printed in English, phonetic Cantonese, the language of the Nationalist Government, and written Chinese. It is arranged in sections such as Emergency Expressions, Location and Terrain, and Landing a Plane. The cloth-covered 275 page *TM 30-273 Cantonese Phrase Book*, right center, is an extremely comprehensive Phrase Book in English, Chinese, and phonetic Cantonese. Instructions contained in *TM 30-273* indicate that the booklet was to be carried by combat personnel. Inside the front cover is a section printed in red entitled "These Words May Save A Life", detailed in Fig. 7.62, that would have been useful in first contacts with local inhabitants. The remainder of the book would have been used for detailed communication once the airman's identity and immediate needs were established.

The 120 page *TM 30-362 Burmese Phrase Book*, bottom left, is similar in form to *TM 30-634* directly above it. This and Phrase Books in many other languages were produced by the U.S. for use by American servicemen. Although they were not Pointie Talkies, the fact that they include phrases written in two languages side-by-side facilitated their being used in the same manner. The *Soldier's Guide to Hindustani*, bottom right, was written by D. T. Shahani in English and phonetic Hindustani. The difficulty with this format is that even if one is able to communicate by speaking what is written phonetically, it is unlikely that one would understand the spoken answer. No written Hindustani is included, so it cannot be used as would a Phrase Card or Pointie Talkie. It is one of at least thirteen language booklets produced by the Educational Publishing Co. in Karachi, India, for sale to allied personnel in India, that included the *American's Guide to Hindustani*, the *Soldier's Guide to Arabic*, and the *Guide to Military Urdu. International Blood Chit Museum*

4. The Chinese should help the bearer by giving him food and medical aid.
5. The Chinese should take the bearer to the nearest American or Chinese headquarters.
6. The Chinese will be rewarded.

The War Diary of MIS-X Technical Section at Fort Hunt for August and September 1945 indicates that *Pointie Talkie No. 5* was not completed in time for distribution to units before the war ended, and describes another PT under development for use in Asia that was cancelled after the cessation of hostilities:

> Pointie-Talkie No. 5. Of the 150,000 copies on procurement, 80,000 of which were near completion will be received. The remainder have been cancelled.
>
> Pointie Talkie No. 7 (Korean), the copy of which was completed, was cancelled.[105]

As with MIS-X chits and flags, it is important to remember that the majority of the Pointie Talkie booklets produced by MIS-X in the U.S. went to units in the Pacific. A total of 74,085 Pointie Talkies were shipped by PO Box 1142 during the war, and of these 57,835 went to POA and AFPAC units, 10,094 went to India, and only 6,018 went to China. The remaining 138 went to units in Europe and the Mediterranean, possibly as examples of what was being produced for airmen the Pacific.[106] The Pointie Talkie booklets made in the U.S. are described in this chapter rather than in the chapter on the Pacific to graphically illustrate the direct connection between MIS-X evasion aids and the theater-made AGAS, OWI and 21st Photo Recon Squadron items from which they were developed.

American servicemen were provided a variety of language material in addition to Blood Chits, Pointie Talkies and Phrase Cards. Some of this material, such as guides to individual countries and geographical areas as well as language guides, was designed to help

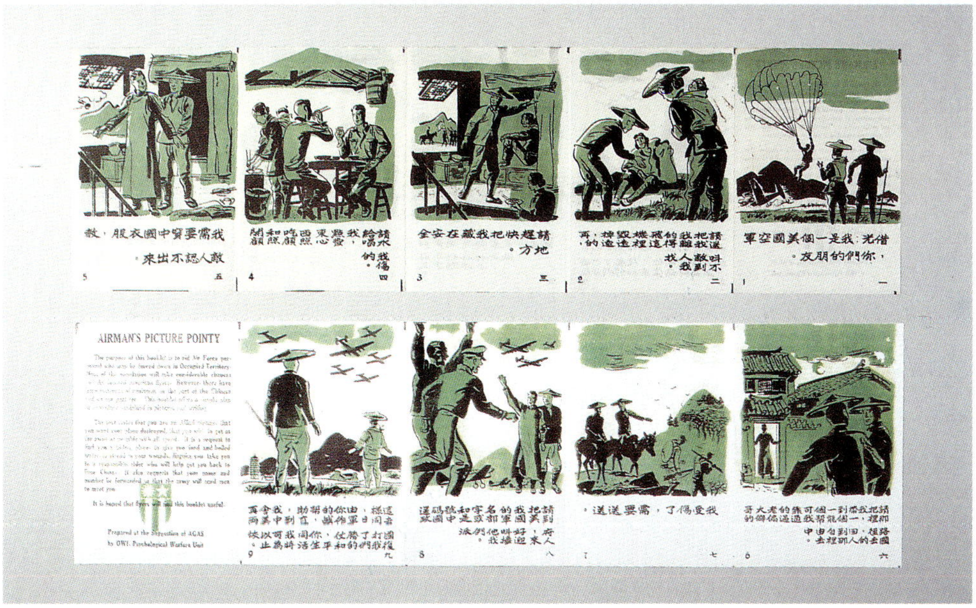

Fig. 7.63: This pictorial chit, entitled *Airman's Picture Pointie*, was produced by the OWI Psychological Warfare Unit at the suggestion of AGAS. It has nine illustrations that instruct Chinese who could not read how they could assist the downed airman. It was to be carried by airmen and used as a Blood Chit when non-literate people were encountered. (REB 3924) *International Blood Chit Museum*

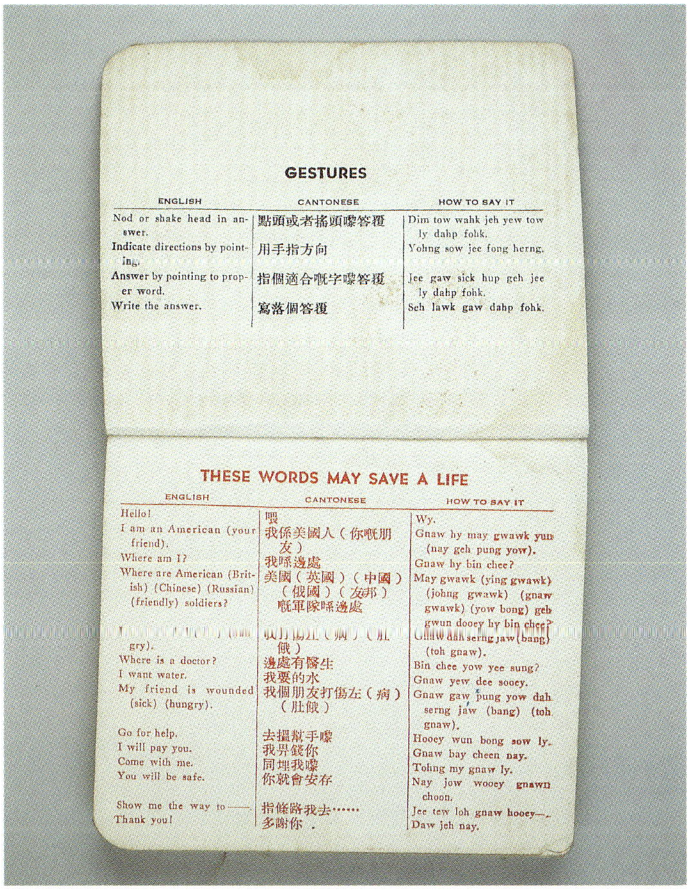

GESTURES

ENGLISH	CANTONESE	HOW TO SAY IT
Nod or shake head in answer.	點頭或者搖頭嚟答覆	Dim tow wahk jeh yew tow ly dahp fohk.
Indicate directions by pointing.	用手指方向	Yohng sow jee fong herng.
Answer by pointing to proper word.	指個適合嘅字嚟答覆	Jee gaw sick hup geh jee ly dahp fohk.
Write the answer.	寫落個答覆	Seh lawk gaw dahp fohk.

THESE WORDS MAY SAVE A LIFE

ENGLISH	CANTONESE	HOW TO SAY IT
Hello!	喂	Wy.
I am an American (your friend).	我係美國人（你嘅朋友）	Gnaw hy may gwawk yun (nay geh pung yow).
Where am I?	我喺邊處	Gnaw hy bin chee?
Where are American (British) (Chinese) (Russian) (friendly) soldiers?	美國（英國）（中國）（俄國）（友邦）嘅軍隊喺邊處	May gwawk (ying gwawk) (johng gwawk) (gnaw gwawk) (yow bong) geh gwun dooey hy bin chee?
[illegible] gry).	[illegible] 餓）	[illegible] serng jaw (bang) (toh gnaw).
Where is a doctor?	邊處有醫生	Bin chee yow yee sung?
I want water.	我要的水	Gnaw yew dee sooey.
My friend is wounded (sick) (hungry).	我個朋友打傷左（病）（肚餓）	Gnaw gaw pung yow dah serng jaw (bang) (toh gnaw).
Go for help.	去攞幫手嚟	Hooey wun bong sow ly.
I will pay you.	我畀錢你	Gnaw bay cheen nay.
Come with me.	同埋我嚟	Tohng my gnaw ly.
You will be safe.	你就會安存	Nay jow wooey gnawn choon.
Show me the way to ——	指條路我去……	Jee tew loh gnaw hooey—.
Thank you!	多謝你．	Daw jeh nay.

Fig. 7.62: Inside the front cover of the *Cantonese Phrase Book*, TM 30-273, is a special section printed in red that could introduce the downed airman to the local people and get him the help he needed. Downed airmen would have used this comprehensive Phrase Book in a similar manner to Pointie Talkies and Phrase Cards. Instructions contained in the book indicate that it was to be carried on operations, telling the serviceman to "put the book in your pocket ... you may need a foreign phrase in a hurry." *International Blood Chit Museum*

them learn about the peoples, customs and cultures of areas to which they would be posted and to teach them some basic language skills of the area. These guides also contained a list of useful words and phrases in English and in phonetic versions of foreign languages. Other material, such as Phrase Books, were to be carried by servicemen to assist them in communicating with the local people.

Shown in Fig. 7.61 are a few examples of the many types of language booklets available to American servicemen in Asia. While language guides were to be used primarily in concert with phonograph records in learning the spoken language and were written in English and their phonetic equivalents, they also contained other useful material. Pocket Guides introduced servicemen to the people, customs, and cultures of an area, and Phrase Books were carried to facilitate communications with local populations. The cloth covered *TM 30-273 Cantonese Phrase Book* contains a special section just inside the front cover entitled "These Words May Save A Life", detailed in Fig. 7.62, that would have been useful in first contacts with local inhabitants.

The pictorial chit added to *Pointie Talkie No. 5* (Fig. 7.60) was not the only such item produced for use in China. The OWI Psychological Warfare Unit, at the suggestion of AGAS, produced a pictorial chit entitled *Airman's Picture Pointie*. This chit, shown in Fig. 7.63, had nine illustrations in addition to text so that the needs of downed airmen could be communicated to Chinese who could not read. The leaflet explains that "the purpose of this booklet is to aid Air Force personnel who may be forced down in Occupied Territory" and that it "offers a simple plan of procedure explained in pictures and writing." That it was intended to be carried by airmen rather than dropped from the air is disclosed by the statement, "It is hoped that flyers will find this booklet useful."

The Psychological Warfare Unit of OWI also produced uncounted thousands of leaflets instructing the indigenous peoples of Asia how to assist Allied air crew members. These were dropped

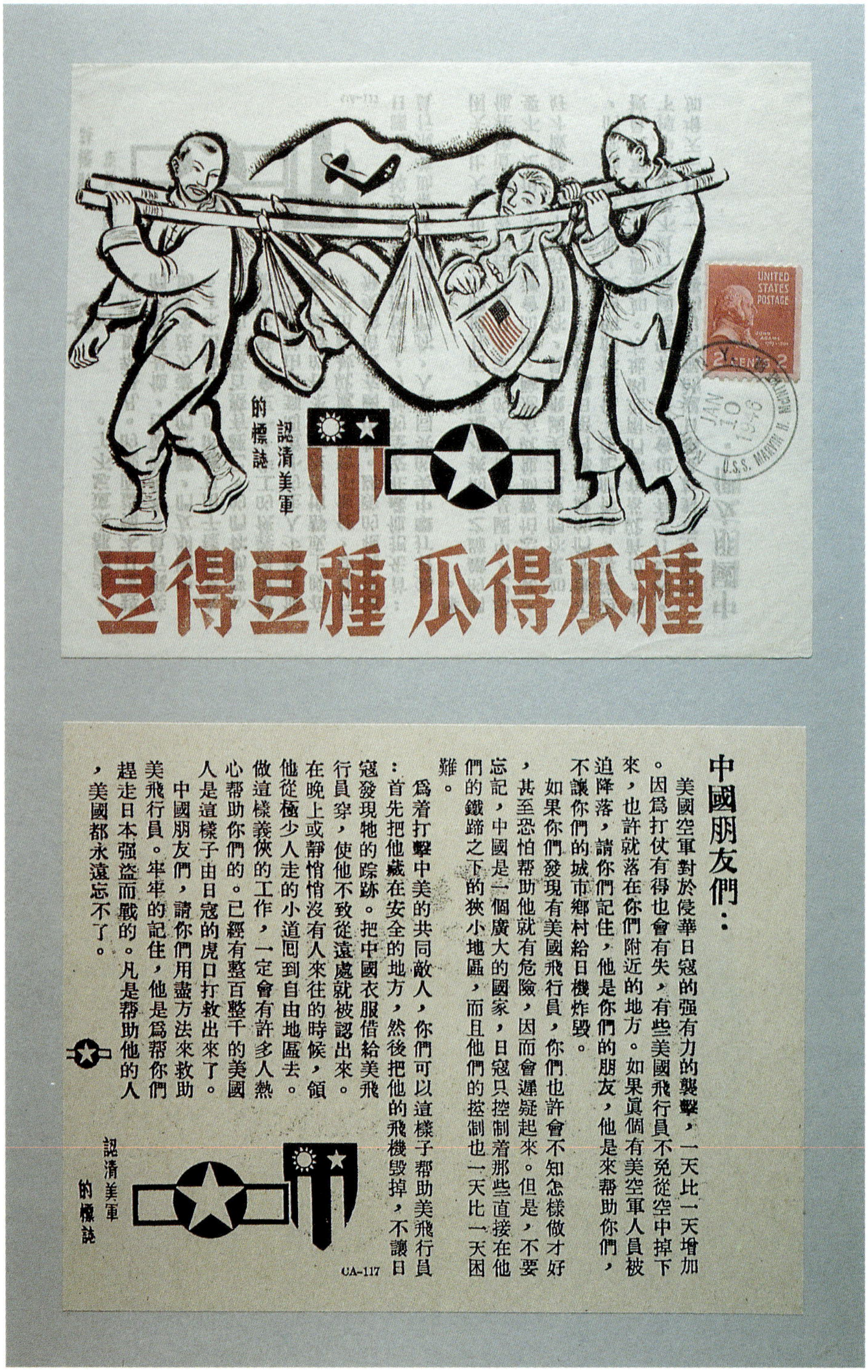

Fig. 7.64: The American airman on this two-sided leaflet, CA-117, displays his Blood Chit to two Chinese carrying him. This leaflet is a souvenir from *USS Marvin H. McIntyre*, as are many of the leaflets that are available on the collecting market. The large red characters on the front read, "Plant melons and harvest melons, plant peas and harvest peas", equivalent to our proverb "As ye sow so shall ye reap." American insignia were commonly included on leaflets in order to instruct the local people that they were the symbols of Americans who had come to help them. The small characters next to the CBI shield and star-and-bars on this leaflet read, "Identify clearly these American military insignia." *International Blood Chit Museum*

by Allied aircraft over occupied and contested territory. One walk-out report stated that "Allied planes had dropped illustrated leaflets in occupied territory instructing Chinese civilians to help downed Allied airmen. Although all Chinese could not read their language, it was hoped they could find people who could."[107]

These instructional leaflets, a clear indication of the advanced development of the Allied evasion and rescue system, almost always carried an illustration of a flyer showing a Blood Chit or flag to identify himself. Often illustrated with the American and/or Chinese nationalist flag, with the shield-shaped red-white-and-blue China-India-Burma patch, and/or with the star-and-bars aircraft insignia, the leaflets told the people to learn to recognize these insignia as symbols of their allies and friends.

The leaflets served several purposes. First, they were written in a simplistic propaganda jargon that reassured the people that the Allies were winning the war and would free them from the Japanese. Secondly, they gave useful instructions, using both text and simple, appropriate illustrations, on how to assist downed Allied aviators. Finally, the leaflets and the efforts to distribute them over areas where Allied aircraft operated and could be forced down prepared local residents for the sudden appearance of an Allied airman seeking help. The leaflets illustrated in this work are but a few of the many produced, yet in spite of the seeming variety they are essentially variations on a common theme. For this reason, and because of space considerations, complete translations for only some of them are given. Most are shown next to a reproduction of the reverse side.

Shown in Fig. 7.64 is a two-sided leaflet for China, CA-117, showing an airman displaying his Blood Chit to two Chinese carrying him in a stretcher. This leaflet is a souvenir from *USS Marvin H. McIntyre*, given to crewmen after the war as evidenced by the stamp and postmark. The large red characters on the front tell a Chinese Proverb: "Plant melons and harvest melons, plant peas and harvest peas." The Chinese have a single word for what we call both melons and squash, likewise a single word for what we call both peas and beans. Regardless of the exact translation, the meaning of the proverb is clear, akin to our proverb "As ye sow so shall ye reap." The small characters next to the CBI shield and star-and-bars read, "Identify clearly these American military insignia". The reverse side of leaflet reads:

To all the Chinese Friends:

To counter the Japanese aggressor who invades China, the mighty United States Air Force is fighting back. The fighting is intensifying every day. In any warfare there are losses, and unfortunately some American airplanes and flyers are being shot down from the air. They may fall and crash where you live. If an American Air Force flyer is forced down to land, you should remember that he is your friend. He is here to help you so that your city and village will not be destroyed by Japanese airplanes.

If you discover an American flyer, you may not know what to do or may even be afraid to give him assistance if it might endanger yourself. For these reasons you may hesitate and delay action. But do not forget that China is a large country, and the Japanese aggressor only controls the small area which they have invaded. Even then, they have a hard time controlling and suppressing these occupied areas on a daily basis.

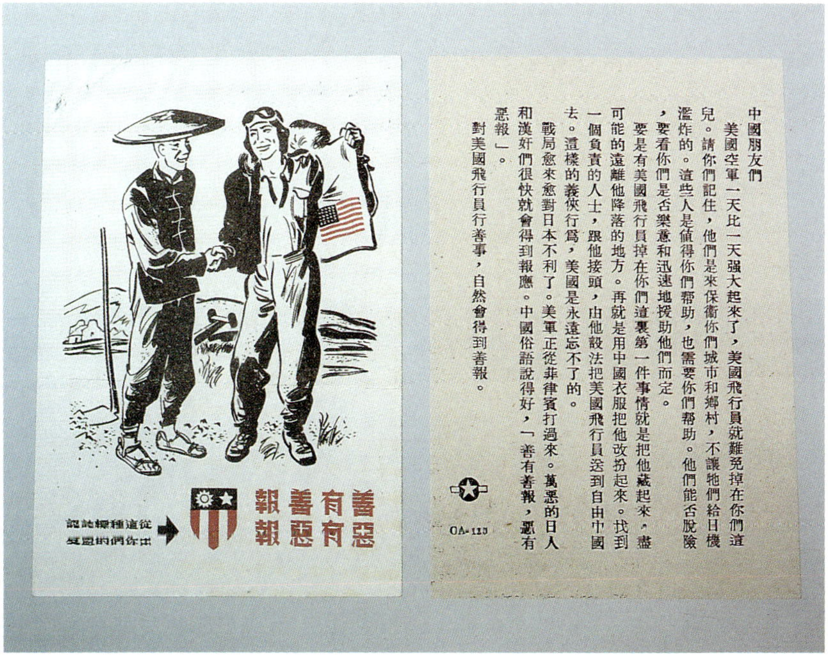

Fig. 7.65: Leaflet CA-125 features a proverb in large red characters: "A good deed has its reward; so has the bad deed." The flyer has an American flag stitched inside his jacket. Like most leaflets, it admonishes the Chinese to learn to recognize the CBI patch and the star-and-bars aircraft insignia. *International Blood Chit Museum*

In order to strike back at the common enemy of China and the United States of America, you can help the American flyer in the following ways: First put him in a safe place, then destroy and get rid of the airplane wreckage so that the Japanese cannot find traces of it. Loan him your Chinese style clothes and outfits so that he may not be recognized from a distance. At night or when no one is around, take him on a seldom travelled trail and get him to freedom. This kind of heroic and patriotic act will eventually result in the reciprocal help and assistance from a lot of people. There are already hundreds and thousands of Americans being saved from the jaws of the Japanese aggressor.

Therefore, to all our Chinese friends, you must try your best to help the American flyers. Remember with all your hearts, they are here to help you drive out the Japanese pirates in this war, so those of you who help the flyers will always be remembered by the United States of America.[108]

Leaflet CA-125 is illustrated in Fig. 7.65, and shows a flyer with an American flag sewn inside his jacket. In the final paragraph the implication is made that a reward for assisting downed airmen may be forthcoming following the war. The large red characters paraphrase a Chinese Proverb: "A good deed has its reward; so has the bad deed." The small characters next to the CBI shield read, "Recognize this emblem to know your ally and friend".[109]

A leaflet with an airman showing his chit while greeting a farmer, CA-115B, is shown in Fig. 7.66. The title on the front is identical to that for leaflet CA-117, previously illustrated: "Plant Melons and Harvest Melons, Plant Peas and Harvest Peas." The smaller writing at the bottom of the front reads, "Please notice this symbol", referring to the CBI shield. The writing on the back of the leaflet by the shield reads, "Recognize clearly this symbol."[110]

Fig. 7.66: This leaflet, CA-115B, is another illustrated with an airman showing his chit to a farmer. The title on the front reads, "Plant Melons and Harvest Melons, Plant Peas and Harvest Peas." The writing at the bottom of the front reads, "Please notice this symbol", referring to the CBI shield. The writing on the back of the leaflet by the shield reads, "Recognize clearly this symbol." *International Blood Chit Museum*

Fig. 7.67: This leaflet, CA-132, combines illustrations with text to graphically show how to rescue Allied airmen. The text between the American and Chinese airmen, each displaying their respective flags, advises "These are symbols of friends!" *International Blood Chit Museum*

A leaflet for China combining text with illustrations to explain the rescue of an airman, CA-132, is illustrated in Fig. 7.67. The text is read, in typical Chinese fashion, from right to left. At the right side of the front of the leaflet is a drawing of two airmen showing the Chinese and American flags sewn into the inside left front of their flight jackets. Referring to the flags, the text in between them reads, "These are symbols of friends!" The title at the left of the American flyer reads, "Notice to the Brave Chinese Friends."[111]

Leaflet CA-19B, illustrated in Fig. 7.68, shows a farmer assisting a pilot who is pointing to a Chinese flag Blood Chit sewn to the inside lining of his jacket. The title at the top of the front side reads, "Give assistance to Allied Pilots." The title on the reverse reads, "Help Save Allied Pilots - Speed Up Your Liberation."[112]

Leaflet CA-118, illustrated in Fig. 7.69, shows a pilot displaying his Blood Chit in front of a downed shark-nosed aircraft on the front. The title on the front reads, "Chinese Friends - Please Help Our Pilots." The smaller text below the aircraft pointedly refers to the Blood Chit in stating, "This flag identifies them as allied pilots." On the back, the title at the upper right reads, "Japan is the Common Enemy of China and the United States of America." The two silhouette views of a P-51B are accompanied by text simply reading "Top View" and "Side View". A portion of the text refers specifically to the aircraft, reading, "If you see an airplane such as this P-51 model land in your vicinity, proceed immediately to the airplane and help the pilot get away from the airplane."[113]

Another leaflet, CA-114, shown in Fig. 7.70, illustrates a Chinese national motioning to an American airman to come out of hiding as Japanese troops leave his home. Food is being placed on the table for the pilot, who has a blood chit sewn inside his flight jacket. A detail of a stylized blood chit is inset at the lower left corner of the leaflet. The title on the right side of the front reads, "Americans will never forget the people who helped them." The smaller writing next to the inset of the blood chit, clearly aimed at familiarizing the Chinese people with the Blood Chit as the symbol of a friend, reads, "Recognize the American Flag."[114]

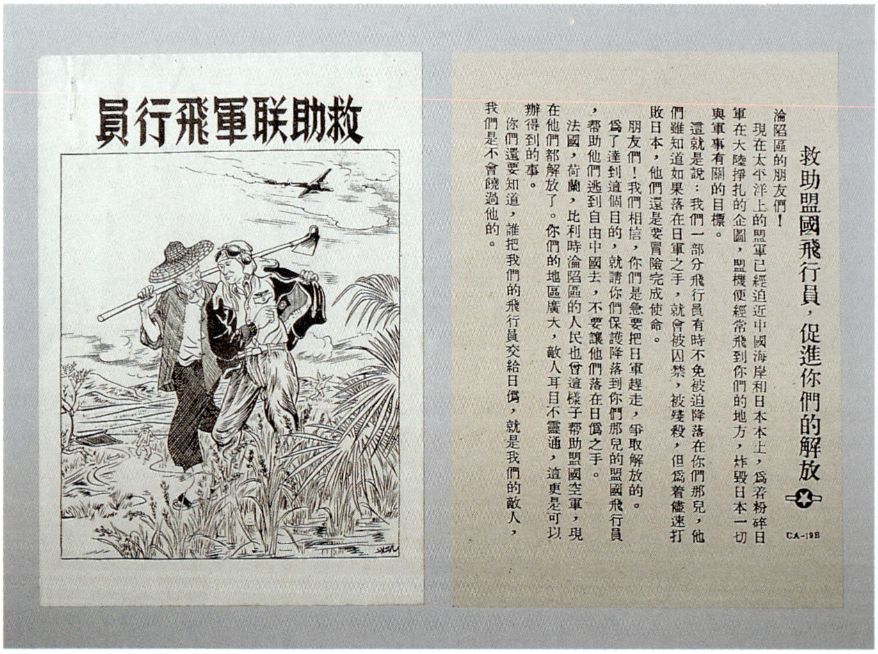

Fig. 7.68: The airman on leaflet CA-19B is pointing out the Blood Chit sewn inside his jacket. Another in the series of leaflets instructing the Chinese on air crew rescue, the title on the reverse side reads, "Help Save Allied Pilots - Speed Up Your Liberation." *International Blood Chit Museum*

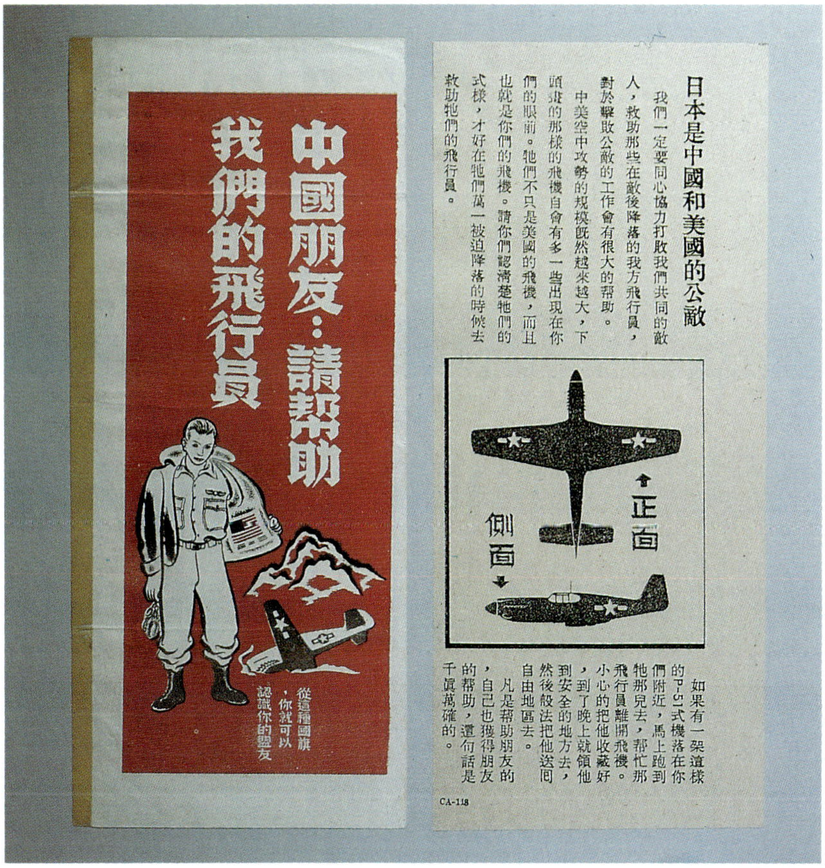

Fig. 7.69: The titles in large characters on this leaflet read, "Chinese Friends - Please Help Our Pilots", and "Japan is the Common Enemy of China and the United States of America." The back of this leaflet, CA-118, shows two views of a P-51B aircraft and advises the reader to learn to identify it. *International Blood Chit Museum*

Leaflet CA-105, shown in Fig. 7.71, combines text with illustrations arranged like a cartoon strip. It is folded like a book, opening in typical Chinese fashion in the opposite direction to our books. The front cover, at the top left of the photo, shows an American airman parachuting to earth. The text accompanying this reads, "This flyer who was forced to parachute (descend) is helping your war effort. Please help him." The illustrations on the inside of the leaflet, shown at the bottom of the photo, are arranged as follows: top right, bottom right, top left, bottom left, on each of the two pages. The writing at the top of the right page reads, "Help American flyers who are helping China fight against Japan," while the text at the top of the left page reads, "All our compatriots with a conscience should courageously help them!"[115] The back cover contains a longer message in the propaganda jargon of the day.

A variant of CA-105 was produced using only the pictorial portion, with modified drawings. This new leaflet, shown in Fig. 7.72, was numbered CA-105B. It is a souvenir from *USS Marvin MacIntyre*. The title in the right border reads, "Helping China Fight the Japanese: the American Pilots." The illustrations are arranged in the same order as those on CA-105, and the text accompanying them is as follows:

1. A white parachute is dropping from the sky.
2. It must be the pilot from the damaged American airplane.
3. Hurry up and take him home, and let him rest.
4. He is a good friend of the Chinese people.
5. Think of a way to avoid letting the Chinese enemy agent know about this.
6. Go tell the underground partisan or the Chinese Army right away, but do it secretly.
7. Hurry up and transport him safely from the front lines.
8. This is the proper way to treat a friend of our country.[116]

Another item, consisting of several pages held together with a staple, was a further development of the cartoon strip idea. Shown in Fig. 7.73, it was a comic book that undertook to explain to the Chinese that Americans were helping them, and that they should in turn help the Americans if they were forced down. Like the leaflets, it is written in a simplistic propaganda jar-

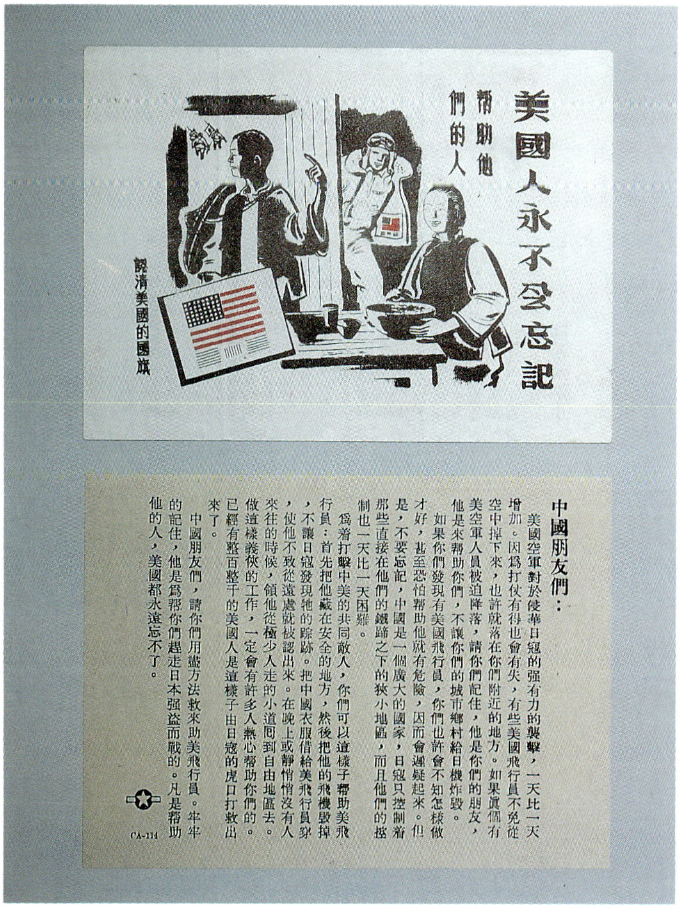

Fig. 7.70: Leaflet CA-114 uses a simple illustration to convey several important messages. The Chinese national has hidden the American airman and has directed Japanese troops away from him. Food has been prepared for the Airman, who is identified by the Blood Chit sewn inside his flight jacket. The title reads, "Americans will never forget the people who helped them." An enlarged, stylized Blood Chit is inset into the illustration, and is accompanied by the words "Recognize the American Flag." *International Blood Chit Museum*

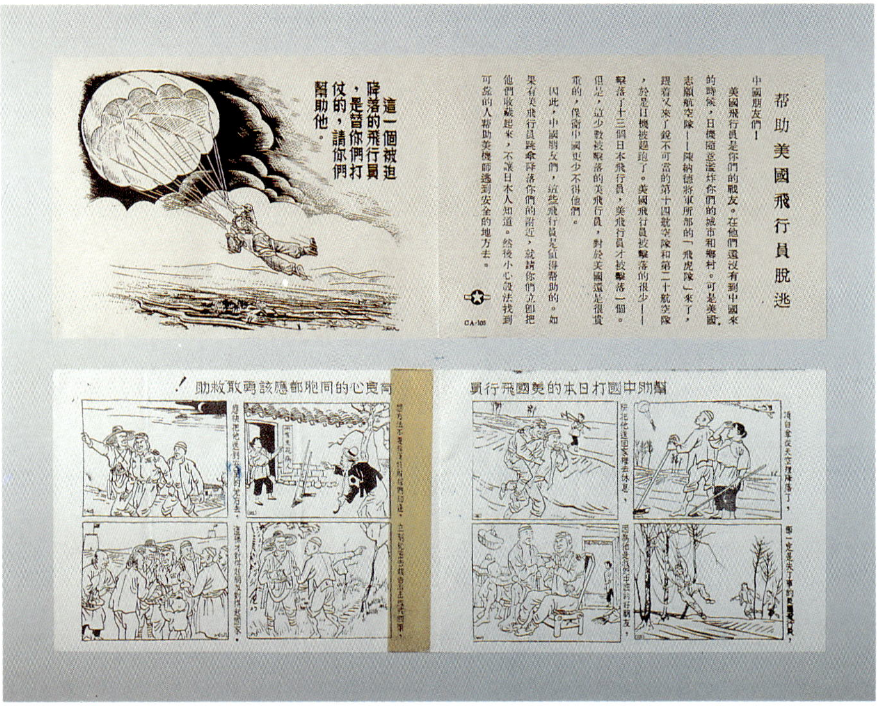

Fig. 7.71: This leaflet, CA-105, combines text with illustrations arranged like a cartoon strip. It is folded like a book, and opens in the opposite direction to books in the English language. The front showing the flyer in his parachute, in the upper left of the photo, reads, "This flyer who was forced to parachute (descend) is helping your war effort. Please help him." The panels inside the leaflet are arranged as follows: top right, bottom right, top left, bottom left, on each of the two pages. The words "Help American flyers who are helping China fight against Japan" appear at the top of the right page, and "All our compatriots with a conscience should courageously help them!" appears at the top of the left page. *International Blood Chit Museum*

gon. The book reads "backwards" to Europeans, that is the front cover is on what would be the back cover of a book in Europe and when open the first page is on the right side rather than the left. As on many leaflets, one of the illustrations in this booklet is of the pilot pointing to a Blood Chit sewn to the lining of his jacket. The text of the booklet reads:

1. (Front cover) This is an American Flyer.
2. These are his parents.
3. This is his wife.
4. He gave up his happy life and a good job to come to China. He was happy to come. (Meaning that he volunteered, he was not drafted)
5. He has a mission. That is to destroy the Japanese robbers, their looting and savagery.
6. He feels this is the most valuable enterprise of his life. If he does not do it this time he will never have the chance again.
7. This young flyer bade his parents farewell and said goodbye to his hometown to come to China to help the Chinese people fight against Japan. He has courage, is professionally trained, and is flying the best planes in the world. But he would need the help of Chinese friends. When he is shot down, he needs

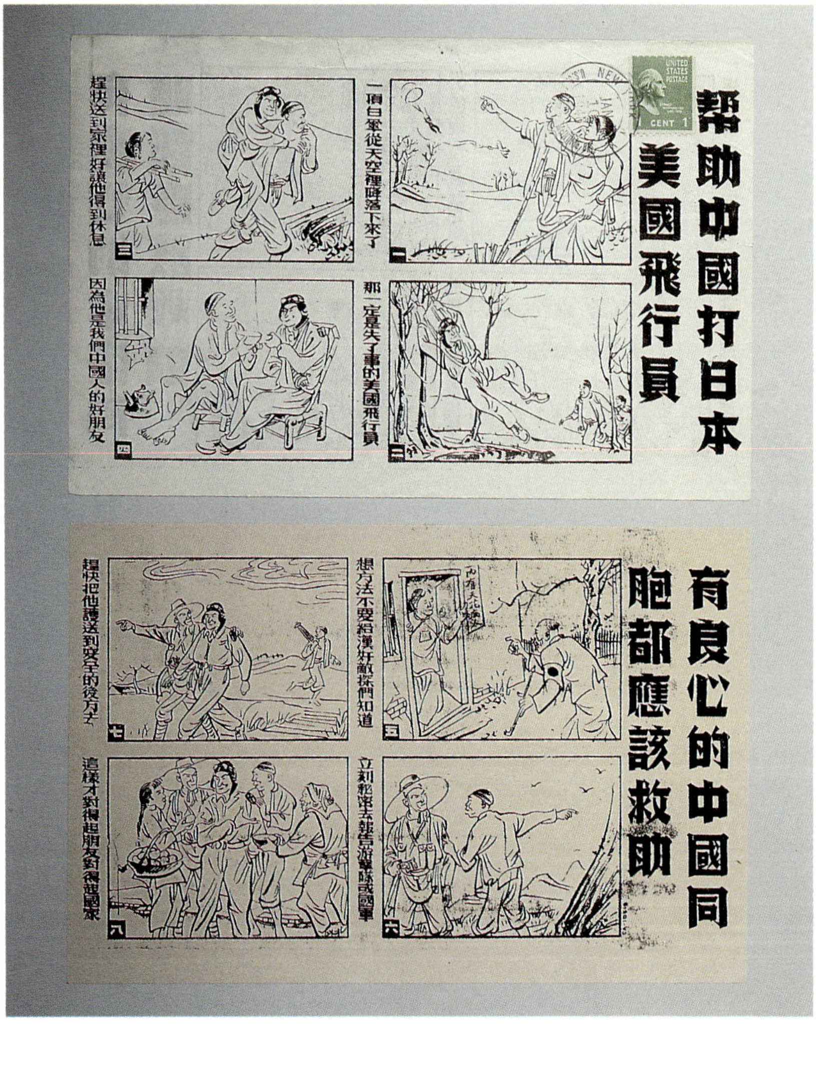

Fig. 7.72: This leaflet, CA-105B, a variation of that shown in Fig. 7.71, is also arranged in panels like a cartoon strip. It may have been useful in instructing persons who could not read. Amplifying text accompanies each panel, and the title reads, "Helping China Fight the Japanese: the American Pilots." *International Blood Chit Museum*

Fig. 7.73: This is essentially a leaflet in the form of a comic book. It explains that the American flyer left his parents, family and hometown to come to China to fight the Japanese, that he was happy to do so (i.e., that he volunteered to do this), and that if shot down he will need the help of the Chinese to return safely to his own people. *Anonymous*

Fig. 7.74: Leaflets were also produced for areas outside of China. This leaflet, CBA-41, was made for use in Burma and is folded like a book. The outside front, shown at the top of the photo, tells the Burmese that Allied pilots are their friends, and asks for assistance in rescuing them. The back explains that American and British aircraft can be identified by their insignia. The inside, at the bottom of the photo, gives detailed instructions for pilot rescue, written in the typical propaganda jargon of the day. *International Blood Chit Museum*

your wisdom and courage to get him to a safe place. Then, he can go back to fight again. He needs your warmth and understanding because he is a young man and he is in a strange place. In order for China and the United States to work together well, you should help him as you would one of your own. Then China and the United States can easily win the war.

8. (Back cover) Then he can return home and every Chinese and American will be able to live happily and peacefully in their homelands.[117]

Leaflets were also produced for deployment over Asian countries outside of China. Leaflet CBA-41, shown in Fig. 7.74, instructed the people of Burma of how to assist downed Allied airmen. The leaflet is folded like a book, and the front and back covers are shown at the top of the photo. A letter from the OWI Psychological Warfare Team that developed the leaflet provides details of its message. The front cover reads, "Allied Pilots Are the Burman's Friends" and "Please help them if they are in danger." The back cover shows the USAAF insignia with the message "American planes bear this marking", and the RAF insignia with the message "British planes bear this Marking". The message on the inside is similar to those on Chinese leaflets.

Leaflets were also dropped over French Indo-China. One of these, AFA-110, is illustrated in Fig. 7.75. It shows a flyer displaying an American flag sewn inside his flight jacket to two Vietnamese. The large text at the top of the back reads, "To All Vietnamese with Courage." The text of the message reads:

> American pilots flying high in the sky are your friends. These people risk their own lives to save you and your children from the oppression of the Japanese. Just like you and your family, these pilots love and cherish freedom. In this kind of love, friendship and brotherhood begin to develop.
>
> Once in a while, there are quite a few airplanes crashing at places far away from their bases. If you are a courageous and loyal Vietnamese, please take in the pilots who survive, save them from the hands of Japanese soldiers, and treat them like your own blood. Run to the site of the crash, help them out of their airplanes and try to destroy the airplanes so that you can cover their traces. Try to fix their wounds, give them food, water, civilian clothing, and help them to get across the border to China where they will be free. Please take them away from the claws of the bloodthirsty Japanese soldiers.
>
> If you are able to accomplish all these tasks, in the future your children can proudly say: "My parents helped the American pilots in the darkest hour of Vietnamese history. They have given me pride and joy in my heritage." The act of giving and the enthusiastic person will always be remembered in history. Bravo Vietnamese![118]

Another leaflet for French Indo-China, AFA-103 shown in Fig. 7.76, has illustrations on both sides, and text in both Vietnamese and French. The illustrations show an airman with a blood chit sewn inside his jacket. The title on the back of the leaflet, in French, reads, "Friends of Indo-China, Please Help." Beneath the illustration of an airman holding his parachute harness and displaying his chit is a message telling people to stay alert for the flag of the United States. Beneath a propaganda message about how the Allies are winning the war and need the valiant aid of the Vietnamese, instructions are given for assisting airmen. These tell the people to

Fig. 7.75: This leaflet, AFA-110, is for use in French Indo-China. The flyer depicted on this leaflet is displaying an American flag sewn inside his jacket to two Vietnamese. Below the title "To All Vietnamese with Courage" are instructions for rescuing downed airmen. *International Blood Chit Museum*

hide and disguise the airman, and when conditions are right to conduct him to the border by way of unused routes. It ends with a message that U.S. will never forget their assistance.[119]

The front of the large leaflet for Indo-China shown in Fig. 7.77 has three photos illustrating airmen using evasion aids while being aided by Vietnamese nationals. A letter from OWI dated 7 November 1944 that accompanies the leaflet states that its purpose is "To urge people in occupied countries to rescue allied fliers that may come down in their land".

The message in French and Annamese (Vietnamese) at the top front of the leaflet reads, "Save the Allied Aviators." The upper, center photo is of an airman displaying the Blood Chit sewn to the lining of his jacket to three people; the middle, left photo shows an airman and Vietnamese man using a Pointie Talkie to communicate; and the lower, right photo shows an airman holding a Pointie Talkie being directed to safety by a Vietnamese man. The title on the reverse side reads, "Speed your Liberation – Rescue Allied Aviators."

In addition to the language-related evasion aids such as the Blood Chits, Phrase Lists, and Pointie Talkies that were officially issued to airmen, a number of souvenir patches patterned after Blood Chits were produced in Asia for sale to

Fig. 7.76: This leaflet, AFA-103, has illustrations on both front and back showing an airman with a blood chit sewn inside his jacket, and messages in Vietnamese and French. The title reads, "Friends of Indo-China, Please Help." Beneath the illustration is a message referring to the chit telling people to stay alert for the flag of the United States. Instructions ask the Vietnamese to hide and disguise the airman, and when conditions are right to conduct him to the border by way of relatively unused routes. It ends with a message that U.S. will never forget their assistance. *International Blood Chit Museum.*

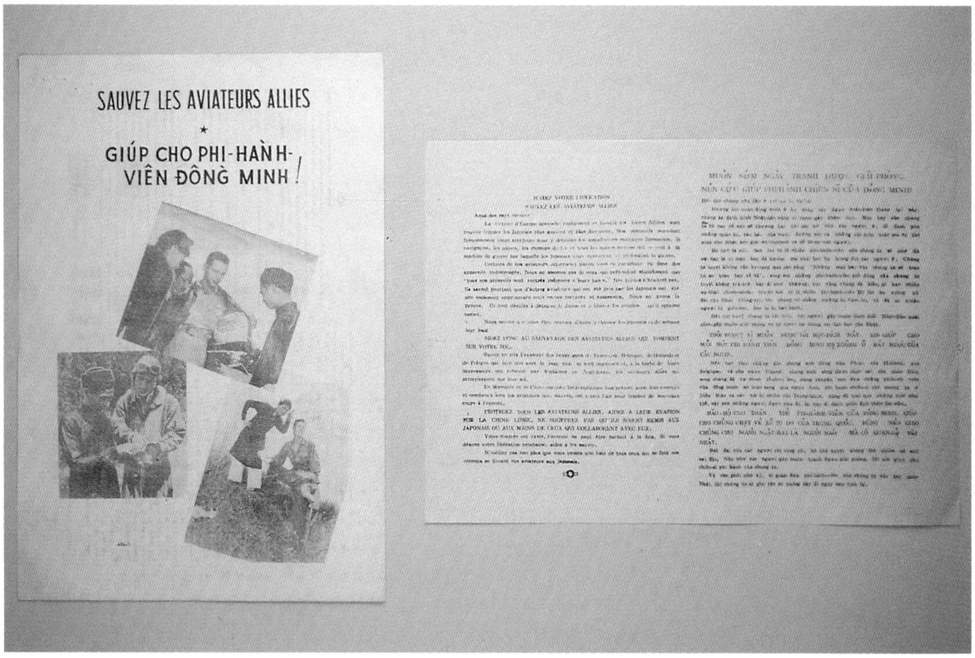

Fig. 7.77: This large leaflet for French Indo China, AFA-5, uses photos showing an airman using his Blood Chit and Pointie Talkie to identify himself and obtain aid. It is written in both French and Vietnamese, and a letter from OWI/CBI explains that is purpose is "To urge people in occupied countries to rescue allied fliers that may come down in their land." *International Blood Chit Museum*

military personnel. Never before nor since has such a variety of hand-made souvenirs patterned after language aids been privately produced for sale to military personnel. These items were made by local entrepreneurs and sold to individual servicemen, who sometimes used them to decorate their flight gear just as they did with other patches. When sewn to a garment worn in combat, it is possible that they they could have been of some use to the evader, although this was not their primary function, and indeed most of these patches available on the collector market today were never sewn to a garment.

Also, as detailed elsewhere in this Chapter, regulations forbade wearing Blood Chits, and by extension patches in the form of Blood Chits, on the outside of flight gear, and most of the clothing decorated with these patches was not worn on operations.

Some of the officially issued Blood Chits are far more rare than many of the theater-made souvenirs, but the cost of souvenirs on the antique and collector market is often far greater than the rarer genuine Blood Chits. This because there are virtually no military artifacts that can match these patches for quality hand-craftsmanship and sheer beauty. These handmade items may be as close to "art" as military collectibles get. Fig. 7.78 shows a storefront in India with several handmade leather patches patterned after Blood Chits displayed for sale among the other patches and related items.

The variety of theater-made patches patterned after Blood Chits is virtually endless, making an accounting of all types and styles impossible. Although the purpose of this work is to detail the history of Blood Chits, some of the souvenir patches are also illustrated. Garments with souvenir patches attached to them are not

Fig. 7.78: This photo of an Indian street scene shows several leather patches patterned after Blood Chits displayed for sale along with other styles of patches and related items. It was from such local entrepreneurs that the many varieties of cloth and leather souvenir patches in the form of Blood Chits were purchased by servicemen. *From* CBI - The Hump, *reprinted by permission of the CBI Hump Pilots Association*

Fig. 7.79: Shown here is an example of a printed cotton souvenier patch. The standard Aero Commission chit text message, rendered in white characters, appears on the red field of the Nationalist flag rather than in a separate text area below it. (REB 9011) *Courtesy of the U.S. Air Force Museum*

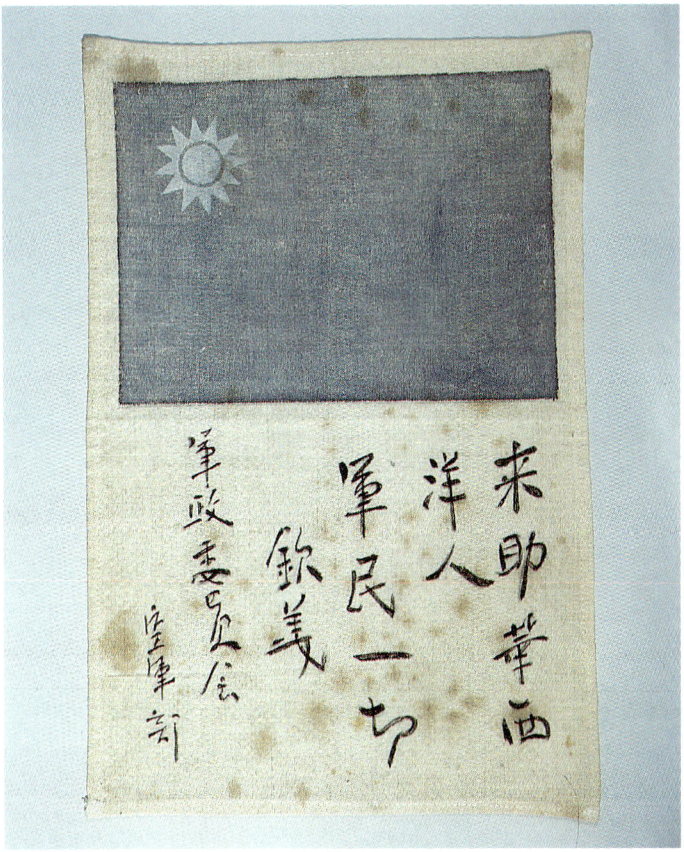

Fig. 7.80: The most striking feature of this unusual cotton souvenir patch is the flag, which is entirely blue and lacks the red field of the Nationalist Chinese flag. The flag was printed on it simply as a blue rectangle, then the sun was stamped in white over the blue field.The text was written by hand. (REB 9012) *International Blood Chit Museum*

Fig. 7.81: This patch, like that shown in Fig. 7.80, is also printed on cotton fabric. The flag and CBI shield were printed on the fabric, and the text was written by hand. (REB 9013) *International Blood Chit Museum*

included in this work, but they would make an excellent subject for a book on decorated flight clothing.

Silk and cotton were among the most common materials used for crafting fabric patches, and these were worked in a variety of ways. Diverse traditional methods for working fabric such as drawing and lettering by hand, printing, appliqué, and embroidery were used singly and in combination to create strikingly beautiful patches.

Printing directly on fabric is the simplest method of mass-producing patches. Figs. 7.79 through 7.81 show examples of patches produced by printing, and by printing combined with hand lettering.

One traditional method of working fabric is called appliqué. This is the sewing of one piece of cut fabric to another in order to form a design or pattern. An application of this technique familiar to most readers would be traditional quilts and coverlets made from many small pieces of fabric. Figs. 7.82, 7.83, and 7.84 show silk patches employing a combination of appliqué, printing, and lettering by hand.

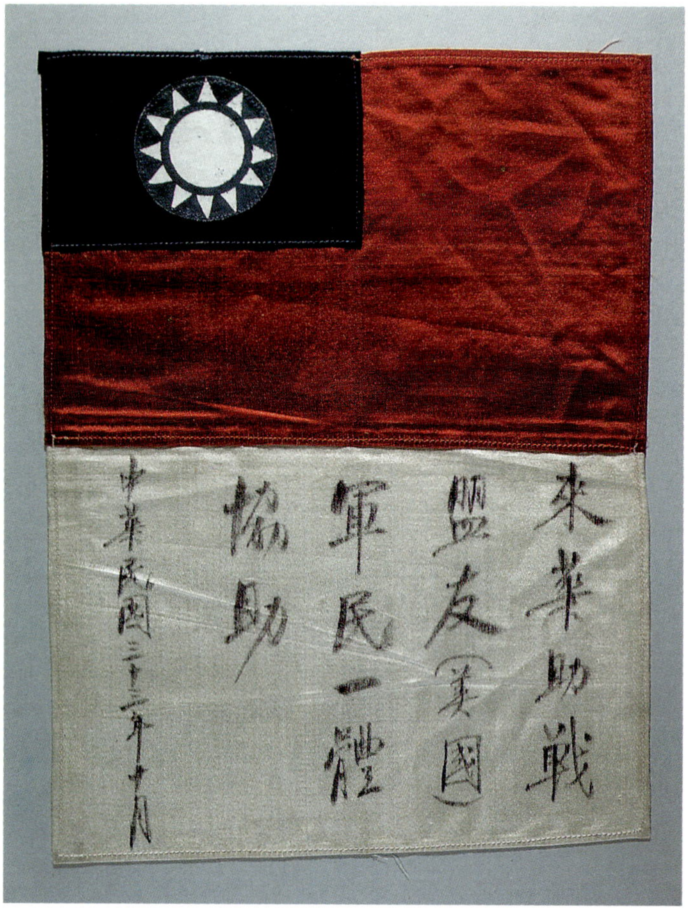

Fig. 7.82: This silk patch combines three traditional methods of working fabric: printing, appliqué, and hand lettering. It is made from four pieces of silk sewn together. The sun on the flag was printed, perhaps with a wood block, on one piece of silk and then sewn to the blue field of the flag. The inked text was hand-written, and is interesting in its own right because of the style in which it was done. (REB 9014) *International Blood Chit Museum*

Fig. 7.83: This simple but elegant five-piece cotton patch combines appliqué and hand-lettering. It is an excellent example of how a few pieces of common fabric can be combined to create a beautiful patch. The inked Chinese characters were added in a particularly fine hand, and are clean and appealing to the eye. (REB 9015) *International Blood Chit Museum*

Embroidery, the decoration of cloth with needlework, was also used extensively in the creation of souvenir patches. Exceptionally beautiful patches were produced by combining appliqué and embroidery on silk, and these items, probably more than any others, illustrate why theater-made souvenirs are so highly valued by collectors. One such patch, illustrated in Fig. 7.85, is made from 14 pieces of silk, and the stars and sun on both flags and the CBI shield are embroidered. The character at the bottom of the third column (from the right) is not identical to that on the Aero Commission chit. It is, however, a simplified character for the same word, "body", so the message is identical even though the characters are not. Simplified characters were sometimes used on souvenir pieces because they were written in the vernacular, whereas official Chinese Blood Chits were always written in classical Chinese.[120] Two columns of additional text have been added beneath the American flag on this patch. The first (right-hand) column of this additional text reads, "Republic of China" and the second (left-hand) column reads "34th Year". This is an example of how years are reckoned by the Chinese, an extension of the ancient method of numbering years during the rule of each Dynasty. The revolution led by Dr. Sun Yat-sen against the Manchu rulers of the Ching Dynasty resulted in the creation of the Republic of China in February 1912. That year, 1912, became the "First year of the Republic of China", and all years during its rule are reckoned from that time. The date on this chit, the "34th year of the Republic of China", is therefore 1945.[121]

Another beautiful example of an item combining appliqué and embroidery is the piece is shown in Fig. 7.86. It is made from 14 pieces of silk with embroidered stars and suns. Although in the form of a Blood Chit, it is actually a commemorative souvenir reading, "Came to China to participate in the war effort. In commemoration, 1945".[122]

While patches combining embroidery and appliqué can be stunningly beautiful, not all are so. Two rather crude looking patches of this type are shown in Figs. 7.87 and 7.88.

The characters on the 3-piece appliqué and embroidered patch shown in Fig. 7.89 are so nicely done that it has been suggested that it was done by machine. A careful inspection, however, indicates that it could also have been done by hand, as was the sun on the flag. Quality hand-work was – and is still – done in many countries, China and India among them. Russell A. "Rusty" Bales, an

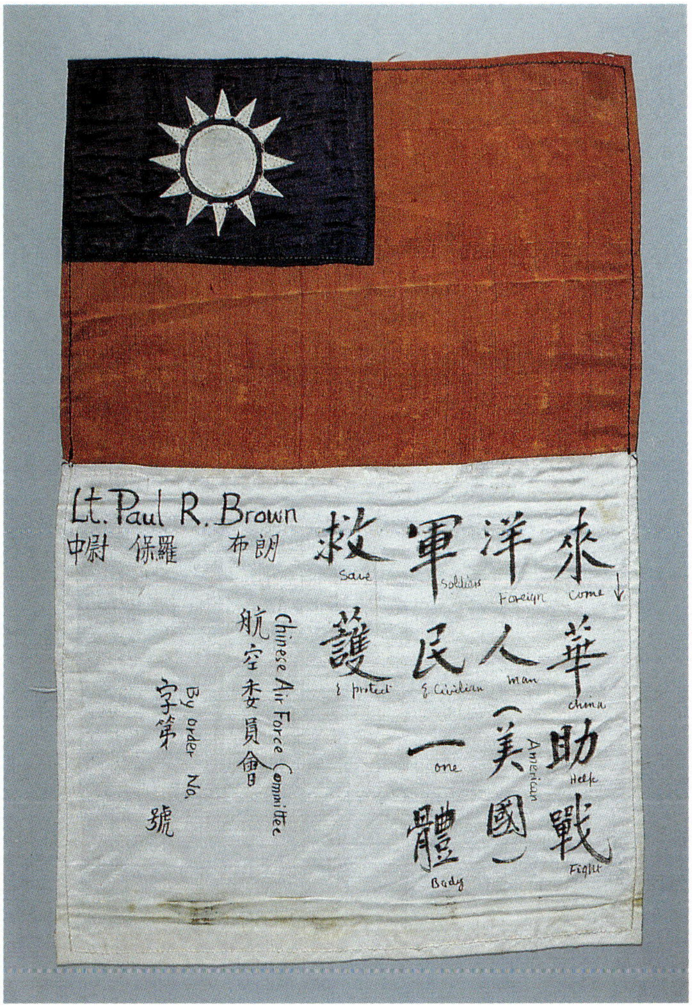

Fig. 7.84: This silk patch, combining appliqué and hand-lettering, belonged to Lt. Paul R. Brown. Each of the Chinese characters is translated into English on the patch, and Lt. Brown's name, transliterated into Chinese characters, appears in the upper left corner of the text area. The patch is made from 16 separate pieces of silk, 13 of which make up the sun on the flag. (REB 9016) *International Blood Chit Museum*

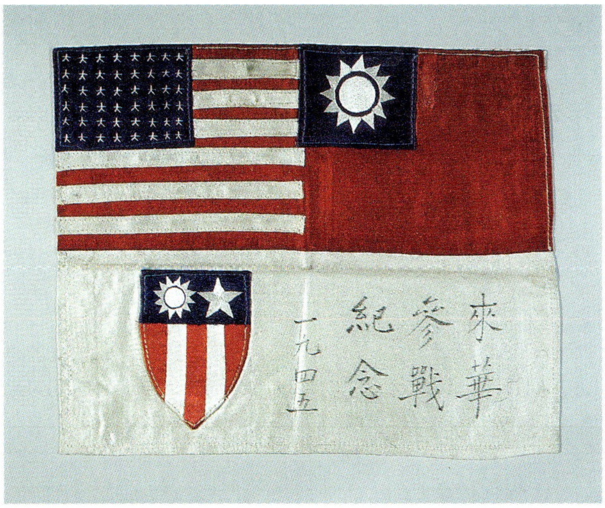

Fig. 7.86: This souvenier patch is another example of the beauty of combining appliqué and embroidery on silk. This patch is in the form of a Blood Chit but its message indicates that it is actually a commemorative piece. The text reads, "Came to China to participate in the war effort. In commemoration, 1945." The four characters next to the CBI shield are the numbers of the year according to the Gregorian calendar: from top to bottom, 1, 9, 4, and 5. This is unusual because at that time in China 1945 was called the 34th year of the Republic of China. (REB 9018) *International Blood Chit Museum*

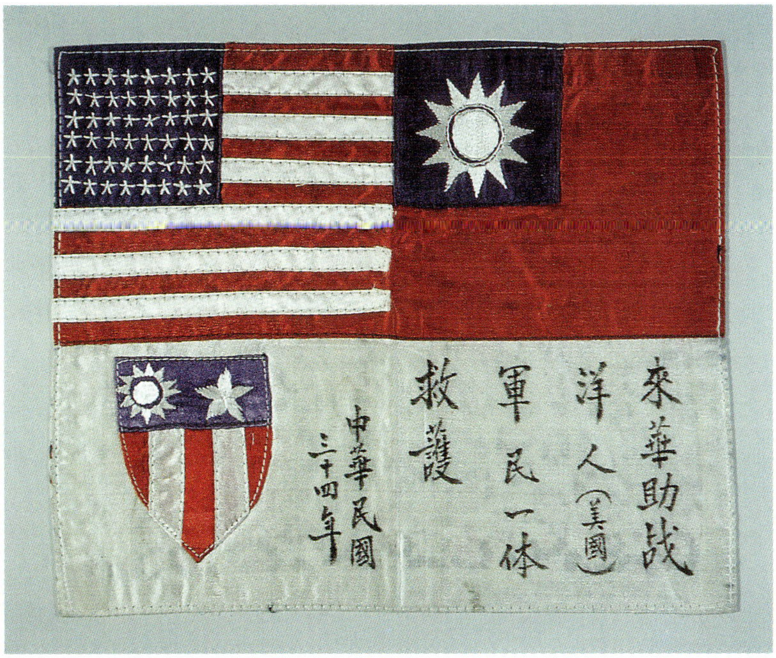

Fig. 7.85: This 14-piece patch illustrates how the age-old techniques of appliqué and embroidery can be combined to create an exceptionally beautiful military souvenir. Exquisite pieces such as this emphasize why theater-made patches patterned after Blood Chits are so highly valued by collectors. The two columns of text beneath the American flag read, "34th year of the Republic of China", being 1945 by western reckoning. Although official chits were written in classical Chinese, souvenier patches were sometimes written in the vernacular as illustrated by this item. The character at the bottom of the third column (from the right) is a simplified character for "body", not used in classical Chinese. The message is identical to that on the Aero Commission chits even though the characters are not. (REB 9017) *International Blood Chit Museum*

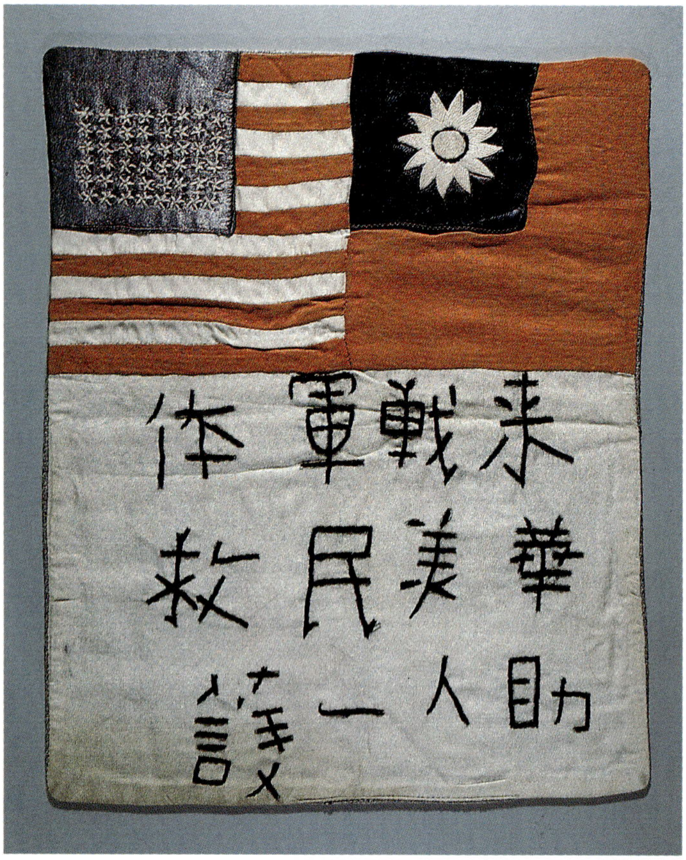

Fig. 7.87: Although the embroidered text is crude, a lot of work went into making the 11-piece silk patch illustrated here. The sun on the Chinese flag is quite stylized, almost appearing as a flower. The embroidered text may have been copied by someone who was not familiar with Chinese, or it may have been done by a novice. The message is that of the first style Aero Commission Rescue Patch without the "may guo" characters. The simplified character for "body" appears on this patch. (REB 9019) *International Blood Chit Museum*

Fig. 7.89: This 3-piece patch combines appliqué with embroidery. The embroidered text is so well done that from a distance it appears to have been done in ink, however a careful inspection indicates that it could have been done by hand. The text is a copy of the standard text used on the first style Aero Commission chits, with stylized, squared-off parentheses surrounding the "may guo" characters. (REB 9021) *International Blood Chit Museum*

ATC C-46 pilot, had a patch of this type sewn to the lining of his jacket. He recalls that he obtained it at Sookerating, in the Assam Valley, India. He writes, "All of the pilots that I knew at Sookerating had the silk Chinese Blood Chit at the time that I was based there. I arrived at Sookerating to begin my tour on August 25, 1943 . . . I could have possibly purchased it at the Post Exchange there."[123]

Some souvenir patches were fabricated using embroidery exclusively. Four examples are shown in Figs. 7.90 through 7.94. Embroidered patches such as these are sometimes erroneously referred to as being "quilted". The patch shown in Fig. 7.89 is in the midst of fabrication, as the edges have not yet been folded back and sewn down. The text of this patch, and of that shown in Fig. 7.91, is a copy of the text on Aero Commission Rescue Patches.

Many types of embroidered patches were made, and while the flags showed some minor variations in size, shape and color, the texts sometimes displayed major differences. The message on the

OPPOSITE: Fig. 7.88: The embroidered text on the 17-piece silk patch shown in here is copied from the standard message on Chinese chits. The sun looks like a flower and the text is highly stylized. The name of the owner, Simms, is written vertically in ink. It is assumed that this patch was not produced in China because of the stylized text. (REB 9020) *International Blood Chit Museum*

Fig. 7.90: This patch, although resembling that in Fig. 7.88, was entirely embroidered. It is in the midst of fabrication as the embroidery is complete and the silk has been sewn to a light cotton backing, but the edges have not been folded and and sewn. Like the patch shown in Fig. 7.88, the embroidered text on this item is so nicely done that from a distance it appears to have been inked rather than done with needlework. Embroidered patches are sometimes erroneously referred to as being "quilted." (REB 9022) *International Blood Chit Museum*

Fig. 7.91: The item shown here is essentially a fully embroidered version of the official Chinese Aero Commission chit, including the standard text, the five "Aero Commission" characters, and the two characters that accompany a serial number. It is the most common text found on theater-made souvenier patches. (REB 9023) *International Blood Chit Museum*

patch shown in Fig. 7.92 reads, "The Chinese Army and people should one and all rescue, protect and provide medical care to foreigners who have come to china to help in the war effort."[124] The character at the top of the fourth column (from the right) is incomplete, but should be identical to one at the bottom of the first (right-hand) column on the patch shown in Fig. 7.87.

The patch shown in Fig. 7.93 also demonstrates the variety of texts found on souvenir items. This reads, "The Americans have come to China to help in the war effort. When these foreign people are wounded or lost, hope our soldiers and civilians one and all will rescue, protect and provide them medical care."[125]

The text on the patch illustrated in Fig. 7.94 indicates that it was intended for sale to Army rather than Air Force personnel. The text reads, "The American Army has come to China to help in the war effort, foreign persons. Soldiers and civilians one and all should rescue, protect and provide medical care to them." The term "Army" was not used mistakenly, nor was it used because airmen were technically part of the Army Air Force. The term refers to ground forces and its use is intentional.[126]

Copies of chits were not the only embroidered items made in Asia. Although the blue silk has faded on the piece illustrated in Fig. 7.95, it is a nice example of an embroidered U.S. flag from the CBI theater. It belonged to George Brouillette, whose Chinese chit is shown in Fig. 7.3.

Leather was another medium used to fabricate souvenir patches in Asia. A combination of appliqué and hand lettering was generally used on these leather pieces, although embroidery was used in rare instances. As is true of the hand-made silk patches, the leather patches exhibit a beauty and quality of handwork virtually unparalleled in the realm of military collectibles. A variety of leather patches fabricated for Army personnel is shown in Figs. 7.96 through 7.105.

Some leather patches were personalized. William Coneley has written his name on the top white stripe of the American flag on the patch shown in Fig. 7.97. The small characters just beneath the flags on the patch in Fig. 7.98 read, "Lt. Col. Vincent John Conrad", while the row of characters to the right of the CBI shield read, "American Army".[127] Additional text has also been added beneath the American flag on the patch shown in Fig. 7.100. The first (right-

Fig. 7.92: The blue field of the flag on this embroidered patch is purple and the red field has an orange hue. Its text is a variation of the standard chit message, reading "The Chinese Army and people should one and all rescue, protect and provide medical care to foreigners who have come to china to help in the war effort." The character at the top of the fourth column (from the right) is incomplete, as if it were done by someone not familiar with the Chinese language. It should be identical to one at the bottom of the first (right-hand) column on the patch shown in Fig. 7.86. (REB 9024) *International Blood Chit Museum*

Fig. 7.93: The diagonal threads on the flag of this fully embroidered patch are farther apart than on most. The text is another variation on a common theme, reading. "The Americans have come to China to help in the war effort. When these foreign people are wounded or lost, hope our soldiers and civilians one and all will rescue, protect and provide them medical care." (REB 9025) *International Blood Chit Museum*

Fig. 7.95: This embroidered flag belonged to George Brouillette, an airman with the 14th Air Force. Flags and other patches were produced and sold to servicemen by local vendors wherever the opportunity presented itself. (REB 9027) *International Blood Chit Museum*

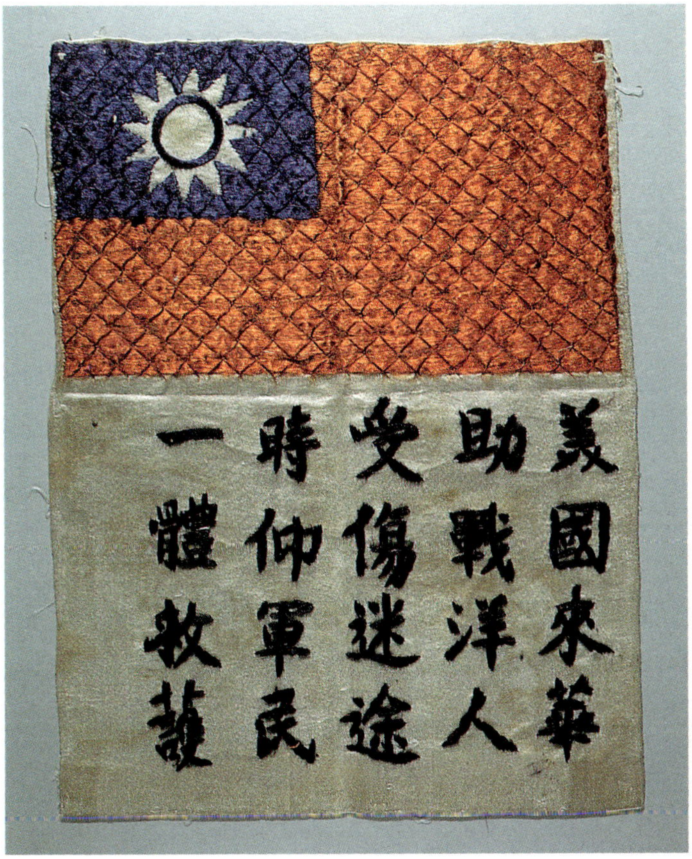

Fig. 7.94: This embroidered patch is a rarity, as it was made for sale to Army rather than Air Force personnel. The text reads, "The American Army has come to China to help in the war effort, foreign persons. Soldiers and civilians one and all should rescue, protect and provide medical care to them." The use of the term "Army" is in the text is intentional, and refers specifically to ground forces. (REB 9026) *International Blood Chit Museum*

Fig. 7.96: This 5-piece leather patch shows the effects of heavy use. Removed from a jacket, it is faded and the text is nearly worn off. While collectors shun such "poor" examples, this serves to remind historians that patches were bought to be used, not as future museum pieces. It also illustrates why leather generally was the material of choice for patches to be sewn on the outside of operational garments, rather than the more delicate silk or cotton pieces. (REB 9028) *International Blood Chit Museum*

Fig. 7.97: This 18-piece leather patch and card belonged to a William Conley, whose name is written on the top white stripe of the flag. The text is stylized, indicating that the patch most likely originated in India. Below the flags, 29 Hump Missions are depicted by stylized mountains. Several symbols are used in conjunction with the mountains to record events that occurred on some of the missions, but the significance of these symbols is not known. (REB 9029) *International Blood Chit Museum*

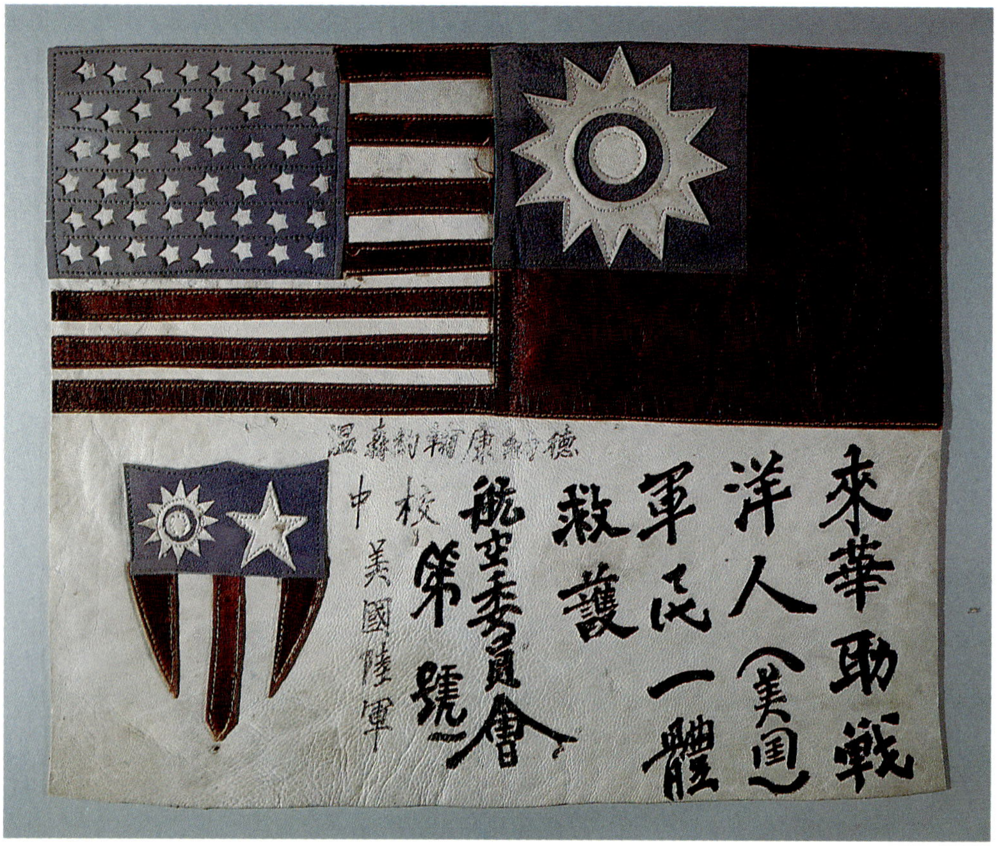

Fig. 7.98: This leather patch is a typical example of an appliqué leather patch that has been personalized. The small characters just beneath the flags read, "Lt. Col. Vincent John Conrad", while the row of characters to the right of the CBI shield read, "American Army." The CBI shield is unusual as the bottom is left open due to its construction from three pieces of leather. Fig. 7.99 shows the back of Lt. Col. Conrad's patch. (REB 9030) *International Blood Chit Museum*

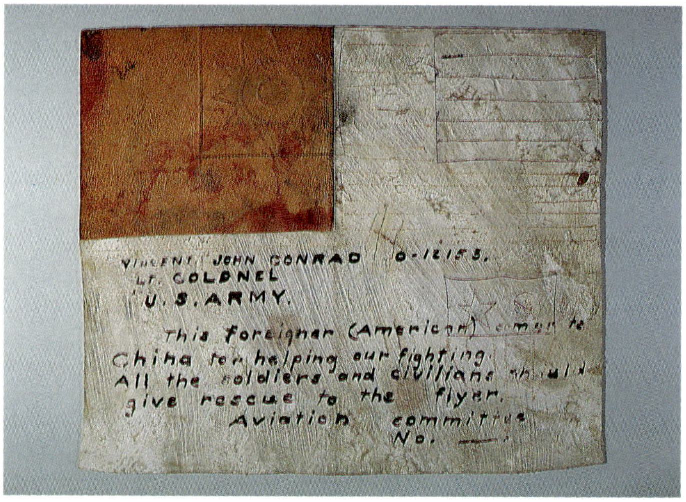

Fig. 7.99: The back of the Lt. Col. Conrad's patch (Fig. 7.98) is shown here. Written on it is a fair translation of the Chinese text that appears on the front of the chit. *International Blood Chit Museum*

Fig. 7.100: The 11-piece leather patch shown here was removed from the flight jacket of Lt. Paul E. Duquette, who is shown in the photo kneeling second from right with his crew in front of their B-25, *Sunday Punch*. This is another variation of the popular two-flag leather patch, with the first style Aero Commission text on the right side beneath the Nationalist flag. The added text beneath the American flag reads, "Person's Name" and "Po Du Keh", a transliteration of the name Paul Duquette. Although faded from being worn, it is still in good condition. (REB 9031) *International Blood Chit Museum*

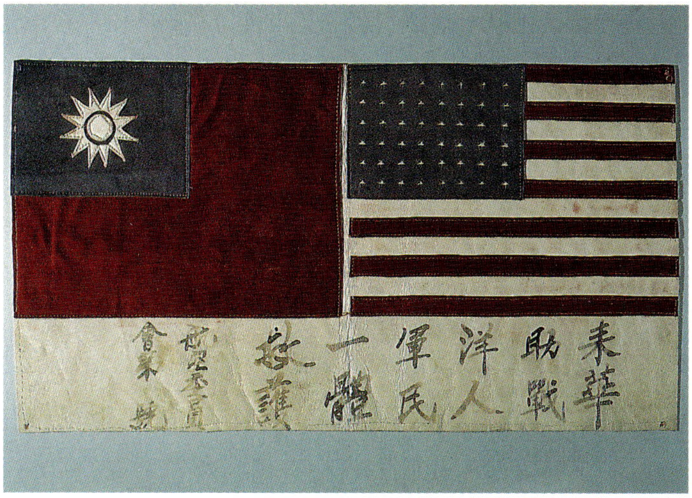

Fig. 7.101: The majority of leather patches feature the American flag on the left side, but in this example the U.S. flag, with inverted stars, is on the right side. The text in this 13-piece patch is the first style Aero Commission message, but arranged in eight columns of two characters instead of in the usual four columns, making the patch wider than most. It is somewhat faded from having been worn on a garment, but the characters are still readable. A patch like this is on display at the EAA Museum in Oshkosh, Wisconsin (August 1991). (REB 9032) *International Blood Chit Museum*

Fig. 7.102: In another of the many variations on the two-flag theme, the entire left side of this 14-piece appliqué leather patch is a stylized U.S. flag within a blue border. The right side is a copy of the first style Chinese Aero Commission chit, with text below the Nationalist flag. Variations of souvenir patches with the two-flag motif seem endless. (REB 9033) *International Blood Chit Museum*

hand) column of this additional text reads, "Person's Name" and the second (left-hand) column reads "Po Du Keh", a transliteration of the name Paul Duquette.[128]

In addition to items for sale to air and ground forces, vendors in China also made a variety of patches for sale to U.S. naval personnel, apparently when the fleet sailed into Shanghai in 1945. Many have the words "U.S. Navy", "U.S.N.", "Shanghai", and "1945", singly or in combination, written on them in English. Four versions of these leather patches for naval personnel are shown in Figs. 7.106 through 7.109.

The text on the Navy patches is often different than that found on those produced for sale to Army personnel. The text on the patch shown in Fig. 7.106 is not well written, but it can be interpreted, with difficulty, to read, "Allied friends from America came to China to participate in the war effort. Soldiers and civilians one and all please protect them."[129]

In another example of a text variation found on leather Navy leather, that on the patch illustrated in Fig. 7.107 reads, "The American ally's army came to China to help in the war effort. All my soldiers and civilians one and all please protect them.".[130]

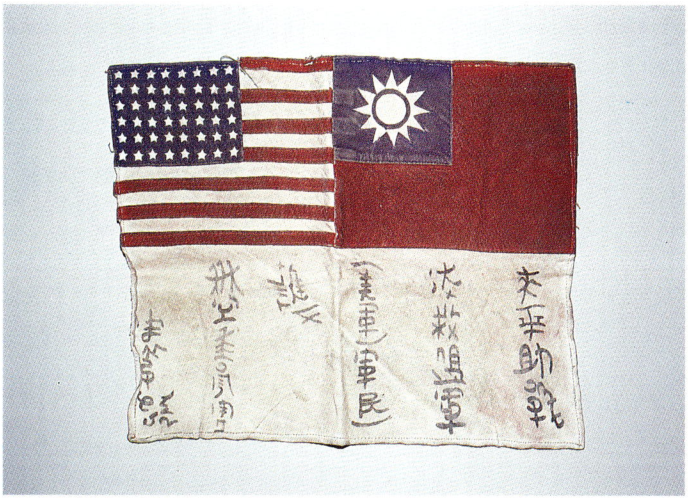

Fig. 7.103: The faded leather patch shown here features an unusual stylized text, again illustrating the endless variety found on theater-made souvenirs patterned after Blood Chits. (REB 9034) *Courtesy of the U.S. Air Force Museum*

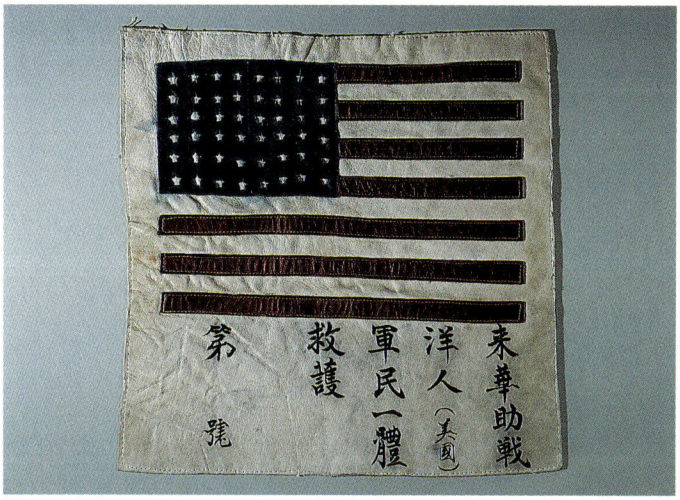

Fig. 7.104: The 9-piece patch illustrated here is still very beautiful in spite of the fact that the blue field on the flag bled onto the white background. It is not unusual for the color on theater made patches to bleed, and it is somewhat surprising how many are actually in good condition considering their age and exposure to the elements over the years. The standard Chinese message is nicely written, and its combination with the American flag rather than the Nationalist Chinese flag is unusual. It has been removed from a flight jacket. (REB 9035) *International Blood Chit Museum*

Fig. 7.105: Illustrated here is a beautiful example of the British chit for Burma rendered as a leather souvenir item. On one of the 8 pieces of leather is a hand-painted peacock. Probably made in India, this patch was available for purchase by both British and American flyers. Like the official issue British chit for use in Burma, this leather souvenir version is one of the most visually stunning to come out of World War II. (REB 9036) *International Blood Chit Museum*

Fig. 7.106: Removed from a flight jacket, this vendor patch for U.S. naval personnel shows a great deal of wear but the text is still readable. It is one of many versions produced when the U.S. Navy sailed into Shanghai in 1945. The message reads, "Allied friends from America came to China to participate in the war effort. Soldiers and civilians one and all please protect them," and the date is given as "34th year of the Republic of China," being 1945. In an unusual touch, the anchor chain is embroidered onto the leather. (REB 9037) *International Blood Chit Museum*

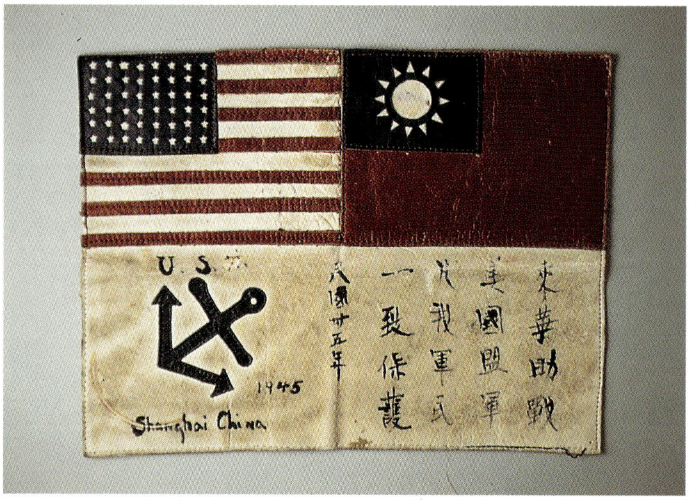

Fig. 7.107: This leather patch is another example of the type made for sale to naval personnel. The message reads, "The American ally's army came to China to help in the war effort. All my soldiers and civilians one and all please protect them." Written in English are "USN" (the "N" is nearly worn off), "Shanghai China", and "1945." (REB 9038) *International Blood Chit Museum*

Fig. 7.108: This patch is printed on a single piece of thin leather. Obviously made for sale to U.S. sailors, it dates from 1945. The message varies from the usual one in several respects, reading, "The American ally's forces have come to China to participate in the war effort. My country's soldiers and civilians one and all should love and treasure them." The left-hand column reads, "34th year of the Republic of China", meaning 1945. The anchor on this patch is aligned vertically. (REB 9039) *International Blood Chit Museum*

Fig. 7.109: Another version of a painted leather Navy patch is this item, with the anchor rotated from the vertical. The message is identical to that on the patch shown in Fig. 7.108 even though two columns of characters have been transposed. (REB 9040) *Courtesy of Randy Graham*

The leather patch shown in Fig. 7.108 features yet another text variation. The message varies from the usual one in several respects. The third character in the first (right-hand) column means "Ally". The first two characters in the third (from the right) column mean "my country". The third character in the fourth (from the right) column by itself means "love", but when combined with the "protect" character below it the combination means "love and treasure". The message therefore reads, "The American ally's forces have come to China to participate in the war effort. My country's soldiers and civilians one and all should love and treasure them."[131] The text on the patch shown in Fig. 7.109 is identical, although the positions of the first two columns of characters have been transposed.

Fig. 7.110: This leather patch in the form of a Nationalist flag was printed on a thin piece of leather. The small characters in the upper right border give the owners name, Simon Gay, along with the notation "American Army." Someone unfamiliar with writing Chinese, possibly Mr. Gay himself, has painted the Blood Chit text onto the flag. (REB 9041) *International Blood Chit Museum*

Fig. 7.111: The small leather patch patterned after a Blood Chit, shown with a CBI shoulder patch to illustrate its size, is identical to the large leather patches illustrated elsewhere. Like its larger relatives, it is a decorative souvenir patch, not a bona-fide evasion aid. (Left: REB 9042) *International Blood Chit Museum*

Fig. 112: Shown here is an example of an American flag, without any accompanying text, crafted in leather. Some, like this, have borders around the flag, and others do not. (REB 9043) *International Blood Chit Museum*

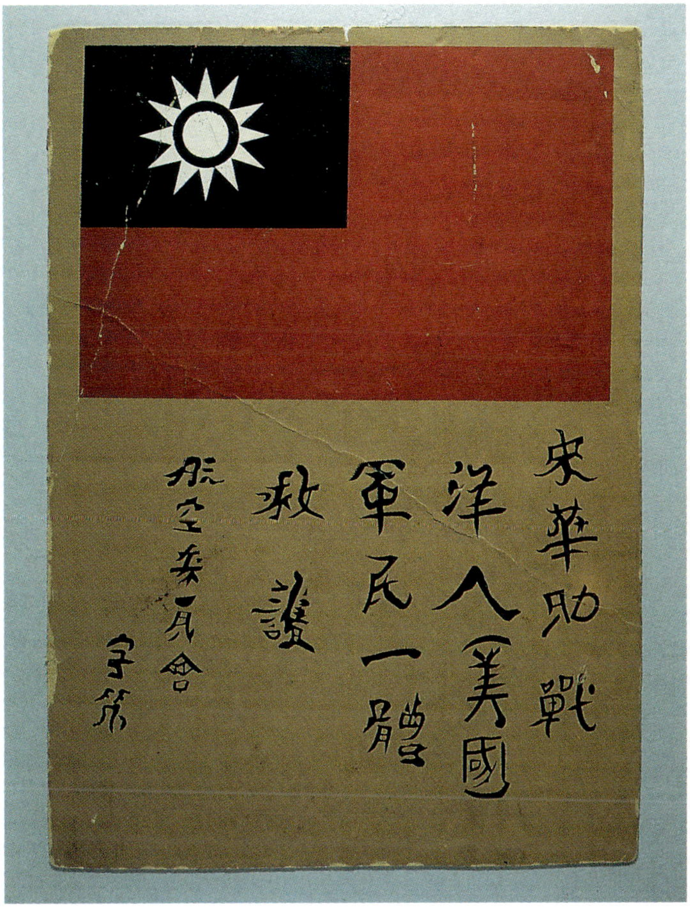

Fig. 7.113: A wartime reproduction, this item is made of a material similar to fiberboard. On the back is a translation reading, "Chinese Air Force Committee. The Westerner has come to China." The fact that items like this were available for sale during the war would seem to preclude their being training aids, and their use other than as souveniers, if any, is unknown. (REB 9044) *International Blood Chit Museum*

Fig. 7.114: A minor variation of the piece shown in Fig. 7.113, this is also printed on what may be fiberboard or possibly a brittle cardboard, although it is smaller in size, the blue is lighter in color, and the characters are "thicker." The translation on the back of this piece reads, "Translation of script, Flying Tigers Rescue Flag. Chinese Airforce Committee. The westerner (Americans) come to China. All the Chinese must help him." (REB 9044.1) *International Blood Chit Museum*

Fig. 7.110 shows a patch in the form of the Nationalist Chinese flag, printed on a single piece of leather, to which writing was added at a later date. The characters in blue in the upper half of the right border were written by a hand accustomed to the language. They read, "Gau Si Mon American Army"[132], the first three being a transliteration of the name of the owner, appearing on the back of the flag as "Simon Gay". The painted black characters on the red field were added by a hand not familiar with writing the language, perhaps that of Mr. Gay himself, in order to make the flag resemble a Blood Chit.

Illustrated in Fig. 7.111 is a small version of an appliqué leather patch in the form of a Blood Chit. It is shown with a CBI shoulder patch done in leather to illustrate its small size. It is identical to the larger leather appliqué patches in every respect except for its size. Like its larger relatives, it is a decorative souvenir patch, not an evasion aid.

In addition to patches created in the image of Blood Chits, patches of the American flag without any accompanying text were also made. Shown in Fig. 7.112 is an example of a flag crafted in leather. Numerous flags patches were made, some with borders as on the example shown, and some without borders.

Several styles of souvenirs were printed on a stiff but brittle cardboard-like material, resembling thin fiberboard. They were available for sale during World War II, but their purpose other than that of souvenirs, if any, is not known. The fact that they could be purchased seems to eliminate the possibility of their being training aids. Three of these items are illustrated in Figs. 7.113 through 7.115.

As previously noted, the variety of designs for theater-made souvenir patches in the form of Blood Chits is virtually endless, and those illustrated are typical examples of the many styles that were produced.

One important consideration in the study of Blood Chits and other language related evasion aids is how well they performed the tasks they were designed for. The responses of local peoples to language aids varied greatly, and the biggest obstacle to their use was the lack of literacy. With the exception of the few pictorial chits, language aids, no matter how well designed, could not be used unless someone was found who could read. A report written more than ten years following the end of World War II summarizes the Air Force's opinion on the effectiveness of Blood Chits in Asia during that conflict:

> Accounts given by evaders show that the effectiveness of the chits varied greatly. The downed flyer frequently made first contact with Chinese or Burmese who could not read. Such illiterate peasants were apparently as helpful to the airmen as more literate contacts. Most flyers felt comfortable with the chits in their possession.
>
> There are some few instances when the evader reported that the chit had been definitely decisive in influencing natives to give aid. Narrative accounts of evasion and escapes are numerous and substantial numbers of these accounts from the CBI make mention of blood chits.[133]

It is important to remember that Blood Chits and related items were never intended to effect the rescue of airmen in and of themselves. Whatever their post-war and contemporary mystique, these evasion aids cannot be analyzed as stand-alone items. They were a small but integral part of the larger evasion and survival systems developed by the Allies, and documented reports of successful evasions and repatriation to military control provides evidence that the entire Allied evasion and survival effort, including the creation and distribution of Blood Chits, Pointie Talkies, Promissory Notes and similar items, paid off in a big way.

Even on a mass, dehumanized scale, war is an individual experience. Nowhere else is the value of Blood Chits, Pointie Talkies, and other evasion aids more clearly shown than in the personal experiences of those individuals who used them to survive, evade, and return from harm's way. Some of the following excerpts of evasion accounts were previously published while others, from declassified documents, are being told in print for the first time here. When reading these accounts, remember that Blood Chits were often referred to by many names, including "patch", "backflag", "rescue flag", or simply "flag".

Despite many shared characteristics, each individual's initial contact with local inhabitants was a unique experience. Although literacy rates varied greatly among local populations so that a written message sometimes could not be read, the initial contact between an airman and the local population often involved, as briefings by E&E officers had instructed, identification using the Blood Chit:

> I took inventory of my survival gear, which I stuffed into pockets of my flying suit and flight jacket. The kit included morphine syrettes, chocolate bars, extra gloves and fatigue hat, matches, water purification pills, compass, fish hooks and line, pocket knife and first aid supplies. I was glad now that I had taken seriously those dull films on survival that were always being foisted on us in training. As usual, in the airplane I had been wearing my filled canteen, machete and .45 automatic on a G.I. web belt. On this day I was very glad to have them, even though they added critical pounds to an already seriously overgrossed parachute.
>
> Shortly two figures approach from below. They turned out to be a Chinese peasant and a small boy of about twelve years. The man armed himself with a large rock and they approached me with great caution and apprehension. I flashed the most disarming smile I could muster, showed them the Chinese/American flags sewed on the inside of my flight jacket and broke out some of my survival chocolate . . . satisfied I was not an enemy spy ready to pounce on them, they approached and accepted the chocolate.[134]

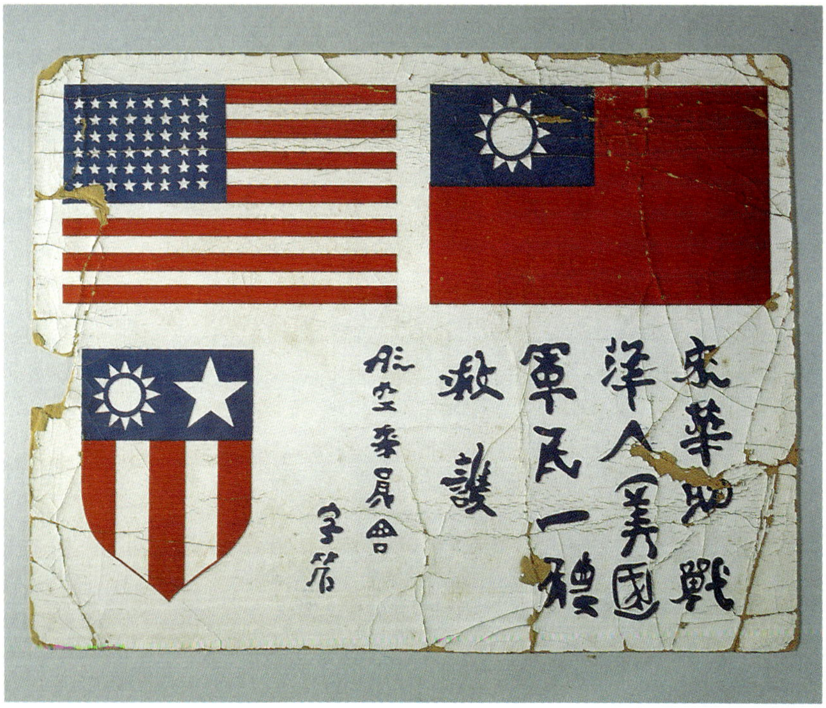

Almost as if from an evasion manual, one pilot put his disabled aircraft down a mere 200 yards from a village, and then approached on foot:

> I walked toward the village and was met by apparently friendly people. They took me into one of their houses where I put my jacket and parachute down and reached for my escape kit. The American flag was all that I needed, and when they saw it, I knew they were friends. I . . . discovered that the people were Chinese since their talk was different from Burmese and they could only read what writing I had that was Chinese. The only Chinese writing I had was the blood chit which I found in my escape kit.[135]

Fig. 7.115: Another example of a patch printed on fiberboard or cardboard is shown here. This one emulates the two-flag patches made in-theater. The entire front is painted, and is badly cracked and peeling in places. What little remains of the paper translation glued to the back of this item reads, "China-Burma-India Air Transport Command Rescue Flag ... ad of the Flying Tiger Rescue Flag ... as we ..." (REB 9045) *International Blood Chit Museum*

Another American fighter pilot downed in China, Capt. Kenneth E. George of the 76th Fighter Group, made full use of the two evasion communication aids he carried as detailed in his walkout report:

> He continued up the stream-bed for what he estimates as two miles. Then, as he was resting, he observed a Chinese walking along nearby and hailed him. The Chinese was startled and frightened at first but understood the situation immediately when he was shown the Chinese Back Flag. The farmer led Capt. George to a village a short distance away and placed him in the rear room of a dwelling.
>
> Then after the Chinese guide had been asked by means of the Pointie-talkie to do so, he set out to find some guerrillas while the rest of the populace inspected the newcomer.[136]

Two members of a Twentieth Bomber Command B-29 lost over China used their Blood Chits and Pointie Talkie to obtain assistance in reaching a guerrilla unit:

> Climbing a steep hill, the fliers came upon several Chinese boys, several of whom apparently understood enough of the writing on the Chinese blood chit the fliers were wearing to point out a direction.
>
> After several hours, a guerilla leader in company with several men appeared. Using a pointie talkie, the Americans inquired if there were any guerrillas about, to which the leader pointed to himself and shook hands with the crew members. Leading the fliers to another small settlement three or four miles away, they entered the back room of one of the houses. Here the two Americans were allowed to wash and rest. All possessions except the American flag blood chit, the Chinese blood chit and maps were taken from them.[137]

A B-24 pilot and his crew who bailed out of their aircraft found that the woods where they had landed were teeming with Chinese:

> A guerrilla first confronted me and I pulled out my little American flag and yelled "American!" I knew we had landed in safe territory but, of course, the Chinese were not so sure who I was.[138]

Another Hump pilot who survived the loss of his aircraft began the long journey home:

> Upon climbing to the top of the mountain, I saw a village far below. After several hours of a slow descent, I ventured into the village very cautiously, not knowing whether they were in sympathy with the Japs or not. I was able to communicate with the use of my flags and China-American Dictionary.[139]

While several accounts mention the use of English-Chinese dictionaries, communication with these was cumbersome since only one word at a time could be translated. Pointie Talkies and Phrase Books greatly facilitated communication when compared to dictionaries, they were issued in quantity, and they were carried by many air crew members.

In some cases, none of the evasion aids were required for rescue. John J. Munro, a U.S. Marine Corps aviator, admits with humor to having "the dubious distinction of having been the last Marine pilot down in north China" where "the two Chinese armies (the Nationalists and the Communists) were contesting the area and we were in sort of a 'water dog' status." Munro did not require the use of his evasion aids, and writes that he "did not release my blood chit since my wingman provided cover until members of the Marine detachment at Tiensen arrived."[140]

The war in China, Burma, and India was often not in the headlines of the newspapers and magazines back home. Even now, it is often overshadowed by the major battles in Europe and in the Pacific. The theater was, however, a crucible of the conflict that led to the eventual defeat of the Japanese Empire's oppression of its Asian neighbors. Even today, after the passage of more than 50 years, the people of Asia acknowledge the tremendous debt they owe to the Allies for their liberation from Japanese tyranny. To the many downed Allied airmen who owe their lives to Asians who risked their own lives, and the lives of their families and their villages, in order to spirit them to safety, the debt was repaid long ago.

In the pacific, the war would be different than in Asia. It would be a carrier war taking the offensive to the enemy, a war of thrust and movement, of battles for fly-speck islands frequently blood-soaked and now largely forgotten. For the airmen it would be a war of long, tedious over-water flights from crushed coral airstrips on small atolls, hot carrier decks, and muddy jungle airstrips, waged against a formidable enemy. The war would change from that of a struggling force of Allied airmen in obsolete airplanes and four American aircraft carriers to the perfection of strategic bombing with the streamlined B-29, and a fleet of more than 100 heavy and escort aircraft carriers.

The war in the Pacific would again raise issues of survival and evasion, and bring Allied airmen into contact with unfamiliar peoples on whom they would depend for safety and survival. The experience of the Allies in Asia would prove vital to their further efforts in the South West Pacific and the Pacific Ocean Areas. Their value confirmed by successful use in Asia, hundreds of thousands of Blood Chits, Pointie Talkies and other language-related evasion aids would be produced for use by both sea and land based air units fighting the final battles of World War II against the Empire of Japan.

CHAPTER 8

Strangers in Paradise World War II in the Pacific

While Japan may have been the first Asian nation to copy the West, significant cultural differences marked the Allies' opponent in the war in the Pacific. Though having adopted Western technology and weapons and studied Western tactics, the Japanese military ethos remained particularly Japanese. Japan had developed an absolutist warrior tradition, a martial culture in which the Emperor was both a political and a spiritual being. Shinto, the Japanese state religion, put forth the belief that the Japanese were unique and of divine origin, spawning a sense of cultural superiority that served as a basis for the Japanese view of themselves as a chosen people, superior to all others, including other Asians. Despite claims that they were liberating fellow Asians from Western imperialism, the Japanese were brutal conquerors.

The Japanese code of *Bushido* formed the essence of the Japanese military ethic and taught virtues for the Japanese, but it had no place for prisoners. Surrender and capture were unthinkable. Taking their own view of capture and imprisonment, the Japanese applied it to Allied POWs. In the Japanese mind, prisoners of war ceased to exist not only as soldiers but also as persons. Allied air crews, often singled out for punishment, faced torture, medical experimentation, vivisection, and death by firing squad, hanging, poisoning, bayonet practice, beheading, and strangulation by hand. The brutality of Japanese captivity is coldly illustrated by raw statistics: about four percent of the British and Americans captured by Germany and Italy died while in their captors' hands, whereas some twenty-seven percent of the Americans and British POWs died while in the hands of the Japanese.[1]

While little was known about the Japanese military ethos regarding prisoners before the war, following the details of Japanese atrocities that became known after the mass surrenders that marked the early months of the war for the Allies, the brutalization of POWs by the Japanese became a strong element in the minds of American and European leaders. It was certainly one of the foremost thoughts of Allied servicemen in action against the Japanese, and fueled the Allies' vigorous survival and evasion efforts in the Pacific.

In 1942 the Pacific theater was divided into two areas, the Southwest Pacific Area (SWPA) and the Pacific Ocean Area (POA). In the SWPA, MacArthur was intent first on stopping the Japanese from invading Australia and then on liberating the Philippines. The POA was dominated by the U.S. Navy, and offensive operations there were targeted at an eventual invasion of Japan.

Under the May 1943 agreement made between MI9 and MIS-X, the United States assumed responsibility for E&E operations in the Pacific. The MIS-X SWPA Section, established in November 1943, had responsibility for all evasion and escape activities in the Pacific, including those for the Far Eastern Air Force, until the activation of MIS-X Section, POA, one year later in November 1944. A U.S. Navy report, explaining the delay in establishing a separate E&E section in the POA, reports that "In the spring and summer of 1943, the War Department, MIS-X Section, sent officers to Australia, India, and China to organize MIS-X sections in these theaters. The POA Theater was not included at this early date because operations had not yet reached land areas within the theater."[2]

At the end of February 1944 the SWPA Section moved into quarters in the Perry House in Brisbane, Australia, which had ample space for workshops, map tables and warehousing.[3]

Although E&E operations were largely an American function, the Australian coast-watchers also rescued downed air crews. The coastwatching network, begun by the Australian Navy in September 1939, eventually included more than 100 stations stretching 2500 miles from Papau Island to the New Hebrides and then to the Solomon and Bismark Islands. Those in the SWPA became part of the Allied Intelligence Bureau (AIB) in July 1942, while the stations in the POA remained attached to Australian Naval Intelligence. Lt. Commander Eric Felt, who established the original network, remained in command of both areas. Before their disbandment in October 1944, coast watchers had assisted in the recovery of at least 188 Allied airmen, a number of downed Japanese air crews, and one future president of the United States.

In a war sometimes marked by strong currents of inter-service rivalry, the MIS-X SWPA Section existed as a rare multi-service and multi-national organization, elements that did much to further its already difficult work. The senior American was an Army officer, Major Paul S. Kraus, with the senior Australian officer being Squadron Leader John Lamb, Royal Australian Air Force. The U.S. Navy was represented by Lt. (Jg.) Thomas R. Drake, USNR, and a Lt. Pierce represented the U.S. Marine Corps. In February 1944 the Section was staffed by a total of "8 officers . . . 9 enlisted men and

Fig. 8.1: Especially developed for use in the Netherlands East Indies (NEI) and Portuguese Timor, this comprehensive evasion purse contained money, chits and vocabulary lists in phonetic Malay and Tetum, letters to village chieftains in phonetic Malay, and Promissory Notes. *Courtesy of David Vincent*

1 WARR (Women's Australian Air Force). . ."[4] The success of MIS-X was also aided by their realistic recognition of the area's political structure as evidenced by their close liaison with other military organizations in the region. These included the Far Eastern Liaison Office (FELO), the office of the Netherlands East Indies (NEI) in Australia, the Allied Intelligence Bureau (AIB) established by the General Douglas MacArthur, the Australian and New Guinea Administrative Unit (ANGAU), and the Allied Translator and Interpreter Section (ATIS).

MIS-X's activities were broad-based since combat operations in the SWPA covered large areas and were conducted by Army, Navy, Marine Corps, Australian, New Zealand and Dutch personnel. For example, from August 1943 through January 1944, MIS-X personnel briefed 6,171 airmen on evasion and survival, including U.S. Fifth Air Force crews, RAAF and Dutch pilots, and Navy Catalina crews. It also trained 104 Intelligence Officers (IOs) during the same period, who would go on to brief crews in widely scattered locations.[5] A memorandum dated 5 July 1944 reports that "Fifty Marine Corps Air Combat Intelligence Officers [ACIOs] have been given MIS-X training. The majority have been ordered to squadrons and air groups."[6] These MIS-X trained ACIOs, and many others before the war's end, would brief thousands of air crew members throughout the Pacific on E&E matters. A secret "Monthly Report" of May 1945 details that during February and March of that year, MIS-X personnel briefed 13,272 American Army, Navy, and Marine airmen and RAAF flying personnel, and reveals that in the SWPA, "MIS-X officers have now briefed, in all areas, a grand total of 95,435 personnel."[7]

Available records of the SWPA Section make specific references to briefings on Mindoro, Luzon, Morotai, Biak, Leyte, Samar, Palawan and Pityilu Island as well as Port Moresby and Nadsab in New Guinea, and Kingaroy, Lowood, and Darwin in Australia. Reports also mention briefings aboard *USS Tangier*, *USS San Carlos,* and *USS Currituck*, and by the war's end ACIOs trained by MIS-X briefed squadrons and air groups on virtually all U.S. Navy aircraft carriers, of which there were 138 of all classes.

The MIS-X SWPA Section was responsible to ensure that "An extensive briefing program was carried out to teach combat airmen and others how best to care for themselves in the event of being forced down in enemy territory . . . Escape kits containing useful emergency equipment were distributed. An evasion method was developed which involved the establishment of safe rendezvous points within enemy occupied territory. These were selected on the basis of availability for rescue facilities, food and medicines, friendly natives, and remoteness from enemy activity."[8]

As a practical matter, survival itself was of primary importance in the Pacific. MIS-X taught that "in the Pacific Theater, survival is a most important factor. The mastery of its techniques will make evasion possible."[9] This relationship between survival and evasion was also articulated by the American Navy, which stated that "while evading capture by the enemy is not regarded as a survival process, survival is certainly a necessary element of evasion."[10]

Preparing for survival at sea was a fact of life in the Pacific, where pilots and crews covered vast expanses of open water to find their bases on small islands or even smaller aircraft carriers. The will to survive was essential to survival at sea, and crews were told, "Don't give in to the sea! Life is worth a gamble. Just when survivors think they can't go on a minute longer, help may be on the way . . . While one still has breath in the body, there's always hope. Men have been adrift without provisions for forty, fifty, sixty and even seventy days, and have lived to tell the tale."[11]

The need for potable water could not be stressed enough, and crews were instructed that "if survivors have enough water on hand and can make a conservative guess as to the time when they will be picked up, they should adopt the eighteen ounce ration . . . If they are going to float a long time, they should hoard their water . . . Make the most of each portion of water by keeping it in the mouth for a long time. Rinse, gargle, then swallow . . . Don't drink seawater! It's suicide! The precious body water will be used up to dilute the extra salt."[12]

In spite of the difficulties, MIS-X briefers gave encouragement to crews, telling them that "survival on the ocean is like survival on the desert; one must have an ample supply of drinking water to live. But there is one great difference – while the arid desert yields practically nothing, survivors can get food and water from the ocean, providing they have proper equipment. With a portable still or vaporizing apparatus one can make sea-water drinkable; with proper lines and fishhooks survivors can supply themselves with nourishment."[13]

For crews downed at sea, fishing was encouraged as means of supplementing or replacing rations, and a variety of fishing gear was provided and expedient methods for use in the field were suggested. Crews were advised that salt water fish could be eaten raw, but that fresh water fish needed to be cooked. Survival manuals gave information about living off of the sea, as well as about fresh water crabs and mussels for crews making landfall.

Survival experts also recognized that natives took full advantage of the coconut tree for food, fluids, and other materials, and gave crews detailed instructions for using and preparing the coconut and other native plants. More familiar fruits like wild banana and breadfruit trees also existed, but were uncommon. The more likely sources of food included pig weed (purslane), Malay apples, taro, and the fruits of the pandanus and morinda trees.

Plants could provide a bountiful platter, and survival manuals were filled with many examples of edible plants. Wild game, however, was difficult to obtain and an unreliable source of food. American crews were advised that "game, other than birds, is in any event scarce and cannot be depended upon. You may be fortunate to get a shot at wild pig with a forty-five, but that is unlikely . . . Most pigs you see will have been domesticated by the natives and will be among their most prized possessions. Do not kill one if you wish to remain on friendly terms with the natives."[14]

The MIS-X SWPA Section instituted special programs, including a Jungle Training School where allotments from various squadrons were taken into the jungle on a two or three day basis for training. Special rendezvous points in enemy territory for pick-up of downed air crews where Cache Kits containing food and medical supplies were buried were also established. The Section also produced a variety of survival and evasion aids including Intelligence Maps "clearly distinguishing enemy territory from friendly territory"[15] and the E3-J Kit ("E" for emergency, "J" for jungle) created by modifying the standard E3 Kit. The E3-J Kit is a good example of the modification of standardized E&E kits to suit in-theater needs, and illustrates the variety found in emergency kits. It should also serve as a warning to collectors that restocking a kit so that it contains what are thought to be standard contents may in fact be destroying an example of a kit that was originally modified in-theater.

The SWPA Section also prepared approximately 10,000 copies of a 50-page survival booklet, the title of which is unfortunately not disclosed in available documents. This was printed on "waterproof light-weight paper" and enclosed a cellophane envelope for distribution to crews.[16] MIS-X also issued "E&E Bulletins" and "E&E Advice Notes", which now provide some of the most detailed records of MIS-X's E&E training efforts. These publications kept combat squadrons current on successful evasions and techniques, listed airfields held by guerrillas available for emergency landings, and introduced new evasion and survival items. Specific information useful to crews flying over sections of enemy-occupied territory was given for Ceram, Timor, Aroe, Kai, the Tanimbar Islands, and other areas of the SWPA.[17]

The Allies recognized the importance of assistance from local inhabitants, noting that "the natives of islands of the South Pacific are in a difficult position. This is not a war of their making, or of their fighting. Yet their attitude is admirable and their help incalculable. The enemy has assisted us by his ill-treatment of the natives, and that word spreads. All reports bear out the fact that the native has learned to hate and fear the Japanese, and all his inclinations are to help us."[18]

MIS-X avoided amazingly well any western cultural biases in dealing with local peoples, and communicated to air crews that the local people were different, not inferior. MIS-X was concerned with saving the lives of Allied servicemen, not evangelizing or colonizing, and recognized that the support and cooperation of local residents was vital to this mission. They advised air crews: "The main thing to remember when dealing with natives is this: – never look down on them. Maybe the evader can fly a plane – all right. But where he stumbles and flounders around in the mud, the natives can run like deer. Where he needs food, they find all sorts of edibles. When the white man is exhausted, the wiry black men are still full of bounce. So treat them with respect. They can do a lot of things a white man cannot do. They can save his life."[19] MIS-X also noted that despite their small stature and limited language, "the native is an ordinary human being. The adult is not a child, but an adult with a grown-up mind, grown-up feelings, and grown-up dignity."[20] The Office of Naval Intelligence echoed this sentiment, when advising crews that "the important thing is to treat them with dignity. Most of them have a strong sense of self-respect . . . They will appreciate being treated as human beings just like your self, neither as inferiors nor as superiors."[21]

The accounts of individual airmen downed in the Pacific confirmed the validity of this advice. Philip R. Corrin, a bombardier on a B-24 with the 13th Air Force shot down over Borneo, writes in an un-copyrighted paper entitled *In Darkest Borneo* detailing his experiences with the Dyak, an Asian people, that neither the British nor the Dutch were liked because they believed that caucasions were superior to native peoples. Corrin found that living among them as an equal created a mutual trust and respect, and that kindness and generosity were a natural part of the Dyak culture that was never shown to those who treated them as inferiors. These people saved his life not out of intimidation or fear, but because he was a fellow human being.

In making first contact with a local people in the South Pacific airmen were advised to be patient:

> They have to be courted. Motion to them to come out. They're very timid and have to be coaxed. Smile at them, offer them a little present; tobacco, a couple of matches, thread, fish hooks, a shiny quarter, or some salt. Swap things with them. Then point to the throat, make signs of eating and say, "I'm hungry." If asked properly, they will take a message to the nearest white man. One should not get boisterous and slap natives on the back. Just be friendly and dignified . . . When encountering natives, try to appear confident but not aggressive . . . One can only win their confidence by appearing open-handed . . . Never make passes at native women . . . because the men will resent it and a person may pay with his life.[22]

A system for obtaining assistance from the natives in the SWPA combining the use of language aids, money, barter kits and Promissory Notes was instituted. A report from SWPA of equipment distributed during April 1945 lists some of these evasion aids, including " Blood Chit, Chinese Flag", "Card, French Language", "Card, Pidgin English", "Kit, Trade", and "Purses, FELO".[23]

Money was an important part of the SWPA evasion system, and purses containing currency and coins were provided to crews as reported in February 1944:[24]

Fig. 8.2: In these plastic pouches were sealed the contents of the NEI purse shown in Fig. 8.1. The pouches, decorated with the flags of the United States, Australia, and the Netherlands, and with the coat of arms of the Dutch monarchy, were to be opened only when the user was forced down and use of the contents was required. (REB 4236) *U.S. Air Force Academy Photo, Courtesy of the Special Collections Branch, U. S. Air Force Academy Library*

Fig. 8.2: The NEI purse, sealed in its plastic pouches, was kept in this leatherette wallet. Like the pouches, the wallet is decorated with three flags and the coat-of-arms of the Dutch monarchy. (REB 4237) *Courtesy of David Vincent*

> One thousand purses containing approximately $63.50 each, are needed for airmen operating over the N.E.I. from North West Area. These funds are to aid them in evading capture if force landed in enemy occupied territory. The purses are to contain the following money:

5 -	$10.00 gold pieces	(U.S.)
10 -	*f* .25d silver pieces	(Java)
25 -	*f* .10d silver pieces	(Java)
2 -	*f* 1.00 silver pieces	(Dutch 1944 date)
5 -	*f* 1.00 paper note	(Java)
3 -	*f* 5.00 paper note	(Java)

Recognizing that some indigenous peoples used a barter economy, MIS-X provided barter kits, and offered advice on adding barter items to existing kits and how to trade. Crews in the SWPA were advised that "Razor blades for trading with the natives are available and at least three dozen should be included in the kit. There is a tin match box for them, but they can also be stored in odd corners. By all means keep a supply of new shillings, or of old shillings brightly shined up. The shine adds to their value. Into the kit, it may be well to place any inconsequential jewelry one happens to have about, and any shiny objects, especially if they have a hole in them and can be strung. Keep pearl buttons from discarded clothing for this purpose."[25]

There were literally hundreds of languages spoken in the Pacific, but MIS-X recommended that communication be attempted using a few widely used languages, recognizing that "Each native group in the South-West Pacific has its own local language. In addition there is generally some language which is commonly used as a means of communications between the natives and the Europeans."[26] In the Netherlands East Indies, which included western New Guinea, the common language was Malay. In eastern New Guinea, New Britain, New Ireland, and most of Melanesia it was various forms of Pidgin English, while in Portuguese Timor it was Tetum. In other areas of the Pacific, such as the Marshall Islands, which had been under Japanese control since 1914, the common language was Japanese. Blood Chits, phrase lists and other language aids were produced in these languages.

One of the more comprehensive evasion kits, shown in Fig. 8.1, was a purse containing money, chits and vocabulary lists in both Malay and Tetum, Promissory Notes, and other material. Issued for use in the Netherlands East Indies (NEI) and Portuguese Timor, it was available by February 1944 and is described in detail in one declassified document:

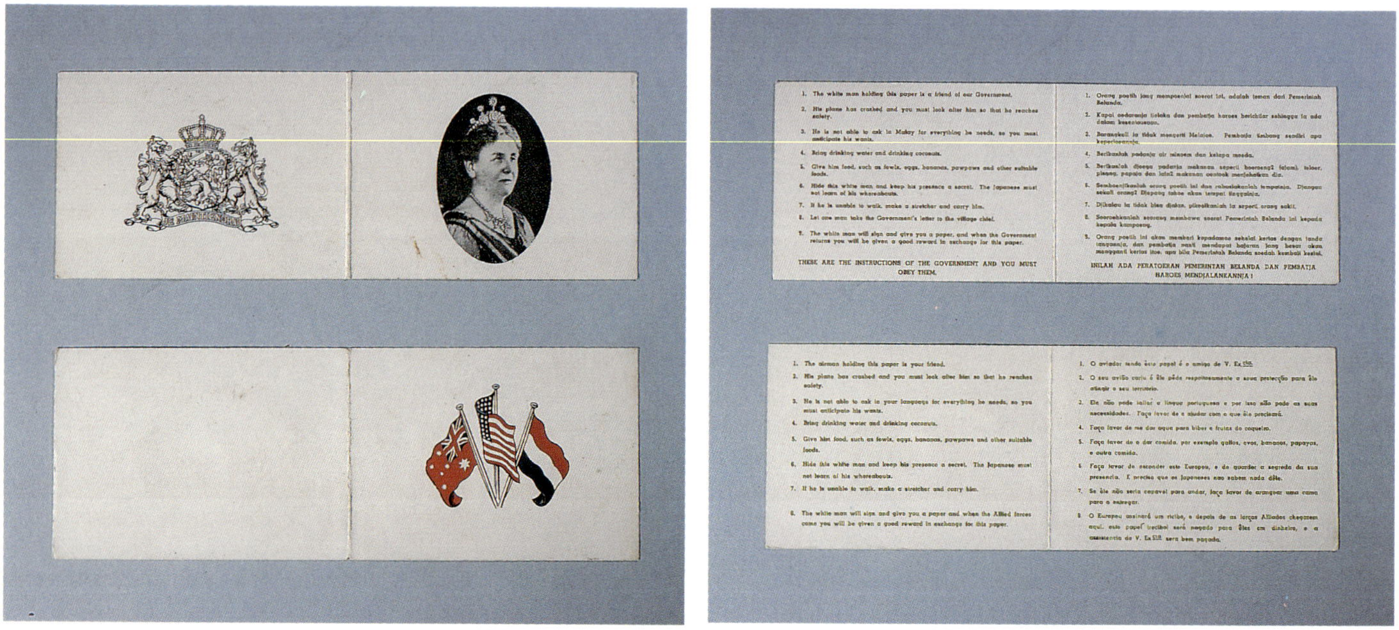

Fig. 8.4: The two folded card chits contained in the NEI purse are shown here. The left photo shows the outside covers of the cards, and the right photo shows their inside messages. The top chit, featuring Queen Wilhelmena of the Netherlands and her coat of arms on the cover, was for use in the NEI and is written in Malay. The bottom chit, featuring the Australian, American and Dutch flags, was for use in Portuguese Timor, and is written in Tetum. (Left: REB 4214, Right: REB 4216) *Courtesy of David Vincent*

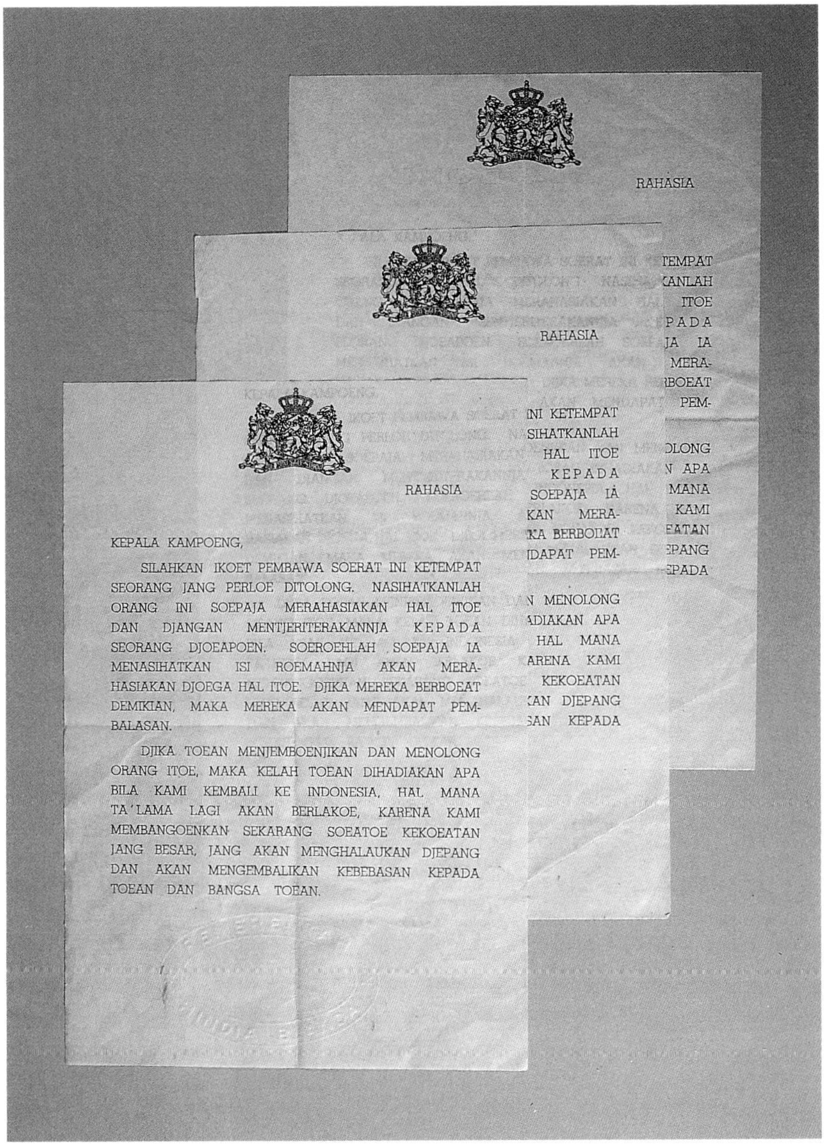

RAHASIA

KEPALA KAMPOENG,

SILAHKAN IKOET PEMBAWA SOERAT INI KETEMPAT SEORANG JANG PERLOE DITOLONG. NASIHATKANLAH ORANG INI SOEPAJA MERAHASIAKAN HAL ITOE DAN DJANGAN MENTJERITERAKANNJA KEPADA SEORANG DJOEAPOEN. SOEROEHLAH SOEPAJA IA MENASIHATKAN ISI ROEMAHNJA AKAN MERAHASIAKAN DJOEGA HAL ITOE. DJIKA MEREKA BERBOEAT DEMIKIAN, MAKA MEREKA AKAN MENDAPAT PEMBALASAN.

DJIKA TOEAN MENJEMBOENJIKAN DAN MENOLONG ORANG ITOE, MAKA KELAH TOEAN DIHADIAKAN APA BILA KAMI KEMBALI KE INDONESIA, HAL MANA TA'LAMA LAGI AKAN BERLAKOE, KARENA KAMI MEMBANGOENKAN SEKARANG SOEATOE KEKOEATAN JANG BESAR, JANG AKAN MENGHALAUKAN DJEPANG DAN AKAN MENGEMBALIKAN KEBEBASAN KEPADA TOEAN DAN BANGSA TOEAN.

Fig. 8.5: Three letters to village chieftains written in Malay were included in each NEI purse. The letters, all identical, feature the coat of arms of the Dutch monarchy at the top and an embossed seal at the bottom. The folded card chit in Malay was used in conjunction with these letters, but the chit was kept by the airman while the letter could be given to the tribal leader. (REB 4218) *Courtesy of David Vincent*

This type of evasion and escape aid is for use of Allied Air Force personnel forced down on missions in the Northwestern Area . . . the purse includes money; an official letter from the Dutch Government; a Glossary of English-Malayan terms; a "Directions and Question" card, and a glossary of English Portuguese terms for use if forced-landed in Portuguese Timor where Tetum is the spoken language of natives . . . The purses will be distributed through Operation Officers of squadrons and in turn to combat personnel. Each receiver of a purse, which is sealed, will sign a Memorandum Receipt for its contents. All will be properly briefed on the use of the purse previous to distribution.[27]

The individual components of the kit are shown in Figs. 8.2 through 8.9.

The contents of the purse were contained in two sealed plastic pouches, shown in Fig. 8.2, that were themselves contained in a leatherette wallet, shown in Fig. 8.3. The pouches and wallet are decorated with the flags of Australia, the United States and the Netherlands, and with the coat-of-arms of the Dutch monarchy.

Fig. 8.4 shows the two folded cards from the purse. The outsides of the cards are shown in the left photo, and the insides are shown in the right photo. These are the "directions and questions" cards referred to in the above quotation, and are the SWPA equivalent of a Blood Chits. The card chit in Old Malay was used in conjunction with letters for village chieftains, also in Old Malay, illustrated in Fig. 8.5. The language is called "Old Malay" because, although still written in the Roman alphabet, the spellings of the words in modern Indonesia have been changed from those shown here.

The following translation of the letters to village chieftans shown in Fig. 8.5 is provided by Peter B. Adeney:

> Secret
>
> Head of the village,
>
> Please take the holder of this letter to the place of someone who will help. Advise that person to keep this a secret and not to tell anyone who is for the Japanese. Tell him to advise those in his house to also keep this a secret. If they do this they will get a reward.
>
> If you alert and help that person, you will be rewarded if we return to Indonesia, which is something that will not be long in happening, because we have now built a massive force, which will drive out the Japanese and will return freedom to you and your nation.

Shown in Fig. 8.6 is a glossary of useful terms and phrases for airmen in Malay, and in Fig. 8.7 is a similar glossary in Tetum. Both glossaries, printed on onion-skin paper, contain hints on pronunciation, plus texts in English and their phonetic equivalents in the other languages.

The two items from the NEI Purse that may have been most persuasive in convincing natives to offer assistance to downed airmen are shown in Figs. 8.8 and 8.9. These are Promissory Notes and an envelope containing 25 NEI Guilders.

Jan Madarasz, now living in Germany, received one of the folded Malay card chits when he was in the NEI during the war. He received it in an envelope, shown in Fig. 8.10,[28] featuring a Bristol *Blenheim* twin-engined bomber with both the Australian and Ameri-

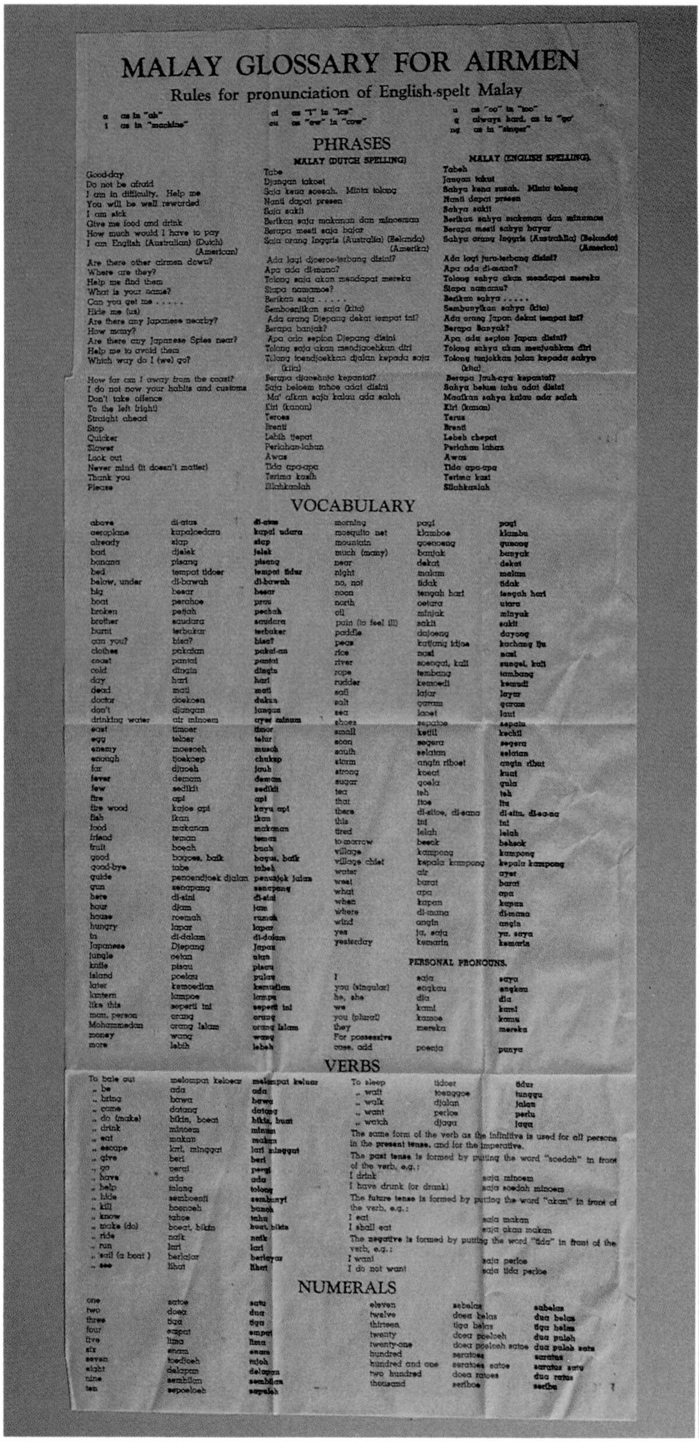

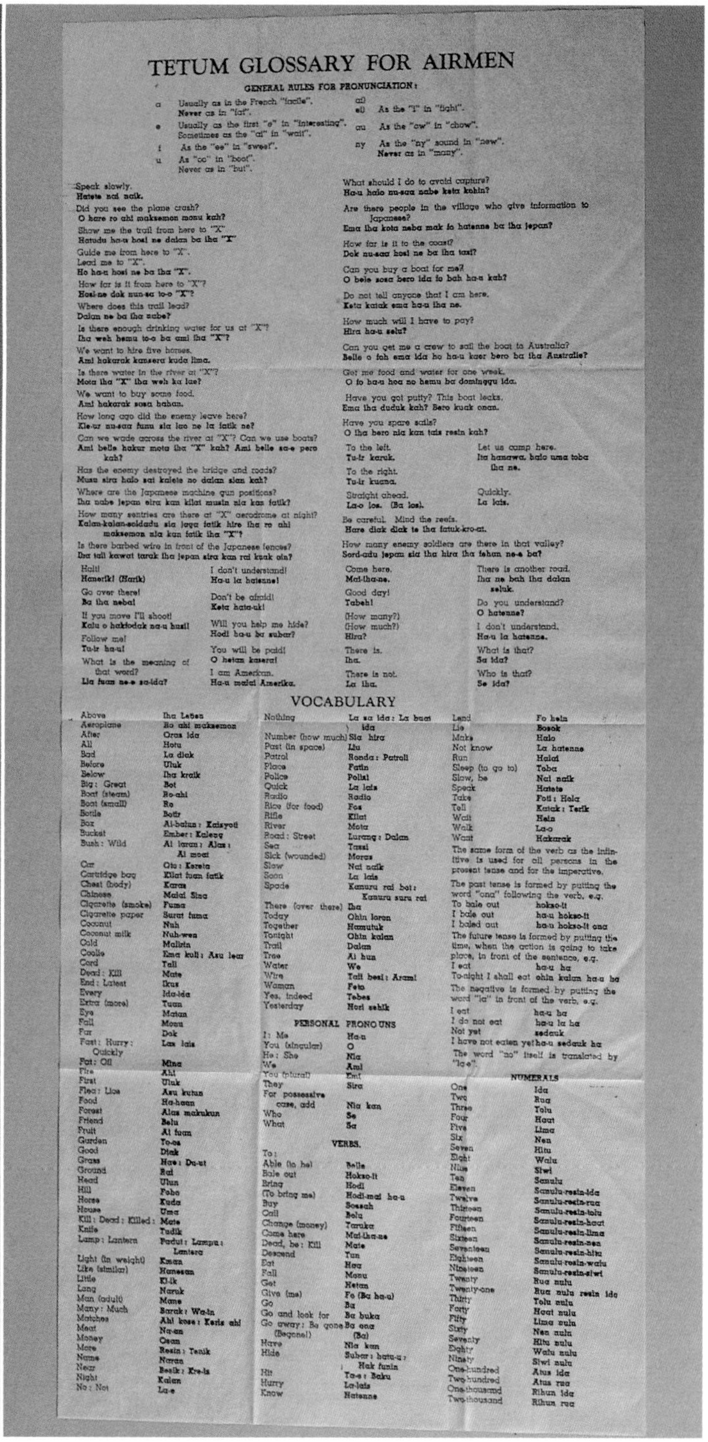

Fig. 8.6: Two glossaries were included in each NEI wallet, this one in Malay, and another, shown in Fig. 8.7, in Tetum. This glossary, printed on onion skin paper, was intended for use in the NEI and allowed airmen to communicate their needs in greater detail than Blood Chits, which simply gave a list of instructions to native peoples. (REB 4220) *Courtesy of David Vincent*

Fig. 8.7: This glossary is a companion to the one shown in Fig. 8.6. It is written in Tetum and is for use in Portuguese Timor. Like the other, it is printed on onion skin paper and gives pronunciation instructions as well as a long list of useful phrases. (REB 4222) *Courtesy of David Vincent*

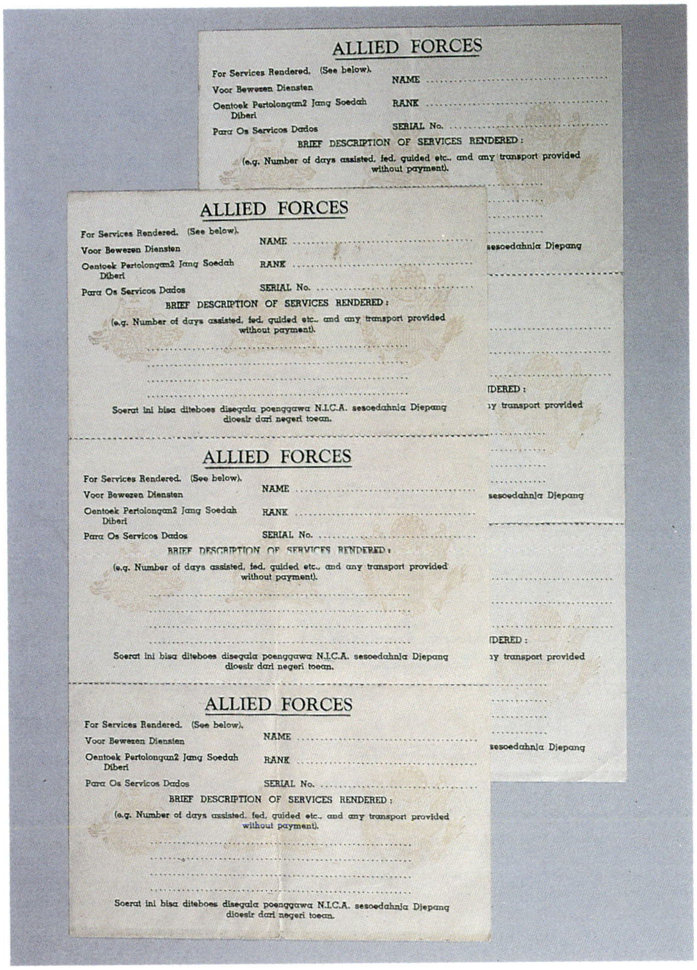

Fig. 8.8: Several sheets of Promissory Notes from the NEI purse are shown here. There are spaces for the airman to fill in his name, rank, serial number, and a brief description of the assistance rendered to him. These notes would be honored for a reward when presented to the authorities. (REB 4224) *Courtesy of David Vincent*

can flags on the front, and the Australian coat of arms and a single-engine fighter aircraft on the back. He writes:

> I received the chit sometime in 1944 through another person, whose name I have forgotten. He asked me whether I would be able to give shelter to a downed allied airman. He had obviously been informed by other members of the underground that I might have such a possibility. In the beginning of the Japanese occupation of the former Dutch East Indies I had given shelter to an escaped officer of the Dutch Army. He was later caught again and was lucky enough not to get killed on arrest, which was the practice of the Japanese Army. Regrettably I was not able to help the downed airman at that time because I was already under suspicion of being involved in resistance activities and was therefore being closely watched by the police. Therefore I don't know what finally happened to him, whether he was rescued and survived. The Japanese used to treat downed airmen as war criminals which meant certain death. The more reason to give any allied servicemen all the help possible under the given circumstances, even though that meant extreme risks for the helpers too.[29]

One of the more interesting languages of the SWPA is Pidgin English. Although using some basic English words in addition to native words, they are pronounced as the natives pronounce them, and are used in a native syntax. Pidgin English was used for communication between Europeans and the islanders.

Fig. 8.9: This paper envelope containing 25 NEI Guilders was a part of the NEI purse. The money, consisting of 24 Guilders in currency and 1 in coin, was to be used by the airman as needed to augment the Promissory Notes as both a reward and an incentive for natives to provide him assistance. *Courtesy of David Vincent*

Fig. 8.10: This colorful envelope contained a folded phonetic Malay card chit received by Jan Madarasz when he was a resident of the NEI during the war. The front features a Bristol *Blenheim* twin-engined bomber with both the Australian and American flags, and the back features a single-engine fighter with the Australian coat of arms. (REB 4238) *Courtesy of Jan Madarasz*

Details of the system used to solicit assistance from native peoples speaking Pidgin English are described in detail in official material:

> Added to their desire to help, certain natives have been instructed by ANGAU to help evaders, and have been promised payment which may be made either by the evader or through "chits", which will be honored by ANGAU. A system has been instituted whereby ANGAU has distributed to natives and Intelligence has distributed to pilots and combat crews a sheet giving instructions to the natives – on one side in English and on the other in pidgin. They go into considerable detail; the evader is a friend of the government; his plane has crashed, they must look after him, feed him, lodge him, nurse him, if necessary carry him by stretcher, guide him and carry his effects, and if needed hide him from the Japs. The native can give the evader information, can frequently send information for him to the nearest ANGAU office or to a mission, and most particularly can guide him either straight through by the same party or by relays of guides. Some natives prefer to give directions, rather than act as guides; but this is a substitute not to be accepted. Remember, however, that the evader will not turn a lazy native into a good guide by turning over all his wealth before the services are performed.[30]

The "chits" referred to in the above quotation are Promissory Notes, while the two-sided "sheet giving instructions to the natives" is actually the SWPA equivalent of a Blood Chit in Pidgin English.

Fig. 8.11 shows some of the evasion material that was issued to air crews in the SWPA. Included are chits in Pidgin English, trade items, RAAF silk maps, a metal signaling mirror, a survival booklet prepared in the SWPA, and a vocabulary book for languages used in the SWPA.

One of the Pidgin English chits from Australia is detailed in Fig. 8.12, along with the envelope in which it was issued. This card was issued to a S/Sgt. Nordin, and the side of the card shown features a message in Pidgin English. Below it is an identical card showing the message in English that appears on the reverse side.

Maj. Donald J. Gibb (USAF, Ret), who served with the 500th BS (M), 254th BG (M) in New Guinea from June 1943 until March 1944, received one of these chits. He writes that the "chit was probably issued to all flying personnel operating in the Eastern half of the island of New Guinea which was administered by the Australian Government." He estimates that the cards were issued from "early 1942 when air operations began over New Guinea to July 1944 when 5th Air Force units moved out of New Guinea for combat operations." Regarding their origins, Gibb states, "The chits were probably provided by the Australian liaison officer assigned to my squadron and issued to our combat crews by squadron intelligence personnel. The chits were provided by the Australian Government whose crest appears on both sides of the card . . . One side of the card is printed in Pidgin English which was used by the Australians before the war to communicate with the natives of the Eastern half of New Guinea and the island of New Britain."[31]

Australian John F. Russell still has the chit and envelope that he received in New Guinea in 1943 where he was "flying Beaufort aircraft with No. 6 GR/B Squadron, firstly from Milne Bay and later from Goodenough Island. Aircrew were issued with the double-sided white card enclosed in the envelope featuring the *Fortress* on one side and the *Beaufighter* on the other. I feel this is the only thing that could be classified as a 'blood chit'. The other papers are more like instructions from the Government to the head tribesmen and any natives working for the Government."[32]

Chits in the SWPA were sometimes issued in envelopes with colorful graphics on both sides, shown in Fig. 8.13. The photo on the left shows the front sides of the envelopes, and the photo on the right shows their back sides.

A chit in Pidgin English, similar to the last described card, was printed on what can best be described as heavy paper or soft cardboard, and was included in what were known as Jungle Kits, issued by the Australian government. The example shown in Fig. 8.14 was reportedly removed from a such a kit.

Another chit, shown in Fig. 8.15, was printed on newsprint, again in Pidgin English and in English. At the top is a variant of the

Fig. 8.11: Illustrated here are aids typical of those carried by air crews in the SWPA. On the American balloon-cloth map are several RAAF maps, an early silk drift chart, silk chits in Pidgin English, the very handy *Pocket Vocabulary* booklet in three major languages used in the Pacific, the survival manual *Castaway's Baedeker to the South Seas* prepared by the Objective Data Section, Intelligence Center, Pacific Ocean Areas, a circular Australian signaling mirror with instructions contained in a cloth case, a compass, and razor blades and Australian shillings to be used for bartering with native islanders. *International Blood Chit Museum*

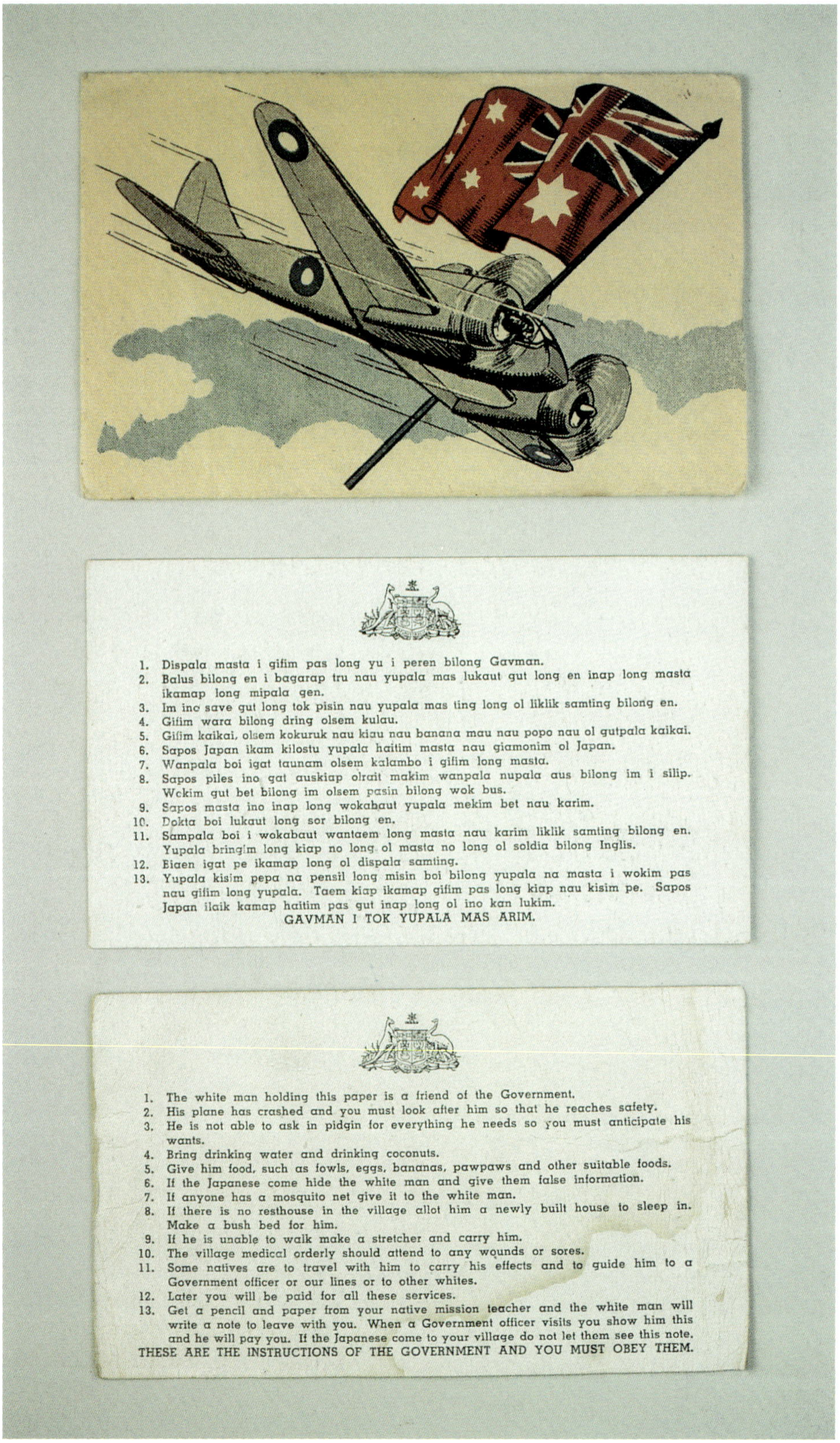

Fig. 8.12: This two-sided chit from Australia, shown with the envelope in which it was issued to a S/Sgt. Nordin, is printed on a stiff cardstock. At the top is the Australian coat of arms, beneath which are instructions in Pidgin English. Another identical card is shown below this, illustrating the reverse side on which the instructions are given in English. This type of chit was issued to Allied aircrews operating over eastern New Guinea and New Britain. (REB 4226) *International Blood Chit Museum*

Fig. 8.13: Chits in the SWPA were sometimes issued in these colorful envelopes featuring the aircraft and flags of the Allies. The fronts of two envelopes are shown the left photo, and the backs of the same envelopes are shown in the right photo. The card chit in phonetic Malay shown at the top of Fig. 8.4 was sometimes issued in the top envelope, which features a B-24 *Liberator* on its front and a Lockheed P-38 *Lightning* on its back. The two-sided cardboard chit in Pidgin English, as shown in Fig. 8.12, was sometimes issued in bottom envelope, with a *Beaufighter* on its front and a B-17 *Flying Fortress* on its back. (Top: REB 4239, Bottom: REB 4240) *Courtesy of Bob Gill*

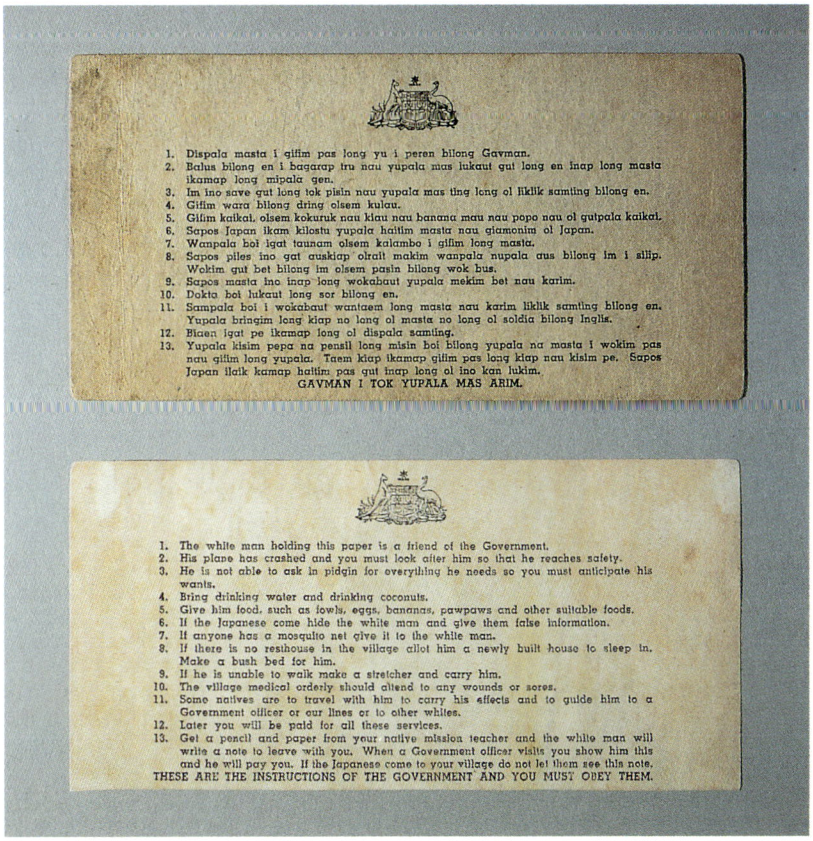

1. Dispala masta i gifim pas long yu i peren bilong Gavman.
2. Balus bilong en i bagarap tru nau yupala mas lukaut gut long en inap long masta ikamap long mipala gen.
3. Im ino save gut long tok pisin nau yupala mas ting long ol liklik samting bilong en.
4. Gifim wara bilong dring olsem kulau.
5. Gifim kaikai, olsem kokuruk nau kiau nau banana mau nau popo nau ol gutpala kaikai.
6. Sapos Japan ikam kilostu yupala haitim masta nau giamonim ol Japan.
7. Wanpala boi igat taunam olsem kalambo i gifim long masta.
8. Sapos piles ino gat auskiap olrait makim wanpala nupala aus bilong im i silip. Wokim gut bet bilong im olsem pasin bilong wok bus.
9. Sapos masta ino inap long wokabaut yupala mekim bet nau karim.
10. Dokta boi lukaut long sor bilong en.
11. Sampala boi i wokabaut wantaem long masta nau karim liklik samting bilong en. Yupala bringim long kiap no long ol masta no long ol soldia bilong Inglis.
12. Biaen igat pe ikamap long ol dispala samting.
13. Yupala kisim pepa na pensil long misin boi bilong yupala na masta i wokim pas nau gifim long yupala. Taem kiap ikamap gifim pas long kiap nau kisim pe. Sapos Japan ilaik kamap haitim pas gut inap long ol ino kan lukim.

GAVMAN I TOK YUPALA MAS ARIM.

1. The white man holding this paper is a friend of the Government.
2. His plane has crashed and you must look after him so that he reaches safety.
3. He is not able to ask in pidgin for everything he needs so you must anticipate his wants.
4. Bring drinking water and drinking coconuts.
5. Give him food, such as fowls, eggs, bananas, pawpaws and other suitable foods.
6. If the Japanese come hide the white man and give them false information.
7. If anyone has a mosquito net give it to the white man.
8. If there is no resthouse in the village allot him a newly built house to sleep in. Make a bush bed for him.
9. If he is unable to walk make a stretcher and carry him.
10. The village medical orderly should attend to any wounds or sores.
11. Some natives are to travel with him to carry his effects and to guide him to a Government officer or our lines or to other whites.
12. Later you will be paid for all these services.
13. Get a pencil and paper from your native mission teacher and the white man will write a note to leave with you. When a Government officer visits you show him this and he will pay you. If the Japanese come to your village do not let them see this note.

THESE ARE THE INSTRUCTIONS OF THE GOVERNMENT AND YOU MUST OBEY THEM.

Fig. 8.14: This is another Australian chit featuring the coat of arms and messages in Pidgin and English. It was printed on a soft, thin cardboard and included in what were called Jungle Kits, issued by the Australian government. (REB 4228) *International Blood Chit Museum*

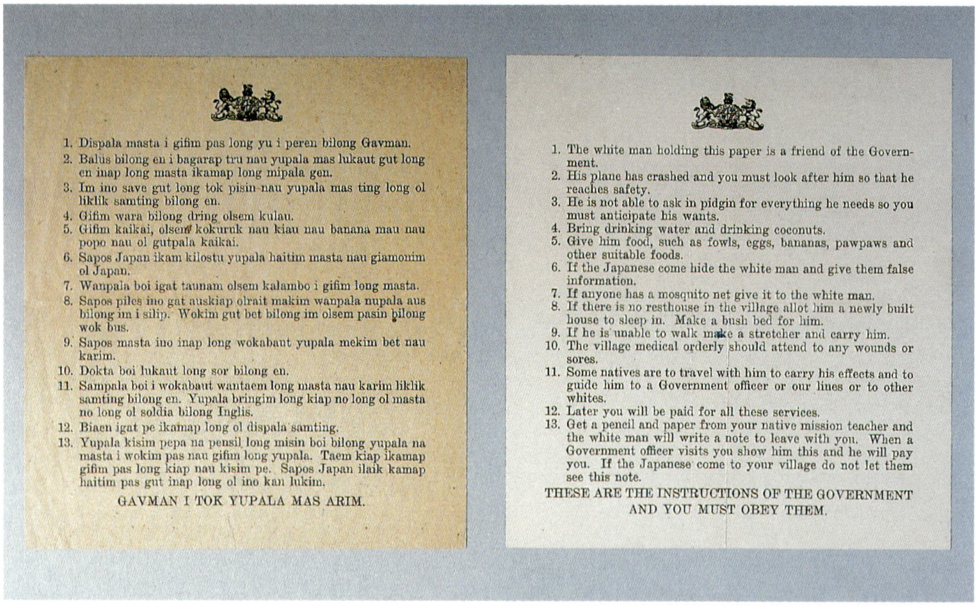

1. Dispala masta i gifim pas long yu i peren bilong Gavman.
2. Balus bilong en i bagarap tru nau yupala mas lukaut gut long en inap long masta ikamap long mipala gen.
3. Im ino save gut long tok pisin nau yupala mas ting long ol liklik samting bilong en.
4. Gifim wara bilong dring olsem kulau.
5. Gifim kaikai, olsem kokuruk nau kiau nau banana mau nau popo nau ol gutpala kaikai.
6. Sapos Japan ikam kilostu yupala haitim masta nau giamonim ol Japan.
7. Wanpala boi igat taunam olsem kalambo i gifim long masta.
8. Sapos piles ino gat auskiap olrait makim wanpala nupala aus bilong im i silip. Wokim gut bet bilong im olsem pasin bilong wok bus.
9. Sapos masta ino inap long wokabaut yupala mekim bet nau karim.
10. Dokta boi lukaut long sor bilong en.
11. Sampala boi i wokabaut wantaem long masta nau karim liklik samting bilong en. Yupala bringim long kiap no long ol masta no long ol soldia bilong Inglis.
12. Biaen igat pe ikamap long ol dispala samting.
13. Yupala kisim pepa na pensil long misin boi bilong yupala na masta i wokim pas nau gifim long yupala. Taem kiap ikamap gifim pas long kiap nau kisim pe. Sapos Japan ilaik kamap haitim pas gut inap long ol ino kan lukim.

GAVMAN I TOK YUPALA MAS ARIM.

1. The white man holding this paper is a friend of the Government.
2. His plane has crashed and you must look after him so that he reaches safety.
3. He is not able to ask in pidgin for everything he needs so you must anticipate his wants.
4. Bring drinking water and drinking coconuts.
5. Give him food, such as fowls, eggs, bananas, pawpaws and other suitable foods.
6. If the Japanese come hide the white man and give them false information.
7. If anyone has a mosquito net give it to the white man.
8. If there is no resthouse in the village allot him a newly built house to sleep in. Make a bush bed for him.
9. If he is unable to walk make a stretcher and carry him.
10. The village medical orderly should attend to any wounds or sores.
11. Some natives are to travel with him to carry his effects and to guide him to a Government officer or our lines or to other whites.
12. Later you will be paid for all these services.
13. Get a pencil and paper from your native mission teacher and the white man will write a note to leave with you. When a Government officer visits you show him this and he will pay you. If the Japanese come to your village do not let them see this note.

THESE ARE THE INSTRUCTIONS OF THE GOVERNMENT AND YOU MUST OBEY THEM.

Fig. 8.15: The two-sided chit shown here is printed on newsprint, and features a variant of the Royal Arms at the top. The message, in Pidgin on one side and English on the other, is very similar to that on the Australian chits shown in Figs. 8.12 and 8.14, but the final instruction promises payment for services rendered. (REB 4230) *International Blood Chit Museum*

Royal Coat of Arms, and the last item on the list of instructions for aiding downed airmen promises payment for assistance.

The chit shown in Fig. 8.16 belonged to Lt. Col. N. W. Pratt, who flew with Marine Air Group 25, stationed on Villa Lavella Island in the central Solomons. This chit is unusual in that two messages appear on it, one to be shown to islanders who could read, and the other to be read aloud by the airman to those who could not read.

Language materials were also produced for use by downed airmen in very small but specific geographical locations. One such aid, issued to a S/Sgt. Nordin, whose card chit is shown in Fig. 8.12, is a Noemfoor Glossary for Airmen, shown in Fig. 8.17. Instructions state that this should be used if the airman is unsuccessful in obtaining assistance using Malay language aids, and that the Glossary is phonetic because natives who do not speak Malay cannot read.

Another language aid for a specific locality was a letter in Pidgin English for use in the Wewak area. Although an example was not available to be photographed, the text of the letter together with its English translation is detailed in the "Report of MIS-X Section, SWPA" issued in February 1944:[32]

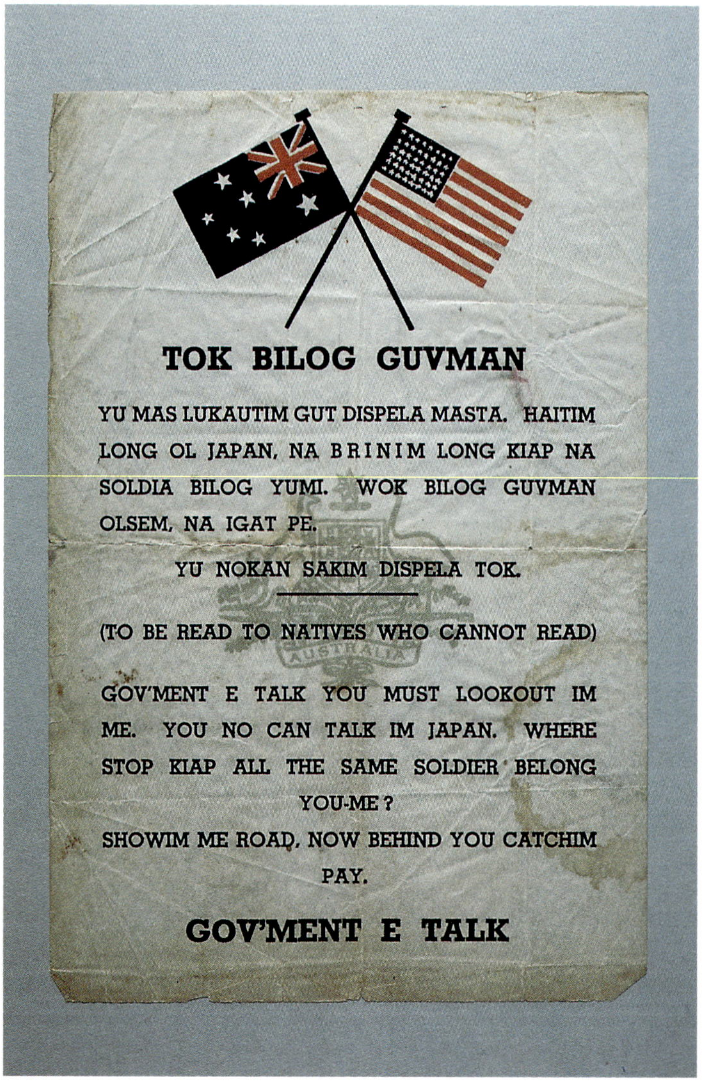

TOK BILOG GUVMAN

YU MAS LUKAUTIM GUT DISPELA MASTA. HAITIM LONG OL JAPAN, NA BRINIM LONG KIAP NA SOLDIA BILOG YUMI. WOK BILOG GUVMAN OLSEM, NA IGAT PE.

YU NOKAN SAKIM DISPELA TOK.

(TO BE READ TO NATIVES WHO CANNOT READ)

GOV'MENT E TALK YOU MUST LOOKOUT IM ME. YOU NO CAN TALK IM JAPAN. WHERE STOP KIAP ALL THE SAME SOLDIER BELONG YOU-ME?

SHOWIM ME ROAD, NOW BEHIND YOU CATCHIM PAY.

GOV'MENT E TALK

Fig. 8.16: This Australian chit, which belonged to Lt. Col. N. W. Pratt, of Marine Air Group 25 stationed on Villa Lavella Island, is printed in color and features the Australian and American flags at the top. Two messages appear on the chit. The top message was to be shown to islanders who could read, and the bottom message was to be read aloud by the airman to those who could not read. The Australian coat of arms appears in the background behind the messages. (REB 4232) *International Blood Chit Museum*

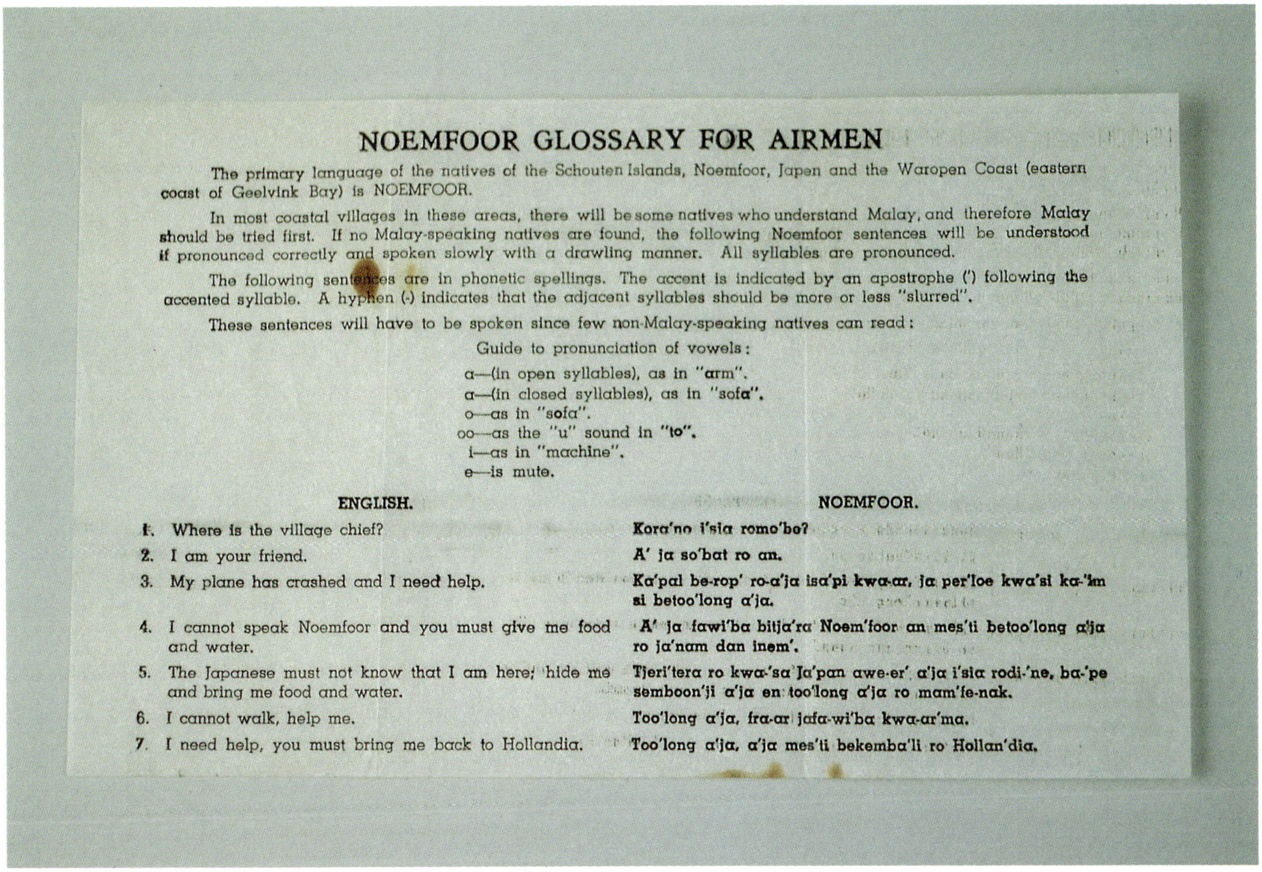

NOEMFOOR GLOSSARY FOR AIRMEN

The primary language of the natives of the Schouten Islands, Noemfoor, Japen and the Waropen Coast (eastern coast of Geelvink Bay) is NOEMFOOR.

In most coastal villages in these areas, there will be some natives who understand Malay, and therefore Malay should be tried first. If no Malay-speaking natives are found, the following Noemfoor sentences will be understood if pronounced correctly and spoken slowly with a drawling manner. All syllables are pronounced.

The following sentences are in phonetic spellings. The accent is indicated by an apostrophe (') following the accented syllable. A hyphen (-) indicates that the adjacent syllables should be more or less "slurred".

These sentences will have to be spoken since few non-Malay-speaking natives can read:

Guide to pronunciation of vowels:

a—(in open syllables), as in "arm".
a—(in closed syllables), as in "sofa".
o—as in "sofa".
oo—as the "u" sound in "to".
i—as in "machine".
e—is mute.

	ENGLISH.	NOEMFOOR.
1.	Where is the village chief?	**Kora'no i'sia romo'bo?**
2.	I am your friend.	**A' ja so'bat ro an.**
3.	My plane has crashed and I need help.	**Ka'pal be-rop' ro-a'ja isa'pi kwa-ar, ja per'loe kwa'si ka-'im si betoo'long a'ja.**
4.	I cannot speak Noemfoor and you must give me food and water.	**A' ja fawi'ba bitja'ra Noem'foor an mes'ti betoo'long a'ja ro ja'nam dan inem'.**
5.	The Japanese must not know that I am here; hide me and bring me food and water.	**Tjeri'tera ro kwa-'sa Ja'pan awe-er' a'ja i'sia rodi-'ne, ba-'pe semboon'ji a'ja en too'long a'ja ro mam'fe-nak.**
6.	I cannot walk, help me.	**Too'long a'ja, fra-ar jafa-wi'ba kwa-ar'ma.**
7.	I need help, you must bring me back to Hollandia.	**Too'long a'ja, a'ja mes'ti bekemba'li ro Hollan'dia.**

Fig. 8.17: Among the language materials produced for use in small, specific locations is this Noemfoor Glossary for Airmen, issued to a S/Sgt. Nordin. Instructions state that this should be used if the airman is unsuccessful in obtaining assistance using Malay language aids. (REB 4234) *International Blood Chit Museum*

Suggested Letter To The Natives Of Wewak Area

Pren bolong mi, yu pella harim tok. Tis pella masta I pren bolong mi tru, alasam barata. Yu pella I savi long long mi, work bolong mi pai im boi. Time mi stop Wewak sipos yu pella I got trubel wusat yet i elp im yu, mi tasal. Na mi laikim yu elpim tis pella masta. Haitim long man bolong yapan, bring im long rot na painim sumpella masta, bring im kaikai long im. Sipos I got moni somting long pai arit sipos no got maski, bihain mi pai im yu. Gib im alasam yu pella I dinau long kongkong bifor. Yu pella lukaut im gut.

Mi pella raus im pinis alla Yapan long salamau na Lae na Pinshaven na mi pella kum raus im long giraun bolong yu. Arait yu lukaut im masta liklik time mi kum luk im yu pella brimg im planti somting bai mi presel long yu. Na yu pella pas im mous, no kan tok.

Masta bolong yu
Masta Eltin.

Interpretation of Letter to the Natives of the Sepic District

Friends of mine, listen to what I have to tell you. This white man is a friend of mine just like a brother. You people know me, I was at Wewak before as a recruiter of native labour. When you were in trouble it was to me you came to straighten things out. When you were in Wewak it was to me you came for food. Now I want you to help this white man hide from the Japanese. Take him along the road and bring to a place where there are other white men. Give him food. If he has money or trade to pay for it very well but if he has not you must give to him and I will pay you when I return. Treat it as a credit sale just as you people used to receive credit from the Chinese traders.

We have already cleared the Japanese out of Lae, Salamau, and Finschaven. Now we are coming to clean them from your land. Very well now, look after this white man. I will see you shortly when I will bring plenty of trade goods and give them to you for nothing. Shut your mouth, do not talk about this.

I am your master
Master Helton.

Fig. 8.18: Shown here are 58 extremely rare Emergency Currency Certificates from the Philippines, sometimes referred to as "Guerilla Currency." The notes shown were issued by the provinces of Bohol, Cagayan, Cebu, Iloilo, Mindanao, Mountain Province, Misamis, and Negros. They were printed with whatever paper and ink were available, and many contained acknowledgements that they would not be redeemed until after the war was over. *International Blood Chit Museum*

By reading both versions of the letter, one can get a feel of how Pidgin English works. The first sentence, for example, can be seen as, "Friend belong me, you fellow hear-um talk", easily translated as "Friends of mine, listen to what I have to tell you". The second sentence, more difficult to interpret than the first, can be seen as "This fellow master I friend belong me true, all the same brother," translated as "This white man is a friend of mine just like a brother." Readers are encouraged to interpret the remainder of the Pidgin English message in order to better understand the situation of Allied crews who learned to communicate in this language. The remnants of western colonialism in the Pacific are evident in Pidgin English, which uses "master" as the term for a Caucasian person. This may serve as a reminder to the reader of the colonial origins of Blood Chits themselves.

As the war moved to different areas, MIS-X shifted the emphasis of its efforts to where the war was being fought. The establishment of the Pacific Ocean Area (POA) Section was prompted in part by the recognition by the U.S. Navy of the need for organized evasion efforts in its other areas of operation, based on their positive experience in the South West Pacific Area. In July 1944 it was recommended, "As soon as they are available, enough officers should be trained by MIS-X in Washington to provide adequate briefing for carrier air groups and replacement groups . . . In order to ensure briefing on areas assigned as targets of opportunity while the carriers are at sea a qualified officer should accompany each Task Force. He would be responsible for the distribution of pertinent information to each ship's ACIO."[34] A postwar Pacific Fleet report gives the following details regarding the establishment of the POA Section:

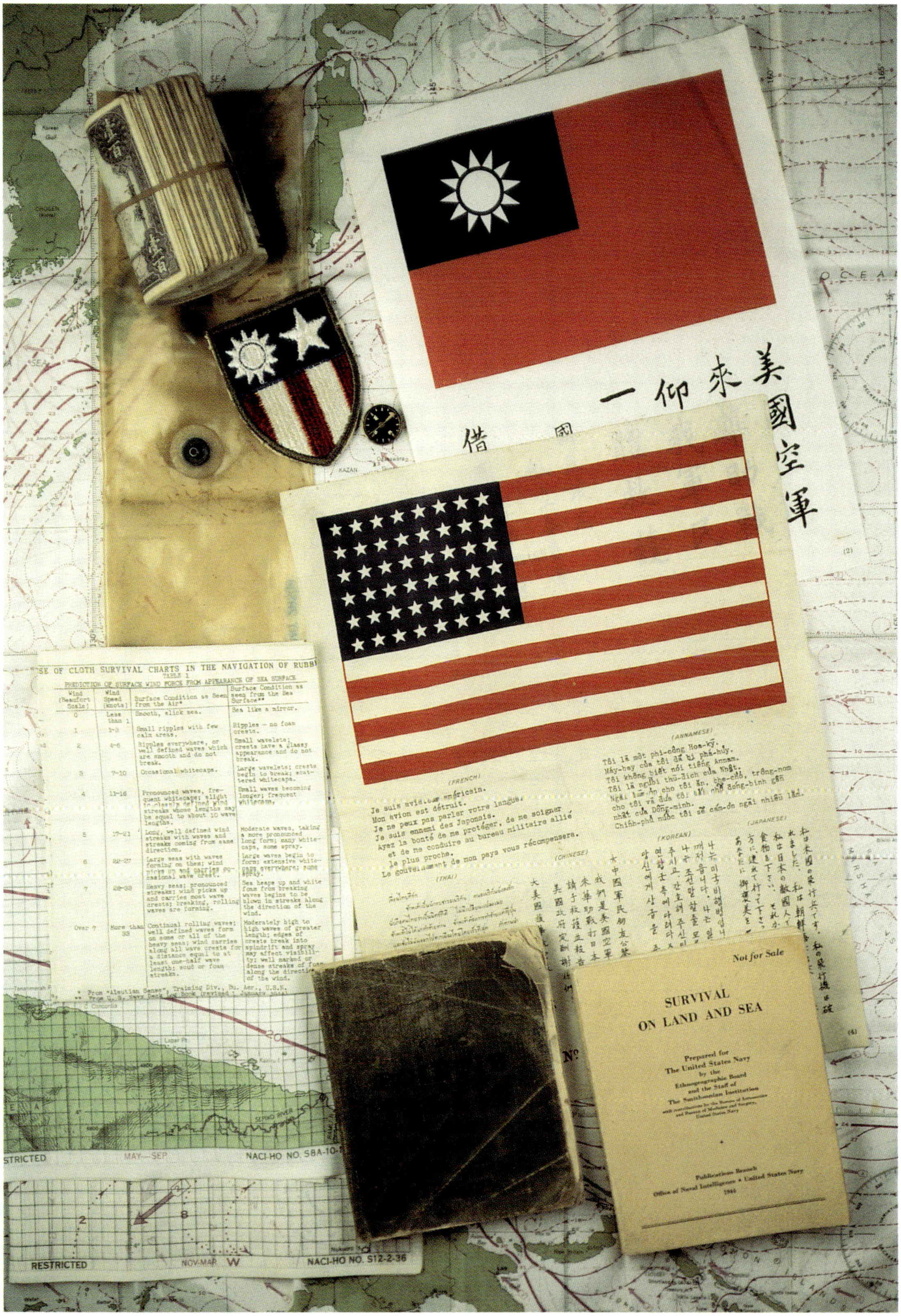

Fig. 8.19: Typical evasion aids issued to American naval aviators operating in the POA are shown here. These include MIS-X type 2 and 4 Blood Chits, NACI-HO rayon drift charts showing wind and water currents, instructions for using drift charts to pilot a raft, a pointie talkie, Evasion Currency (5,000 CN in 100 CN notes), and a CBI shoulder patch. Most of these items were issued in a waterproof pouch except for drift charts which were packaged in their own envelope. *International Blood Chit Museum*

On 30 September 1944, the Chief of Staff, CINCPAC-CINCPOA, authorized that a single theater agency be established under the Officer in Charge, JICPOA. On 8 November 1944, MIS-X, POA was officially established as a section of JICPOA.

Contacts were immediately established with ComAirPac, Operational Intelligence Section; Aircraft, Fleet Marine Force, Pacific; and Army Air Forces, Pacific Ocean Areas, A-2 Sections, through which all operations with the Navy, Marine Corps and Army air units were handled . . . A permanent MIS-X subsection was established at Advance Intelligence Center, Guam, when it was activated about 12 February 1945.

Most of the large commands in the forward areas maintained their own equipment and personnel for air-sea rescue operations, the Navy taking the leading part in advising and coordinating . . . ComAirPac took an active part in organizing rescue work throughout the Pacific and MIS-X, POA had an opportunity to coordinate its program with the closely related functions of the air-sea rescue agencies in the forward areas.

MIS-X, POA's responsibilities extended to all branches of services in the theater. Principle effort was directed to the air components of the Navy (carrier and land-based), Marine (carrier and land-based) and Army . . . The information received by MIS-X, POA and distributed throughout the combat area was used . . . by MIS-X, POA officers as the basis for their briefing of operating units. It was also used by the Air Combat Intelligence and S-2 officers attached to individual units as the basis for briefing in preparation for specific missions.[35]

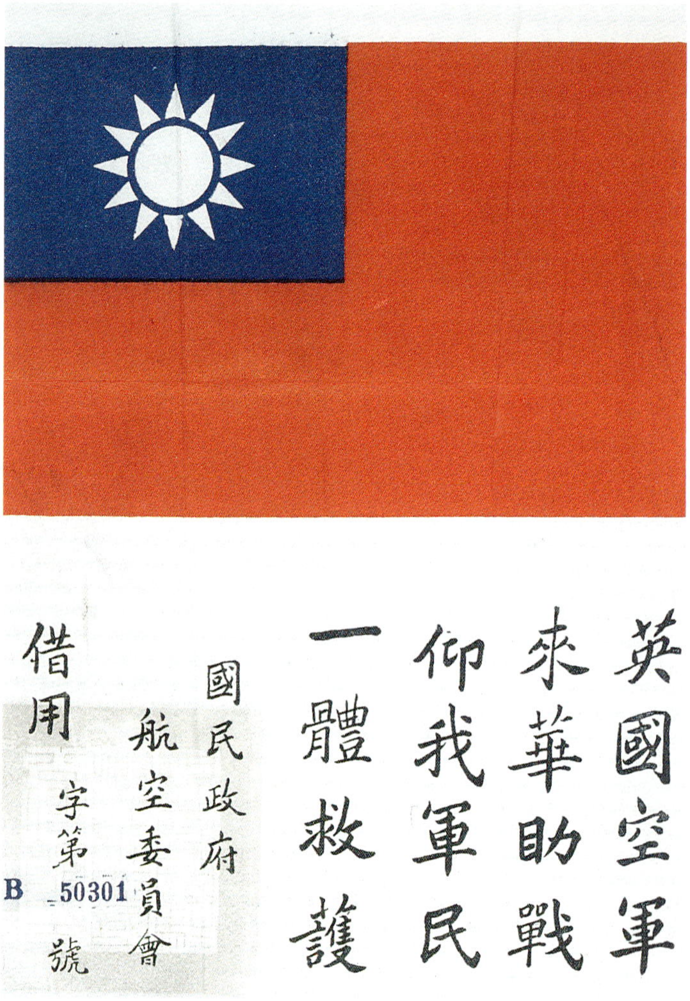

Fig. 8.20: Shown here is an extremely rare type 2 MIS-X chit from the Pacific. It is in the Special Collections library of the United States Air Force Academy, and belonged to Capt. Harry L. Osterweis, a prominent member of MIS-X's SWPA Section. The serial number on this chit is preceded by the letter "B" rather than the letter "W", which may indicate that it was issued in Brisbane, Australia, where headquarters of MIS-X SWPA Section was located. (REB 3246.3) *U.S. Air Force Academy Photo, Courtesy of the Special Collections Branch, U.S. Air Force Academy Library*

The area covered by the MIS-X POA Section was impressive. Navy documents disclose that during June and July 1945 MIS-X officers conducted operations at such widely separated locations as Ie Shima, Okinawa, Iwo Jima, the Marianas, the Philippines, the Hawaiian Islands, the Aleutian island chain in Alaska, and in Sydney, Australia, with the British Pacific Fleet.

In the POA, as in the SWPA, the rescue of downed airmen relied heavily upon the assistance of native peoples. The Office of Naval Intelligence (ONI) advised its pilots, "If you are in a region inhabited by native peoples make every effort to get in touch with them and ask them for help. You run little danger if you approach them in a friendly manner. Never show fear, threaten, or flourish a gun. As a rule it is fear on the part of natives that makes them attack strangers and if you do nothing to cause concern you will be perfectly safe. Go up to them as you would to individuals of your own race and color, smile, offer a cigarette if you have one, and make your want known."[36]

Operations in the Pacific Ocean Area included targets not only in island chains like the Marshalls, Gilberts, Marianas, Bonins, and Ryukyus, but also in Formosa and China, and on the Korean peninsula. Instructions to crews in the POA reflected the diversity of areas over which missions were flown:

Approach an elderly, rather than a young man, preferably an elderly Chinese, Formosan, Korean, or Filipino farmer or poorly dressed person . . . Identify yourself with the identification flag with which you will be provided. (If in China use the Chinese flag which contains characters describing you as an American and promising a reward for your safe return.)

Use your Pointie-Talkie to request assistance. Remember that Chinese characters can be read and their intended meaning understood in China, Formosa, and Korea, although the spoken languages in those countries differ widely. A young educated man in these countries may not be able to make out the characters while an elderly literate man can.

The Chinese, Koreans, Formosans and Filipinos appreciate a smiling, pleasant person. Try to keep in good humor always, even when annoyed by the inevitable delays. Be modest; maintain your dignity. Avoid the slightest familiarity with all women.[37]

Fig. 8.21: Crews of the 5th and 13th Air Forces operating in the Pacific were issued materials somewhat different than those issued to Naval aviators. These included cloth maps and currency for the area of the mission, plus the type 2 and 4 MIS-X chits. Included in kits for missions to China were three paper items: a single-sheet pointie talkie, a letter bearing the CBI shield at the top and the 5th and 13th AF insignia at the bottom, and a leaflet featuring the 5th and 13th AF insignia below an airman showing his chit to a farmer. These were issued to crews in the standard 4 by 11 inch waterproof pouch. *International Blood Chit Museum*

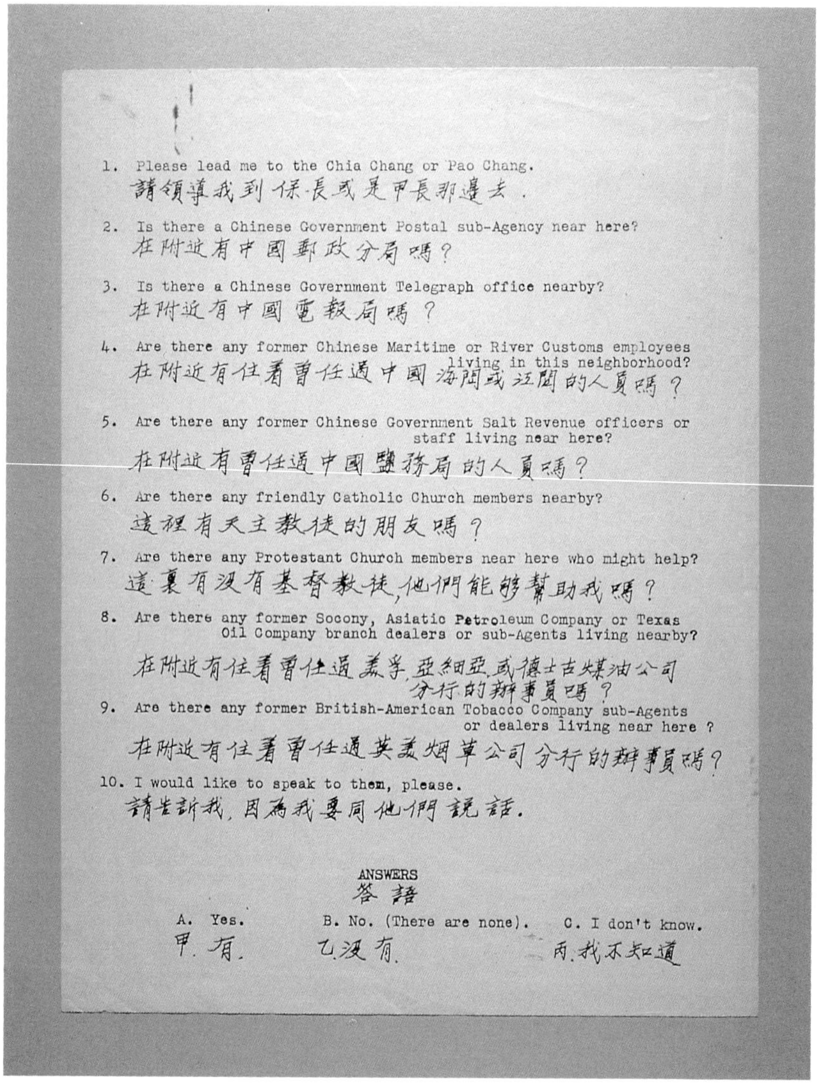
1. Please lead me to the Chia Chang or Pao Chang.
請領導我到保長或是甲長那邊去.
2. Is there a Chinese Government Postal sub-Agency near here?
在附近有中國郵政分局嗎?
3. Is there a Chinese Government Telegraph office nearby?
在附近有中國電報局嗎?
4. Are there any former Chinese Maritime or River Customs employees living in this neighborhood?
在附近有住着曾任過中國海関或江関的人員嗎?
5. Are there any former Chinese Government Salt Revenue officers or staff living near here?
在附近有曾任過中國鹽務局的人員嗎?
6. Are there any friendly Catholic Church members nearby?
這裡有天主教徒的朋友嗎?
7. Are there any Protestant Church members near here who might help?
這裏有沒有基督教徒,他們能够幫助我嗎?
8. Are there any former Socony, Asiatic Petroleum Company or Texas Oil Company branch dealers or sub-Agents living nearby?
在附近有住着曾任過美孚,亞細亞,或德士古煤油公司分行的辦事員嗎?
9. Are there any former British-American Tobacco Company sub-Agents or dealers living near here?
在附近有住着曾任過英美烟草公司分行的辦事員嗎?
10. I would like to speak to them, please.
請告訴我,因為我要同他們說話.

ANSWERS
答語

A. Yes. 甲.有.
B. No. (There are none). 乙.沒有.
C. I don't know. 丙.我不知道

Fig. 8.22: The single-sheet pointie talkie included in 5th and 13th AF kits for missions to China is shown here. Intended for use on the eastern Coast, it includes just ten questions and statements and three possible answers. (REB 3926) *International Blood Chit Museum*

The term "identification flag" used in the above quotation is another term for the Blood Chit.

Air crews operating over the Philippines were advised that if down in rural areas they should remain in the hills and make contact with the various bands of guerrillas. These groups knew the local situation, and could arrange for rescue by boat or aircraft.[38]

The language-related evasion aids and money issued to crews operating in the POA also reflected the diversity of the area. Some of these items are detailed in the *Final Report* of the POA Section, written in September 1945:[39]

The main items issued to POA units were:
- E-17, E-3a, or Mark I (Navy) Emergency Kits
- Chinese Flag, blood chit
- U.S. Flag, blood chit
- C.B.I. shoulder patch
- Pointie-Talkie (several languages)
- Chinese money purse
- Waterproof Compass
- Cloth survival charts.

Identification and Native Contacts
- Dog tags
- Shoulder patch
- Blood chit, local country
- Blood Chit, U.S. flag
- Money of native country (but not enemy country)
- Pointie-Talkie

The CBI shoulder patch included in POA evasion kits, officially an Army Shoulder Sleeve Insignia (SSI), is the shield-shaped sleeve patch designed by Brig. Gen Frank Dorn for American troops in the China-Burma-India theater. It features the star of India and the sun of the Nationalist Chinese in white on a blue field, above red and white stripes. The Burmese peacock was not included in the patch design as Burma had already fallen to the Japanese. The insignia was universally recognized in the China and India-Burma Theaters as the symbol of American forces, and was therefore included in evasion material issued to crews operating over Asia from bases in the POA who did not otherwise have access to it.

As in the SWPA and elsewhere, the POA Section knew the value of Evasion Currency as an evasion aid. Mrs. Walter D. Innis writes, "I have some Chinese blood money which was issued to my late husband, Rear Admiral Walter D. Innis, USNR, in World War II. This was issued to aviators to buy their way out if shot down over enemy territory. Fortunately, my husband did not get shot down so I have this money in mint condition in the original waterproof envelope."[40]

Control of money purses was strict. An untitled report explains that "Packets will be issued to crew members only prior to a mission, and will be recovered from crew members immediately upon return from a mission. Necessary steps will be taken by inspection, receipt or otherwise, to insure return of the purses intact."[41] In order that the maximum benefit be achieved from the money, the Navy briefed its crews, "Do not be over-generous. Display as little of your wealth as possible. Before starting on your mission, learn the value of the money in terms of what the ordinary Chinese or other foreign citizen earns per day."[42]

There were usually two types of money in circulation, Japanese-issued occupation money and native currency and coin. Evasion kits included the local money, since native peoples might logically assume that someone carrying enemy currency was the enemy.

Currency produced in the Philippines, often called "Guerilla Money", was possibly the most difficult to use operationally because of the many local varieties. The banknotes, officially known as Emergency Currency Certificates, were produced in each province or currency district on whatever paper was available, and there was no standard format so bills varied considerably from province to province. Bills were often signed by hand or stamped with rubber stamp signatures. Some had hand-embossed seals on them. Many contained messages acknowledging that their actual value was negligible until the Japanese were defeated, including "This note is redeemable at face value after the emergency and will not be devaluated or discriminated against" on notes from Mindanao, or "Payable After the War" on some Cagayan notes. Figure 8.18 shows 58 of these bills, issued by the Emergency Currency Boards of Bohol, Cagayan, Cebu, Iloilo, Mindanao, Mountain Province, Misamis, and Negros.

Following the war Guerilla Currency became very collectible, and thousands of notes virtually identical to the wartime currency were produced in the Philippines and sold to collectors. Unless Philippine guerilla currency is obtained from a veteran or other person who can guarantee its authenticity, it should be viewed with suspicion as possibly being of postwar manufacture.[43]

While many of the Blood Chits and other evasion materials used in the SWPA were produced in Australia, their counterparts in the POA were almost exclusively products of the United States. Fig. 8.19 shows some of the evasion aids issued to American naval aviators operating in the POA. This includes MIS-X type 2 and 4 Blood Chits, drift charts showing wind and water currents, instructions for using drift charts to pilot rubber rafts, a pointie talkie, a survival book from the Office of Naval Intelligence, Chinese currency, and a CBI shoulder patch.

Clear instructions for the use of Blood Chits was given to Navy crews operating in the POA: "The main use for identification flags has been in China. They should also be useful in Manchuria and Korea. One flag is issued under the sanction of the Chinese Central Government. Underneath the flag of China and over the Chinese Government's seal is a statement to the effect that you are an American airman and are to be conducted safely to the government or American authorities. The serial number on the flag is important because it is registered with your name at U.S. China headquarters and enables you to be positively identified . . . Another identification flag that is useful in many countries is a small American flag under which appears an identification statement in several languages similar to that on the Chinese flag, but also promising a reward for assistance and delivery of the party to American headquarters."[44]

The evolution of the MIS-X type 2 and 4 Blood Chits from their China and India-made predecessors is detailed in Chapter 7. Since the primary area in which these two types of Blood Chits were used was the Pacific, the reader is encouraged to again familiarize himself with these types.

Three versions of the type 2 MIS-X chit, in Chinese, are illustrated in in Figs. 7.34, 7.36 and 7.37 of Chapter 7. These show chits with and without the number "(2)" in the lower right corner, and two possible locations for the serial number. The chits are stamped with the chop of the Chinese Ambassador to the United States, and serial numbers begin with the letter "W", denoting that they were issued by authorities in Washington, D.C.

One of the most unusual type 2 chits seems to appear only in the Pacific. The example illustrated in Fig. 8.20 is in the Special Collections Library of the United States Air Force Academy, and belonged to Capt. (later Col.) Harry L. Osterweis, of MIS-X's SWPA Section. The serial number on this chit is preceded by the letter "B" rather than the letter "W". MIS-X did supply Blood Chits to the RAAF, and it is possible that, just as the letter "W" denotes that a

通告

誠摯的中國朋友們:

凡在左袖上帶有這些徽章的,都是諸位的朋友。他們就是美國航空員,前來幫助你們把那可恨的日寇驅逐到中國國以外。

諸位或者曾經看過這張紙上端的徽章,過去的五年之間在 蔣委員長指揮之下作戰的美國航空員,就是佩帶着這種徽章。

其他的美國航空隊々員現在也來 幫助 貴國抗戰。他們在左袖上所佩帶的,就是下面兩種徽章。對於佩帶這三種之中任何一種徽章的外國人,於有必要時,請求諸位儘量地加以援助。他們就是諸位的朋友。諸位因援助他們而用掉銀錢,不論多少,美國政府必將全額償還。

美軍總司令佈

Fig. 8.23: This letter, included in E&E kits for missions over China, told the Chinese that persons who wore the 5th and 13th AF insignia on their left sleeves were American airmen come to help rid them of the Japanese, just as those who wore the more familiar CBI insignia. (REB 3928) *International Blood Chit Museum*

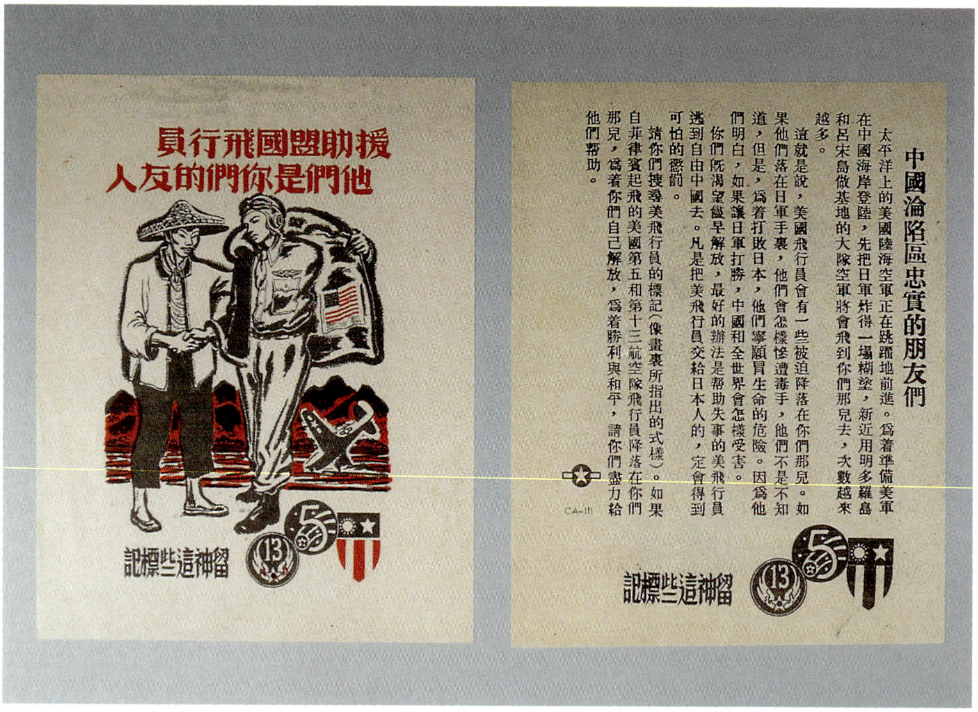

Fig. 8.24: This leaflet was included in 5th and 13th AF E&E kits for missions over China. It features an airman showing his Blood Chit to a Chinese farmer, and tells the Chinese that the bearer is an American flying from bases in the Philippines. To the left of the 13th AF, 5th AF and CBI insignia is the admonition, "Pay attention to these patches." (REB 3930) *International Blood Chit Museum*

chit was issued by authorities in Washington, D.C., the letter "B" may denote that this chit was issued by the authorities in Brisbane, Australia, where headquarters of MIS-X SWPA Section was located.

Two variations of the type 4 chit are illustrated in Figs. 7.39 and 7.40 in Chapter 7. They are identical except for the addition of the notation "№" in front of the serial number on one version. Serial numbers are printed in red at the bottom of the chits, and the message is given in French, Thai, Lao, Annamese (Vietnamese), Chinese, Korean and Japanese. As previously noted in Chapter 7, the inclusion of a message in Japanese on the type 4 chit was extremely important to air crews forced down in areas such as Korea or the Marshall Islands, where Japanese was the official language and the speaking of native tongues had been discouraged since 1910 and 1914 respectively.

The issue of MIS-X chits to Naval and Marine aviators is well documented. Lt. Cdr. H. Paul Brehm, who received the Navy Cross for the sinking of the battleship/carrier *Hyuga* on 24 July 1945, received his MIS-X type 2 chit, numbered W 47235, during the summer of 1945 aboard *USS Ticonderoga* where he served with Airgroup 87, Bombing 87.[45] Marine Aviator Roger Perreault flew Curtiss Helldivers with VMSB-244, Marine Air Group 24, 1st Marine Air Wing, in the Pacific, as well as F-4U Corsairs as high cover for dive bombers and for ground support missions. He was issued his MIS-X type 4 chit, № 62555, "in 1944 while I was attached to VMSB-244. We were on the island of Mindanao on a dirt strip called Malabang. As I'm sure you are aware, the islands of the Pacific are held by many different nations and we never know where we might be shot down, thus all the different languages."[46]

Evasion aids produced in the U.S. were also issued to Army Air Force crews operating in the Pacific. Some of these aids, illustrated in Fig. 8.21, include MIS-X Blood Chits, cloth maps, Chinese currency, an Army Air Forces survival booklet, and three special paper items. Robert S. McCarter, who flew P-51 Mustangs with the 341st Fighter Squadron, 348th Group, 5th AF, still has his evasion kit containing many items identical to those shown. He writes, "The Escape Kit was only carried on long missions and was given back to the personal equipment officer afterwards. I first used the kit in early 1945 when we were based in Luzon. This was for missions to Hong Kong or to Formosa. These were long missions lasting over seven hours. The last time I used the kit was when we were based on Ie Shima, a little island off Okinawa, near the end of the war. This was for a mission over Shanghai."

McCarter provides details of the kit's contents: "The kit contained a silk Blood Chit with the American flag and another silk chit with the Chinese flag. There were three paper items: one is a picture of a downed flier facing a Chinese coolie and showing his open flight jacket with the chit of the American Flag; another shows the insignia of the fifth, thirteenth, and CBI air forces surrounding a paragraph written in Chinese; the third is a typewritten sheet of ten questions in English and Chinese. These items were folded and placed into a clear plastic packet. In addition to the above, my packet contained two cloth maps of the area we were to cover . . . there was some Chinese paper money included. With all of this, the packet became quite bulky and too large to carry in a pocket. We mostly wore a summer khaki shirt and trousers rather than a flight suit. I carried the kit in my B-2 Survival Kit. The B-2 replaced the back pad of the seat type parachute."[47] Note that "silk" was the term in general use during World War II for the rayon-acetate fabric on which most U.S. Blood Chits and cloth charts were printed.

The currency shown is unusual as it is valued as "10 dollars", not Chinese national yuan (CN). The bill was issued by the Central

Bank of China in Shanghai in 1928, and may be valued in this manner because of the American and other settlements there before the war. The banknote in McCarter's kit is also a Central Bank of China 10 dollar note, dated 1923.[48]

The three paper language aids from the 5th and 13th Air Force evasion kits are detailed in Figs. 8.22 through 8.24. Shown in Fig. 8.22 is an example of the simple pointie talkie printed on one side of a single sheet of paper that was included in 5th and 13th AF evasion kits. It has only ten questions and three possible answers, and was apparently intended for use on the eastern Coast of China, an area where European oil and other interests were located before the war.

The letter illustrated in Fig. 8.23 is intended to familiarize the Chinese with insignia worn by American air crews, and to encourage them to aid anyone wearing these insignia. By this time the CBI insignia had become recognizable to the Chinese as symbols of the Americans, and the intent of this document was to familiarize the Chinese with the 5th and 13th AF patches as well. The translation is as follows:[49]

To All The Chinese Friends:

The people who wear these patches on their left sleeve are your friends. They are the American pilots, here to help you drive the comtemptible Japanese bandits out of China.

You may have seen this patch [the CBI patch] before. American pilots who fought for Generalissimo Chiang during the past 5 years wore this patch. Other American Air Force pilots are here to assist your honorable country in its war of resistance. They are wearing the two patches shown below on their left sleeve. Please generously assist the foreigners wearing any of these three patches. They are your friends.

All your expenses will be paid by the government of the United States.

Commander-In-Chief
American Forces

The leaflet shown in Fig. 8.24, was included in evasion kits of the 5th and 13th Air Forces. It was unusual for aerial leaflets to be

Fig. 8.25: A packet of five Army Air Force Life Raft Charts is shown here, displayed on the charts for the South Pacific Ocean. The charts are printed on a plastic material, and the packet is contained in a brown paper cover with "AAF Life Raft Charts" printed on it. Like the Navy's drift charts, these show summer conditions on one side and winter conditions on the other. *International Blood Chit Museum*

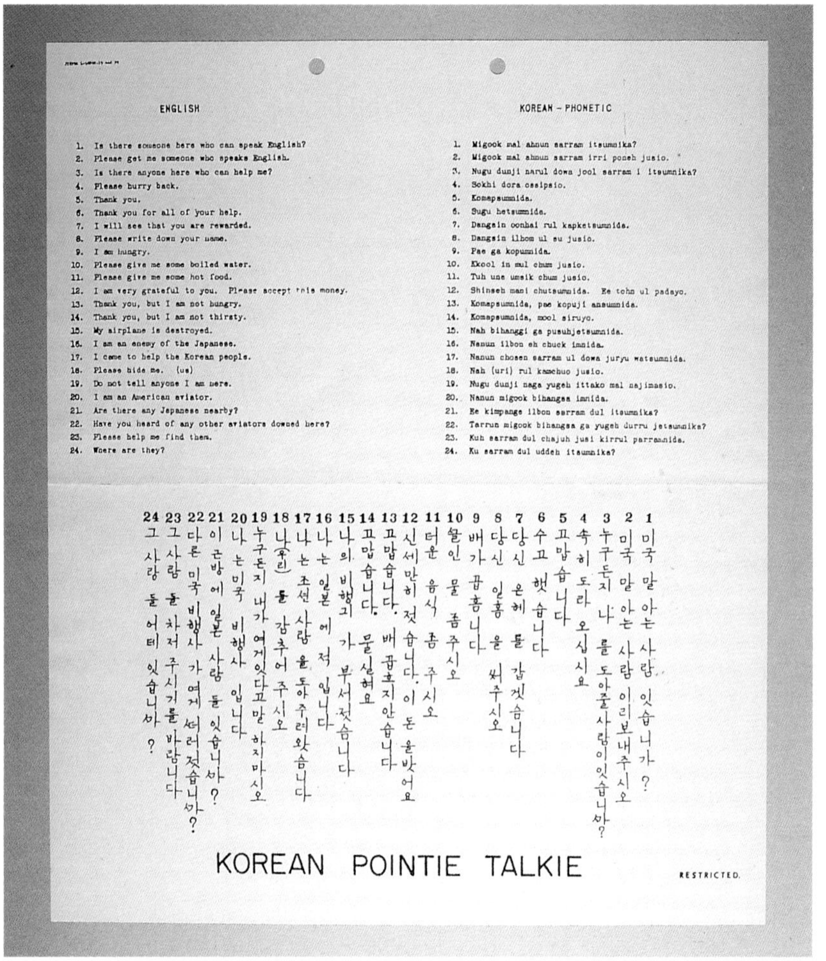

	ENGLISH		KOREAN - PHONETIC
1.	Is there someone here who can speak English?	1.	Migook mal ahnun sarram itsumnika?
2.	Please get me someone who speaks English.	2.	Migook mal ahnun sarram irri poneh jusio.
3.	Is there anyone here who can help me?	3.	Nugu dunji narul dowa jool sarram i itsumnika?
4.	Please hurry back.	4.	Sokhi dora ossipsio.
5.	Thank you.	5.	Komapsumnida.
6.	Thank you for all of your help.	6.	Sugu hetsumnida.
7.	I will see that you are rewarded.	7.	Dangsin oonhai rul kapketsumnida.
8.	Please write down your name.	8.	Dangsin ilhom ul su jusio.
9.	I am hungry.	9.	Pae ga kopumnida.
10.	Please give me some boiled water.	10.	Kkool in mul chum jusio.
11.	Please give me some hot food.	11.	Tuh une umsik chum jusio.
12.	I am very grateful to you. Please accept this money.	12.	Shinseh mani chutsumnida. Ee tohn ul padayo.
13.	Thank you, but I am not hungry.	13.	Komapsumnida, pae kopuji ansumnida.
14.	Thank you, but I am not thirsty.	14.	Komapsumnida, mool siruyo.
15.	My airplane is destroyed.	15.	Nah bihanggi ga pusuhjetsumnida.
16.	I am an enemy of the Japanese.	16.	Nanun ilbon eh chuck imnida.
17.	I came to help the Korean people.	17.	Nanun chosen sarram ul dowa juryu watsumnida.
18.	Please hide me. (us)	18.	Nah (uri) rul kamchuo jusio.
19.	Do not tell anyone I am here.	19.	Nugu dunji naga yugeh ittako mal najimasio.
20.	I am an American aviator.	20.	Nanun migook bihangsa imnida.
21.	Are there any Japanese nearby?	21.	Ee kimpange ilbon sarram dul itsumnika?
22.	Have you heard of any other aviators downed here?	22.	Tarrun migook bihangsa ga yugeh durru jetsumnika?
23.	Please help me find them.	23.	Kuh sarram dul chajuh jusi kirrul parramnida.
24.	Where are they?	24.	Ku sarram dul uddeh itsumnika?

KOREAN POINTIE TALKIE

RESTRICTED

Fig. 8.26: This pointie talkie from JICPOA appears in CINCPAC-CINCPOA E&E Supplement No. 15, issued on 5 August 1945, too late for its use in combat since the invasion of Korea, like that of the Japanese home islands, became unnecessary. This pointie talkie was later copied, with only a few changes, for use during the Korean War. (REB 3932) *From Evasion and Escape Supplement No. 15, CINCPAC-CINCPOA, 5 August 1945, SECRET*

included in evasion kits, but this one was. It shows an airman displaying a Blood Chit to a Chinese farmer, and bears the insignia of both the 5th and 13th Air Forces. The front and reverse of two original leaflets are illustrated.

The text at the top of the front of the leaflet reads, "Help the Allied aviators, they are your friends." The small text to the left of the insignia at the bottom of both the front and back reads, "Pay attention to these patches." The back of the leaflet reads:

> To our loyal friends in the Chinese occupied territory: Currently in the Pacific Ocean, the Army, Navy and Air Force of the United States of America are rapidly hopping and jumping forward. In preparation for the American forces to land on the beaches in China, the Japanese forces must be bombed and destroyed. Lately, sorties from air bases such as those on Mindanao and Luzon Islands have reached your area. These sorties will increase day by day.
>
> Therefore, some of these American Flyers will be forced down and land in your area. If they end up in the hands of the Japanese, they are definitely going to face torture and execution. Even though these dangers exist, these flyers risk their lives because they wish to defeat the Japanese. They know that if the Japanese win this war, China and the rest of the world will surely suffer.
>
> Since all of you are expecting an early liberation, the best way is to help these American flyers who have been forced down to escape to Free China. For those of you who turn them in to the Japanese, you will surely receive severe retribution and punishment.
>
> You are to search for and locate these patches (just like the designs shown here) on the American flyers. If you find the patches of the 5th and 13th air groups, they must be of the flyers who took off from the Philippine Islands.
>
> For your liberation, victory and peace, you should try your best to help them.[50]

The Army Air Force issued its own version of drift charts, known variously as Pilot Charts and Life Raft Charts. These were printed on a plastic material much like the organisol that would later be used for EVC charts during the Vietnam conflict. The charts were issued in a packet of five in a brown paper cover with "AAF Life Raft Charts" printed on it, and an example is displayed on a chart of the South Pacific Ocean in Fig. 8.25.

The issue of Blood Chits to AAF personnel is well documented. AAF veteran, T/Sgt Eldon Ray Ford, a Radio Operator/Waist Gunner on an 18-gun B-25J with the 47th Bomb Squadron, 41st Bomb Group, indicates that he was issued the two MIS-X chits on Okinawa. He received his type 2 chit, number W 24018, on 1 July 1945, and his type 4 chit on 17 July 1945. Ford adds that the squadron was the only AAF unit in the Pacific that used torpedoes in combat.[51]

The Mexican Air Force also flew missions in the Pacific from bases in the Philippine Islands (PI). The Escuadron Aereo de Pelea 201, Fuerze Area Expedicionaria Mexicana (201st Fighter Squadron of the Mexican Expeditionary Air Force), was attached to the 58th Fighter Group of the V (Fifth) Fighter Command, and received their orientation between 7 May and 12 June 1945. On 8 May the 201st was scheduled to receive Air Sea Rescue training from a Captain R. C. Suehr, and E&E training from a Captain F. Futz.[52] Santiago A. Flores writes that MIS-X Blood Chits "were issued to some of the pilots that participated in the Long Range fighter sweeps and later a dive-bombing mission to the Island of Formosa, in July and August 1945. I saw one of them in the home of the late Gra. de Div Fapadema Fernando Hernandez Vega, which was the standard Blood Chit with the American Flag . . . Mexican pilots, if they could not return back to the P.I., they could fly directly to Mainland China, and try to contact the Nationalist or Communist Chinese for help back to the Allied lines."[53] Readers wishing further information regarding Mexican Air Force operations in the Pacific may wish to access File No. K110.40301-1 and other microfilm records at the Albert F. Simpson Historical Research Center at Maxwell Air Force Base.

As in other theaters, chits in the Pacific were worn or carried in various ways. Naval aviator Mark A. Hardisty writes, "The China Blood Chit was sewn on the back of my Navy flight jacket and the French Indo China Blood Chit was carried in my flight suit pocket."[54] The two Blood Chits referred to were the MIS-X types 2 and 4 respectively.

Former Marine Aviator Roger Perreault relates that, "Originally Blood Chits were sewn to the outside of the intermediate leather flight jacket or to the outside of the flight coveralls. They soon became tattered and dirty from chaffing against the parachute harness. Some pilots had them sewn inside. They were placed high in the middle of the back. I never had mine sewn at all. I carried them folded in my jacket or coverall pockets . . . I just transferred them from pocket to pocket into whatever I was wearing for a particular flight"[55]

Fig. 8.27: Material was developed by various branches of the services from several nations to introduce airmen to the languages, peoples and customs in SWPA. The materials illustrated are in the three major languages used for communications between the natives of the area and non-natives: Malay, Pidgin English, and Japanese. Details of two of the most useful booklets, the *Melanesian Pidgin English Language Guide* and the Pidgen section of the *Pocket Vocabulary*, are shown in Fig. 8.28. *International Blood Chit Museum*

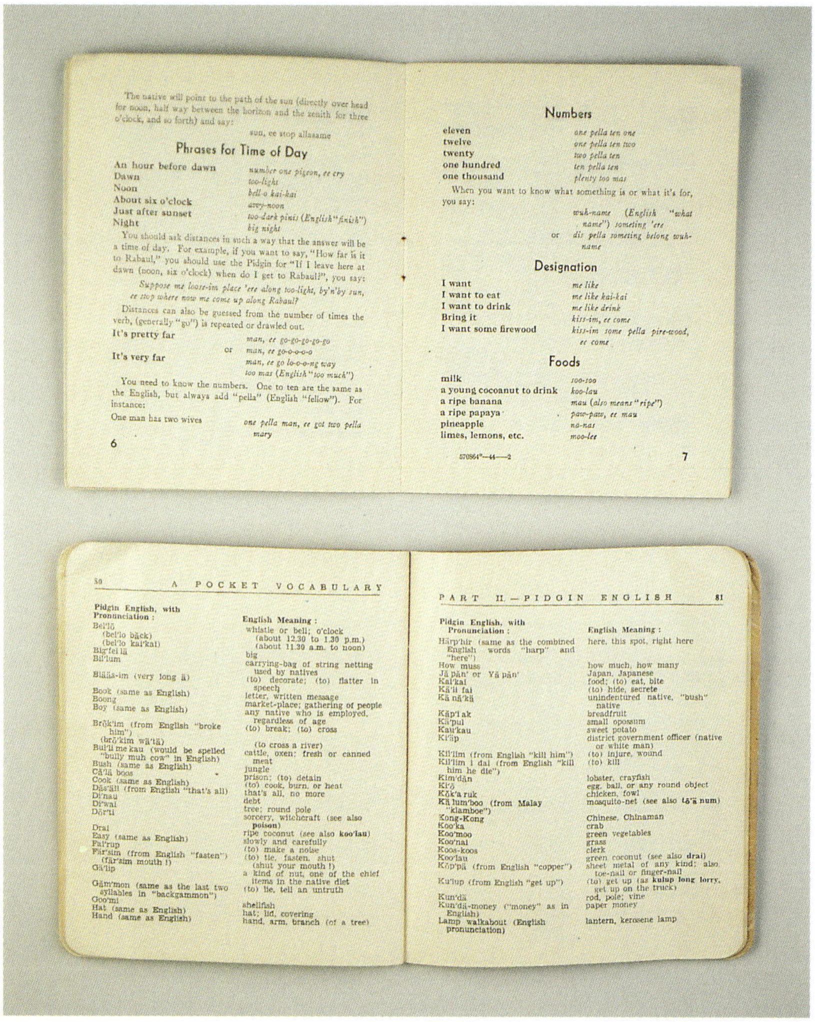

Fig. 8.28: Some of the Pidgin English language material used by servicemen in the SWPA is shown here. At the top are pages from the *Melanesian Pidgin English Language Guide*, which was prepared by the Army Education Branch, Morale Services Division, Army Service Forces for issue to Naval personnel. Althongh a teaching tool, this booklet could have been useful in the field as well. At the bottom of the photo are pages from the *Pocket Vocabulary of Malay, Pidgin English, and Japanese Phrases*, prepared by Army G-2 in the SWPA. It was designed for ease of use in the field, and contained a wealth of language information. *International Blood Chit Museum*

Army Air Force veteran Eldon Ray Ford indicated that "I never saw anyone wear a chit. We were advised to keep them neatly folded and kept in a knee pocket of our flying suit which could be securely buttoned." His type 2 chit "was kept in the plastic pocket that contained the Chinese money. The money had to be turned in immediately after each strike during the debriefing session."[56] The "plastic pocket" referred to is a roll-up waterproof pouch, part number SB209, that was the standard American issue item used in the Pacific for carrying evasion kits, maps and money.

Maj. G. L. Ferguson, USMC (Ret) considers his chits unused because they were never sewn to his jacket. He writes, "I donated two Pacific Theater rescue patches WWII . . . to the Museum at MCAS El Toro . . . They were unused (never sewn on jacket) . . ."[57]

In the Pacific, as elsewhere, Blood Chits were issued by serial number. An untitled Navy manual instructs that "Upon issue to a crew member, a record in quadruplicate, of his name, rank, serial number and organization, together with the register number of the chit assigned to him, will be forwarded to CinCPAC-CinCPOA ADVON, Advance Intelligence Center, Att'n: MIS-X Section, who will forward three (3) copies to the proper agencies in China."[58] The "proper agencies in China" was Naval Group China, headed by Admiral Milton Miles, discussed in Chapter 7. The records were thus on hand in China in the event that an aviator based in the POA went down there. These records still exist today, in the form of nearly 1,500 onion-skin papers on file in the National Archives in Washington, D.C. at "RG-38, Admiral Miles COMNAV Group, China, Box 37, Folders 35.10".[59] Families of Naval aviators who flew over China during the war may wish to access these records. Examples of the format used in keeping records the unit level and at MIS-X HQ for Blood Chits issued by serial number are shown in Appendix IV and V.

One final evasion aid was produced in the Pacific at the very end of the war. *Evasion and Escape Supplement No. 15* issued by CINCPAC-CINCPOA on 5 August 1945 includes a single-sheet Pointie Talkie that was prepared by JICPOA for use in Korea. Although produced too late in the war to be actually used by air crews, the Pointie Talkie, shown in Fig. 8.26, was later copied with only a few changes and used during the Korean War.

In addition to language aids, other material was developed to acquaint servicemen with the peoples and languages of the Pacific area. Much of this was similar to that produced for other theaters, but some of it was unique to the Pacific. Shown in Fig. 8.27 are

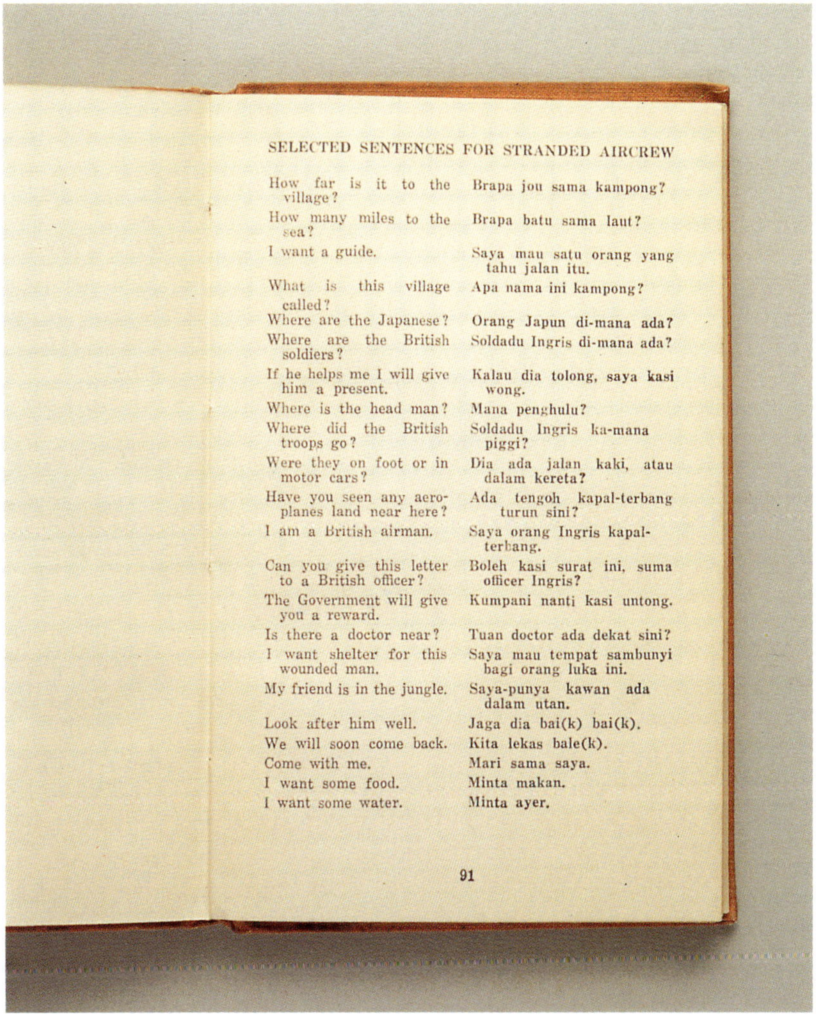

SELECTED SENTENCES FOR STRANDED AIRCREW

How far is it to the village?	Brapa jou sama kampong?
How many miles to the sea?	Brapa batu sama laut?
I want a guide.	Saya mau satu orang yang tahu jalan itu.
What is this village called?	Apa nama ini kampong?
Where are the Japanese?	Orang Japun di-mana ada?
Where are the British soldiers?	Soldadu Ingris di-mana ada?
If he helps me I will give him a present.	Kalau dia tolong, saya kasi wong.
Where is the head man?	Mana penghulu?
Where did the British troops go?	Soldadu Ingris ka-mana piggi?
Were they on foot or in motor cars?	Dia ada jalan kaki, atau dalam kereta?
Have you seen any aeroplanes land near here?	Ada tengoh kapal-terbang turun sini?
I am a British airman.	Saya orang Ingris kapal-terbang.
Can you give this letter to a British officer?	Boleh kasi surat ini, suma officer Ingris?
The Government will give you a reward.	Kumpani nanti kasi untong.
Is there a doctor near?	Tuan doctor ada dekat sini?
I want shelter for this wounded man.	Saya mau tempat sambunyi bagi orang luka ini.
My friend is in the jungle.	Saya-punya kawan ada dalam utan.
Look after him well.	Jaga dia bai(k) bai(k).
We will soon come back.	Kita lekas bale(k).
Come with me.	Mari sama saya.
I want some food.	Minta makan.
I want some water.	Minta ayer.

91

Fig. 8.29: This page of Malay phrases for use by downed airmen is contained in the RAAF booklet *Elementary Malay*, illustrated in Fig. 8.27. The opposing page, entitled "Notes:", is blank. *International Blood Chit Museum.*

examples of some of this material, produced by the governments of Australia and the United States. Details of the two of the booklets, the *Melanesian Pidgin English Language Guide* and the *Pocket Vocabulary* are shown in Fig. 8.28. A detail of the special section for use by downed airmen contained in the RAAF booklet *Elementary Malay* is shown in Fig. 8.29.

Language material was also produced that allowed Allied troops to communicate on a very basic level with Japanese troops. These were not intended for evasion, but are included in this work as examples of another type of language aid for U.S. military personnel. These cards, shown in Figs. 8.30 and 8.31, were primarily intended for use in controlling Japanese prisoners of war.

None of the evasion stories from the Pacific located during research for this work contains information regarding the use of Blood Chits. This is not surprising, considering that Blood Chits and related language aids played only a small part in an evasion and rescue. Moreover, the simple use of a chit pales in memory in comparison to the complex anxieties faced and the formidable dangers overcome during evasion, and this is perhaps especially so when remembering back more than 50 years.

One correspondent, however, did relate his knowledge regarding the success of chits in the Pacific. Marine Aviator Roger Perreault of VMSB-244, Marine Air Group 24, 1st Marine Air Wing, writes, "We lost a number of men and planes, with some being returned to us via Blood Chits. All in all the Blood Chit worked very well and many men were returned from the Pacific Islands and China. Be aware that the Island peoples and the Chinese peasants are extremely poor by our standards. Aside from patriotism, the fact that each chit said that the person who returned a downed pilot in good health would be paid well by the U.S. government made it work."[60]

The exact number of evaders who returned to friendly control is not known. The best analysis of available information from both MI9 and MIS-X appears in Appendix I of *MI9, Escape and Evasion 1939-1945* by M.R.D. Foot and J.M. Langley. The authors indicate that MI9 estimated as of 30 June 1945 that some 10,350 Allied personnel had successfully evaded capture through their efforts and the efforts of other British E&E agencies such as BAAG. This figure does not include evaders from the Pacific, and does not include evaders recovered by American E&E agencies. They also caution readers that American and British estimates do not always

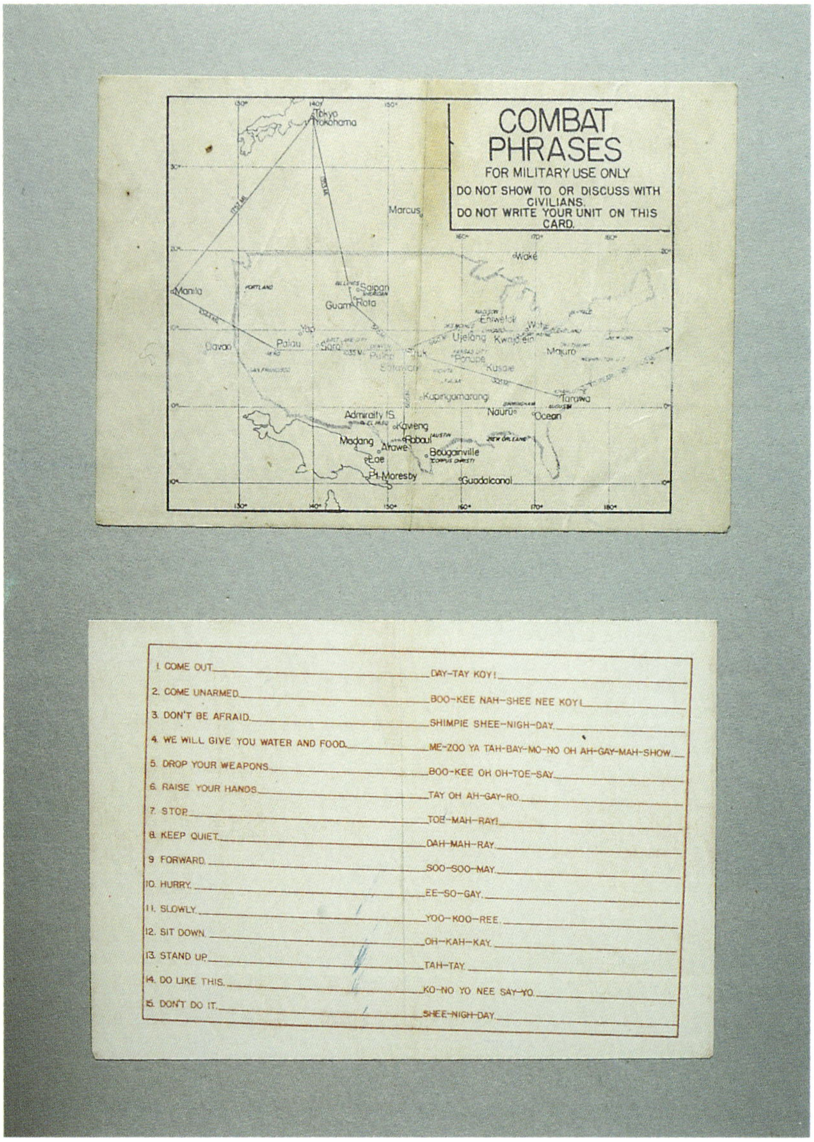

Fig. 8.30: This two-sided Phrase Card, not intended for evasion, gives phrases that might be used for communicating with surrendering Japanese soldiers and for controlling POWs. The front of the card shows a map of the Pacific with a map of the United States to show the relative size of the theater, and the back gives the phrases in English and phonetic Japanese. (REB 3934) *International Blood Chit Museum*

support each other, and that information given by MIS-X does not give separate figures for evaders and escapers. In the final analysis, Foot and Langley estimate that a total of nearly 35,200 Allied evaders and escapers were recovered by the various E&E agencies during the course of World War II. For an endeavor where success was measured in human lives, this was no small accomplishment.

America's commitment to its airmen during World War II is exemplified by the heroic efforts of the U.S. Navy in the Pacific, which often risked aircraft and submarines, together with their crews, to rescue a single downed pilot. These rescue operations were the foundation upon which America's present rescue efforts are based. Modern efforts routinely employ large numbers of fixed and rotary wing aircraft, utilize huge amounts of fuel, supplies, and ammunition, and often operate in severe danger when attempting to recover downed airmen, Rescuers have been injured and even killed in their efforts to retrieve a fallen comrade, and those who serve today in this role "That Others May Live" are among the most highly respected members of America's armed forces.

While the tradition of making extraordinary efforts to rescue downed airmen survived the hasty and haphazard demobilization that followed VJ Day, the American evasion and escape organization MIS-X, whose unheralded yet heroic efforts resulted in the recovery of uncounted Allied airmen, did not. In the myopia brought by the exhilaration of victory, all known records and artifacts of MIS-X on hand at its headquarters at Fort Hunt were burned, and the buildings that housed the agency were razed. Barely three years later Col. E.W. Ridings, chief of the Air Force Intelligence Training Command, could find no useful information about MIS-X when attempting to establish a similar organization.[61] The Evasion and Escape Section of the Directorate of Intelligence, USAF, had to be invented from scratch.

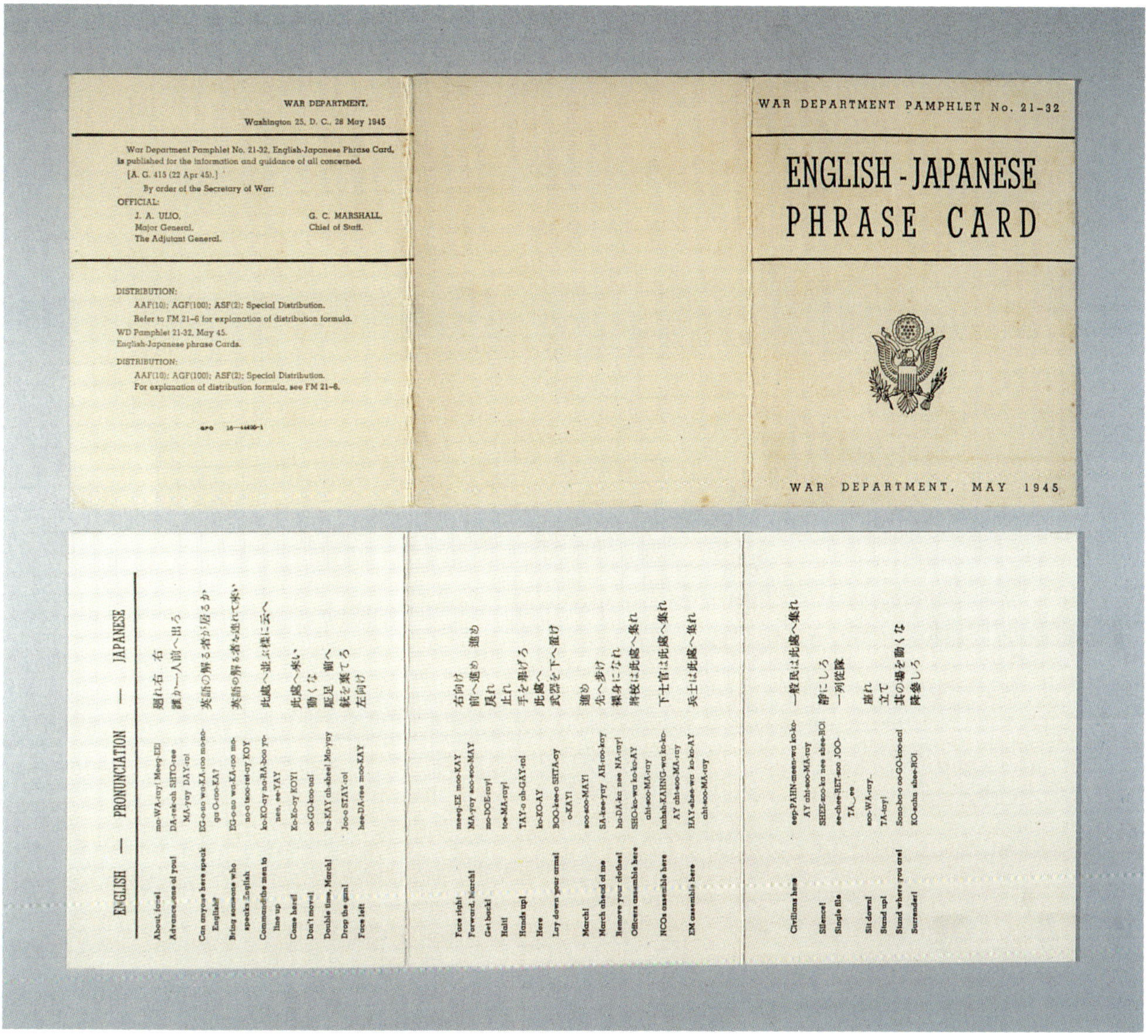

WAR DEPARTMENT,
Washington 25, D. C., 28 May 1945

War Department Pamphlet No. 21-32, English-Japanese Phrase Card, is published for the information and guidance of all concerned.

[A. G. 415 (22 Apr 45).]

By order of the Secretary of War:

OFFICIAL:

J. A. ULIO,
Major General,
The Adjutant General.

G. C. MARSHALL,
Chief of Staff.

DISTRIBUTION:

AAF(10); AGF(100); ASF(2); Special Distribution.

Refer to FM 21-6 for explanation of distribution formula.

WD Pamphlet 21-32, May 45.
English-Japanese phrase Cards.

DISTRIBUTION:

AAF(10); AGF(100); ASF(2); Special Distribution.
For explanation of distribution formula, see FM 21-6.

WAR DEPARTMENT PAMPHLET No. 21-32

ENGLISH - JAPANESE
PHRASE CARD

WAR DEPARTMENT, MAY 1945

ENGLISH — PRONUNCIATION — JAPANESE

Fig. 8.31: Another Phrase Card with phrases useful for dealing with surrendering Japanese or POWs is shown here. This was printed in 1944 by the Government Printing Office (GPO). (REB 3936) *International Blood Chit Museum*

Despite organizational difficulties following World War II, new generations of military airmen would routinely carry Blood Chits, updated versions of the simple printed messages introduced by the Royal Air Force in early 1917, in the skies over Korea, Egypt, Aden, Vietnam, the Middle East, Bosnia, and elsewhere. Perhaps nothing more clearly illustrates the value of Blood Chits than the fact that, despite the vast changes in the world and in warfare during the three-quarters of a century since their genesis, these simple devices continue to perform a vital role in the continuing British and Ameri-can efforts to ensure the safe return of their aerial warriors.

Additional Photos

The artifacts illustrated in this section are from the personal collection of the late John A. Scott. At the time this book was being compiled the artifacts were on display at the Oregon Military Museum, and were not available for photography. The display was dismantled just prior to the publication of this book, and this special section has been added to accomodate the inclusion of photos of these important artifacts.

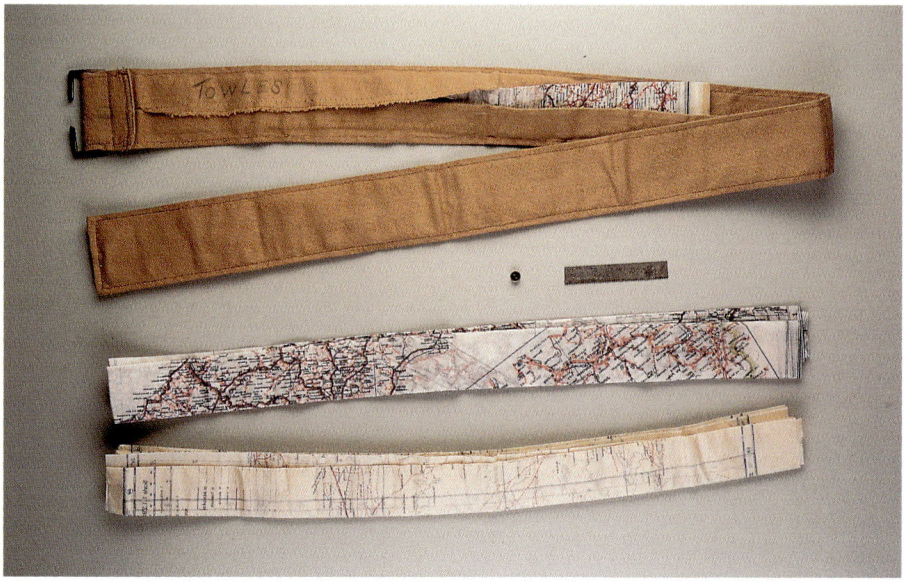

Fig. ADD.1: This RAF flying suit belt was identified by John Scott as having been fabricated in the North African theater. Sewn into the belt at the time of its fabrication were three Tissue Maps, a small compass, and a hacksaw. The maps are J3 and J4 (Northern and Southern Italy) and an un-numbered map of the north coast of eastern Libia and western Egypt, centered around Tobruk. The hacksaw, unlike those included in Evasion Purses, is square on both ends and has teeth on only one side. The tiny compass is illustrated in Fig. ADD2. The airman would have used his evasion kit, not the contents of the belt, during his initial evasion of capture. The real value of the belt was that its contents could have gone undetected if the airman were captured, and would have then been available for use both in POW camp and upon escape. From the maps included, it is thought that the belt was issued during the period that missions over Italy were being flown from bases in North Africa. *International Blood Chit Museum, from the John Scott Collection.*

Fig. ADD.2: Three compasses mentioned elsewhere in this book are shown here. On the left is the tiny compass that was sewn into the RAF belt illustrated in Fig. ADD.1. It contains a small disc with a white dot that points north, and a thin white line that points south. In the center of this photo is a British compass referred to as a "wee compass", or sometimes simply as a "wee". It contains a tiny disc with a red dot that points north, and is notable because it is at least as tall as its circumference. Wee compasses, mentioned in an official British document quoted in Chapter 6, were sewn into the seams of some RAF flying suits. On the right side of the photo is a German evasion compass of the type issued to Luftwaffe pilots during the Battle of Britain, discussed in Chapter 6. It has a compass rose on both the top and bottom of the inner ring. *International Blood Chit Museum, from the John Scott Collection.*

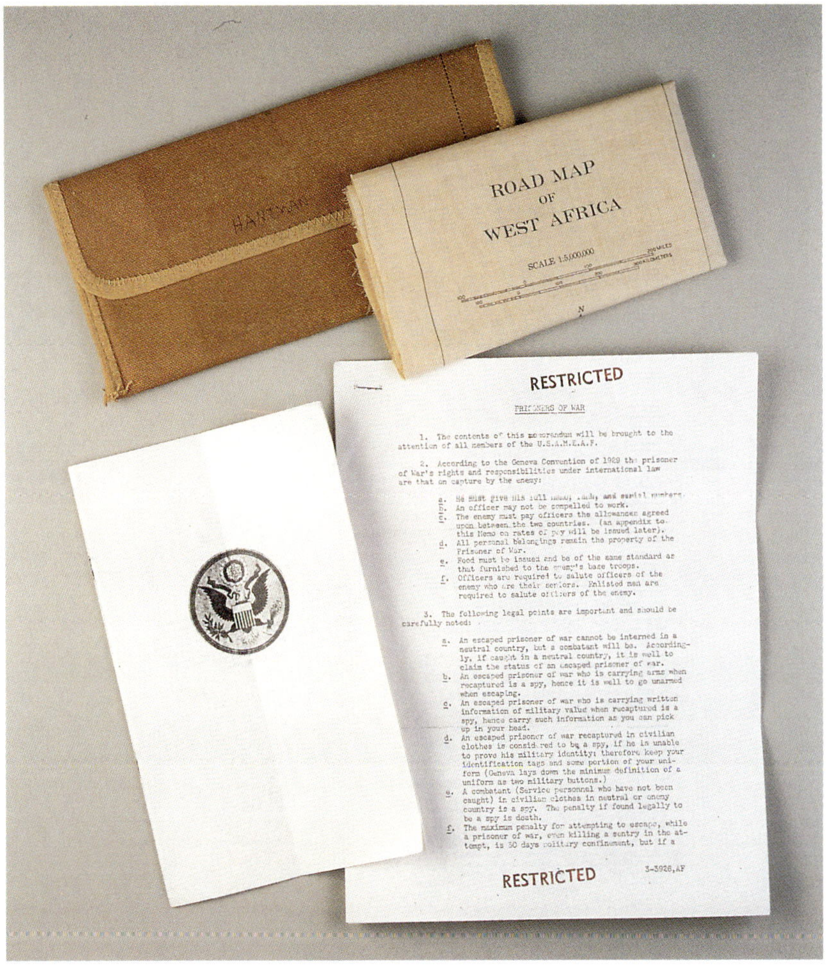

Fig. ADD.3: Shown here is what is identified by John Scott as an example of the first American evasion kit. Prepared for use in North Africa, the cloth pouch contains a booklet of Blood Chits in four languages used in West Africa, also illustrated in Figs. 5.16 and 5.17, a balloon-cloth road map of North Africa, also shown in Fig. 5.18, and capture instructions entitled "Prisoners of War". The booklet of chits is folded once vertically, and the other papers are folded in thirds. The kit may have also contained gold seal currency. *International Blood Chit Museum, from the John Scott Collection.*

Fig. ADD.4: Three photos of an American airman in civilian clothes are illustrated here. Photos like these, discussed in Chapter 6, were carried by airmen in the European Theater. Should the airman have been forced down, the photos would have been used by resistance or partisan organizations in forging identity papers for him. *International Blood Chit Museum, from the John Scott Collection.*

Appendix I:

COPY

TRANSLATION

ORDER GOVERNING REWARDS AND PUNISHMENTS FOR RESCUE CHINESE AND ALLIED FLIERS
YU (諭) #8437 MAY 23, 1945

1. In order to encourage the people's hatred of our enemy, to promote an allied fighting spirit, and to rescue downed Chinese and Allied flyers, we have decided upon the following regulation:

2. Anyone who rescues a flier in the occupied area and escorts him to safety will be given one hundred thousand dollars reward (Chinese currency).
The reward should be approved by the first organization (air and its branches, local government or unit) which receives the escort, and then forwarded to the Aeronautical Commission for settlement.
Should the escort go through one or more occupied areas, the first organization to receive the escort should plan very carefully for the transfer to the next. All expenses incurred will be repaid to an organization on the endorsement of the man rescued.

3. If the rescuer is a unit, organization or social group, a reward of honor will be given, but according to the will of the said unit, organization or social group, it can be changed into a reward in money.

4. If the rescuer is not willing to accept the reward in money, he may appeal for other rewards of honor.

5. If a rescuer because of his efforts in rescue meets with any harm from the Japanese or puppet troops, in addition to being given a reward according to the regulation, he will also be given a pension by the Relief Commission according to his qualifications.

6. If a rescuer escorting a flier to the rear cannot return to his home, in addition to being given a reward according to this regulation, he will be given an appropriate job according to his will and ability. During the time he is waiting for orders he will be taken care of by the air force organization in the vicinity.

7. Any organization (air and its branches, local government or unit) after receiving a downed flier, which does not do its best to escort him to safety, will be punished severely according to the law.

8. Any individual who does harm to downed fliers in the occupied area or hands them over to the enemy will be considered a traitor and will be punished severely according to the law after our territory is restored.

9. This regulation will be carried into effect on the day it is proclaimed.

Commission of Military Affairs

(CHOP) CHIANG KAI SHEK

AGAS/091400/873.
(to accompany AGAS/05100/840/--Clarke's 74/51--evidently detached therefrom).

Appendix II:

The Production of Cloth Charts and Blood Chits in the United States During World War II

Fig. A2.1: Thousands of blood chits were stamped with the chop of the Chinese ambassador by MIS-X personnel under armed guard at the Chinese embassy in Washington, D.C. The signatures of two of the men who performed this duty, Lloyd R. Shoemaker and Carl N. Petersen, appear on one of the chits shown here. *International Blood Chit Museum*

While the essential purpose and messages of Blood Chits have changed little since the paper chits of the fledgling Royal Air Force, the material on which they are produced underwent several changes, particularly during World War II. Blood Chits, like cloth evasion maps and drift charts, were utilitarian objects, designed to be carried as an essential item of air crew combat equipment, not as future museum pieces. They were made from materials which were deemed best able to withstand rugged use under difficult conditions by a downed airman.

During World War II, and for some time afterward, the material of choice was cloth. Particularly when compared to paper, cloth is durable, relatively water-resistant, treatable to resist mold and fungus, and the printing will not be damaged or destroyed when it is folded. Cloth maps have been used for military purposes since before the Christian era, and their use before World War II led directly to their consideration for use as evasion aids. One authoritative monograph on the cloth maps of World War II, written by Barbara A. Bond, a ranking member of the Ministry of Defence (MOD) Mapping and Charting Establishment, reveals:

> The excavation in 1973 of the Han dynasty tomb at Mawangdui, Chjangsha, in China unearthed 3 maps drawn on silk. The subsequent restoration of these maps showed that one of them, now known as the Garrison Map, was produced for military purposes and dates from before 168 B.C. The map is in 3 colours at a scale of approximately 1:90,000. It is considered to be the classic military map showing, as it does, the position and designation of troops and military units, command posts, logistics information, immigrant settlement, the number and disposition of local households, frontier roads, and distance indicators. The map appears to show a state of combat readiness in Changsha state against expected invasion from Nanyue state. Thus, one of the oldest extant maps in the world is a military map on silk.
>
> In more recent times we know of the existence of military maps on silk during the American War of Independence and of the German area in the late eighteenth century. Ordinance survey, which started life as a military mapping organisation, was printing maps on silk in the nineteenth century. The War Office Committee asked to report in 1892 on the form that a military map of the United Kingdom should take, made frequent mention in its report, and throughout the minutes of evidence, to the superior durability of linen over paper as a material on which maps could "be printed for use in the field".
>
> The foldability and durability of silk and linen had important military applications over a considerable time span, particularly for cavalry use prior to World War I. These same characteristics proved to be of particular use more recently in the twentieth century, specifically for escape and evasion purposes during war.[1]

The characteristics which made cloth an ideal medium for maps also recommended it for use in the production of Blood Chits. There is no extant study on the printing of Blood Chits during World War II because this was simply a use of manufacturing and printing methods that had already been created, tested, and used to produce cloth evasion maps. A brief account of the development and methods for printing wartime cloth charts is therefore valuable.

Fig. A2.2: This chit is an excellent example of how the fragile rayon-acetate fabric deteriorated when sewn to a garment. Probably sewn to the inside of a flight jacket, it was used as a pocket as evidenced by the stitching which appears only on the sides and bottom of the chit. The inability of rayon to stand up to this sort of use was one reason that MIS-X was considering switching to cotton fabric near the end of World War II. *International Blood Chit Museum*

When MI9 was formed at the end of 1939, Clayton Hutton, the genius behind its evasion and escape aids, faced new problems in producing cloth maps for evasion use. Maps had evolved into documents featuring very fine print and thin lines, and existing methods for printing on silk were unsuitable for modern maps. The fine details of the maps would bleed into surrounding areas making the map unreadable and useless. The problem was perplexing:

> For days we tried to find a satisfactory solution to our problem, but nothing could induce the colored inks to "sit" on the silk. Then, as a last resort, I wondered if a little pectin - a gummy substance found in certain fruits and roots to which the setting power of jams is due - would serve as an agent to coagulate the ink and prevent it from running. Without any real faith in what we were doing, my friend and I prepared the new mixture and put it to the test. Our spirits soared when we gazed down at the resultant print. It was perfect down to the last detail. Even the smallest topographical feature was sharply defined.[2]

The pectin proved to be the solution to Hutton's dilemma, and evasion maps and Blood Chits could be printed on thin cloth, such as silk. A Blood Chit for use in Ethiopia, where in the early months of World War II the RAF was resisting the Italian advance, was printed on linen, but silk remained the fabric of choice for maps between 1939 and 1941. The silk used for maps was that which had been found to be unsuitable for use in manufacturing parachutes.[3]

Silk was also used in India for printing both maps and Blood Chits, and silk continued to be used in India for production of these items well after it ceased to be used in Britain. By early 1942, however, silk was in short supply and in the West and maps began to be printed on a form of rayon known at the time as Bemberg Silk.[4]

British and Australian maps were in wide-spread use well before similar items were produced by the Americans. Australian maps were printed on silk, and Australian pilots customarily wore them around their necks as scarves, making them immediately available in time of emergency.[5]

While the English and Australians were producing and issuing cloth evasion maps to their forces, American textile manufacturers and chemical companies developed a special cloth which would become the accepted standard material for the printing of cloth maps and Blood Chits in the United States. Once the new fabric was introduced, methods were developed to treat it to withstand the ill-effects of humidity, saltwater, and mildew. The official designation for this material was "cloth, map, rayon acetate", although it is commonly called "calendered taffeta". It was obtained in rolls varying in length from 200 to 250 yards and in widths varying from 42 to 45 inches. The primary manufacturers for the fabric were Beaunit Mills, Inc., Burlington Mills Corporation, and National weaving company, all located in New York.[6]

The film-positives required for printing the maps were produced from the original map "art work", one for each color, and were provided to the printers by the Aeronautical Chart Service (ACS). Although no records exist regarding the production of Blood Chits, it can be reasonably speculated that either MIS-X itself, or ACS at the request of MIS-X, provided the film-positives for Blood Chits.

The production of maps was a slow process because the complexities of printing in seven colors required that some colors be printed in several tones. Each color required a separate pass through the printing press. The maps, unlike Blood Chits, were printed on both sides of the fabric, further complicating and slowing the printing process.

The procedure for producing cloth maps was to cut the fabric to size and treat the edges chemically to keep it from unravelling, then to mount it on a paper base to hold it flat for printing. The sheets were then printed on a sheet-fed offset lithographic press, one run for each color. When one side of the map was finished, the fabric was removed from the paper base and mounted to a second paper base with the blank side up, after which the printing process was repeated for the back side.[7]

The finished maps were folded to approximately 3 x 5 inches, and the folded maps were then packed for shipping, twenty packages to a container, with each package containing twenty five folded maps wrapped in Kraft paper.[8] The containers, designed for overseas shipment, were then routed, possibly through the MIS-X warehouse at PO Box 1142, for distribution to air crews through the various E&E sections. The bulk of the cloth charts and Blood Chits distributed by MIS-X were for Asia and the Pacific, where American evasion and escape organizations had the authority for E&E efforts.

Four companies printed all of the rayon-acetate cloth charts for the American military during World War II as well as other specialized items for the OSS and Britain's War Office and Intelligence Bureau. These companies also produced Blood Chits using essentially the same production methods and the same acetate-rayon fabric. These firms were A. D. Steinbach and Sons, Inc. of New Haven, Connecticut, Kaumograph Co. of Wilmington, Delaware, Sweeney Lithograph Co.of Belleville, New Jersey, and William Kuhlman and Co. of Doyleston, Pennsylvania.[9] Of these companies, only the Kaumograph Co. remains in existence, but all of its records from the war years were disposed of long ago.[10]

An important element of the Chinese flag Blood Chits produced by MIS-X was the inclusion of the chop of the Chinese ambassador to the United States. This made the chit official in the eyes of the Chinese. The chop itself was made of stone and set in a wooden handle, and was applied by MIS-X personnel at the Nationalist Chinese embassy in Washington, D.C. The work was done under armed guard to protect the chop, a valuable seal of a Kuomintang government official. One of the Blood Chits illustrated in Fig. A2.1 bears the signatures of two of the MIS-X personnel, Lloyd R. Shoemaker and Carl N. "Norm" Petersen, who spent many hours under difficult conditions stamping thousands of Blood Chits with the Chinese ambassador's chop.

MIS-X Technical Section routinely evaluated their methods and materials to ensure that they produced the best possible product for use in the field. Particularly useful in this regard were reports and suggestions from E&E sections in the various Theaters of Operation where the chits were in use. One example of this process is illustrated by the following exchange of correspondence. In December of 1944 the following was sent from AGAS Headquarters to the MIS-X Technical Section at Fort Hunt:

> Enclosed is one of the American Identification Flags . . . prepared by Technical Section, C P and M, MIS-X Branch and shipped to AGAS-CHINA for distribution.
>
> This flag, supplied one of the officers, Headquarters, AGAS-CHINA, had been sewn inside a field jacket as recommended, and in the course of time the jacket was sent to the laundry for laundering. Upon its return, it was discovered that the ink used in the "chit" section of the flag had faded almost completely away.
>
> It is recommended by Headquarters, AGAS-CHINA that tests be made on the ink used on these flags and if it is discovered that all the ink used in the chit section does fade through wetting, that a "water-proof" ink be substituted immediately on all subsequent flags turned out for field use.[11]

MIS-X Technical Section, in concert with one of the firms that printed the chits, conducted tests on the inks themselves and on their durability with different types of fabric. Their tests indicated that the fabric was as much a consideration as the inks, as expressed in this July 1945 letter to Lt. Col. A. R. Wichtrich, HQ AGAS, from Col. Sweet, Chief of CPM Branch:

> The matter has been taken up with the concern that prints the chits. After experiments this ink was changed and it is thought that inks used at present are as waterproof and durable as can be gotten. Printing inks are used, not vat dyes as they are not suitable for this kind of work. It is reported that the fabric on which the ink is used is a factor affecting its permanency. The fibers of acetate rayon (fabric used) are smooth and hard like glass rods. They do not offer a favorable surface for ink to adhere to and do not allow ink to penetrate. Cotton fibers are rough and frayed, offering an ideal surface for ink to adhere to, and permits ink to penetrate. For these reasons, the printer thinks that cotton chits will stand up better than rayon, and submitted some sample Blood Chits printed on cotton. These samples have been treated with a filler which stiffens the cloth and makes unnecessary the backing up operation which must be performed to print on rayon. The cotton cloth has been treated so that it is fungus and mildew proof.
>
> Tests by Technical Section show that prolonged soaking (2 weeks) in 5% salt solution and 5% acetic acid (room temperature) has no pronounced effect on either cotton or rayon chits. When laundered by boiling in 1% solution GI soap for 30 minutes, the cotton chit showed no pronounced effect, but the rayon chit faded considerably, though remaining legible.[12]

Samples of both the cotton and rayon chits, printed with the same inks by the same manufacturer, were sent with the above letter to AGAS for evaluation. Following their testing, AGAS informed MIS-X that they preferred the cotton chits for several reasons:

> 1. Use of cotton for blood chits is recommended for the following reasons:
>   a. Durability: Blood chits are frequently used for pockets on the inside of jackets. It is believed cotton chits will endure, where rayon ones will ravel or tear under slight pressure.
>   b. Tests have shown that cotton chits are more nearly waterproof than rayon or silk.
>   c. It is believed that there will be less tendency to send home cotton chits as souvenirs than there has been in the case of silk.[13]

The Blood Chit shown in Fig. A2.2 is a good example how a rayon chit deteriorates when it is sewn inside a garment, exactly as described above. Although scorned by collectors, examples like the one illustrated provide important information to historians, illuminating documents like those above with specific examples. This chit was sewn on the sides and bottom, indicating that it was used as a pocket, as was often the case. The rayon acetate fabric was far too fragile for this sort of use.

It is ironic that cotton was found to be a superior fabric because the first Blood Chit produced in United States by MIS-X, in March 1944, was printed on cotton. Discontinued in favor of the new rayon-acetate fabric, cotton now seemed about to make a comeback as the material of choice for Blood Chits. This was not to be, however, for the date of this last memo was 8 August 1945. The shortsightedness of the military, which did its best to erase even the memory of MIS-X at the end of the war, insured that this information would be lost to the next generation of evasion and escape personnel, and indeed Blood Chits manufactured from 1948 through 1961 continued to be printed on rayon acetate fabric.

The tradition of producing maps and Blood Chits together, which began with four printing companies during World War II, continues today at the Defense Mapping Agency Aerospace Center (DMAAC), the modern successor to the Aeronautical Chart Service (ACS). Here Blood Chits (BLDCHIT) and evasion charts (EVC) are printed today on Tyvek, a nearly indestructible artificial material created by the DuPont Company.

Appendix III

The Payment of Blood Chit Rewards During World War II

The payment of a reward for rescue is an important factor in the success that Blood Chits achieved in securing assistance to downed airmen during World War II. Reward amounts varied from location to location, and theater to theater, as disclosed by a memorandum issued more than a decade after the war ended:

> During World War II, the following values were established in the automatic payment of blood chits . . .
>
> 1. (Western Eurasia) France, Belgium, Holland, Denmark, Norway - $50
> 2. China, Indo-China, India - $100
> 3. Philippine Islands and New Guinea - $50
> 4. Greece and North Africa - $100
>
> In explanation for the reason of varying amounts, this was occasioned by different theater commander policies since no general staff guidance existed in this field. The above amounts did not include additional payments for expenses incurred in lodging, food, and transportation of downed air crewmen.[1]

Payments were sometimes made locally and at the end of the walkout when a helper accompanied an air crew member to safety. In other instances payment of the reward was postponed until the

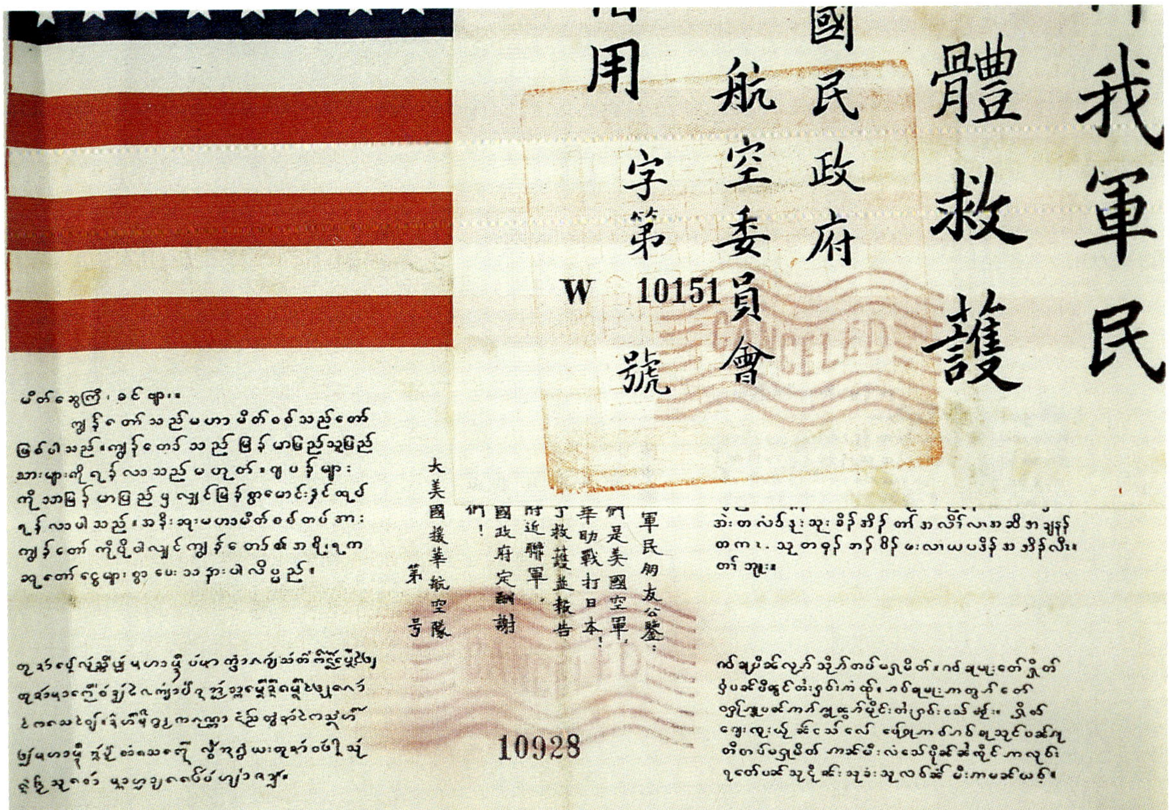

Fig. A3.1: The two chits shown here were cancelled with a special stamp for this purpose. Chits were sometimes cancelled when rewards were paid, and probably in some instances when an airman wanted to take his chit home. The ink seems not to be waterproof and has smudged somewhat on one of the chits. *International Blood Chit Museum*

enemy was defeated or driven from occupied territory, in order to make payments physically possible and to prevent retribution against helpers, their families, and their villages.

Following the end of World War II, official commissions were sent to various parts of the world to keep America's promise to reward those who had risked everything to rescue and assist its airmen:

> U.S. claims commissions were established in these areas and all valid claims were processed in conjunction with the appropriate governments. An example of this is in the Philippine area where Hukbalahups and head hunters existed in Aparri, the Legaspi Peninsula, and Northern Mindanao. The Philippine government screened and rejected any claims from these areas due to the hostile attitude of these natives to US downed air crewmen.
>
> The above is (also) applicable to the French and Benelux areas.[2]

In Europe, claims commissions were sometimes activated as the the enemy was driven from occupied countries. These commissions sometimes included members of former clandestine organizations because of their familiarity with assisting downed airmen:

> the Allied Screening Commission was activated with two main headquarters: one in Rome, the other in Athens. The commission consists of forty military personnel and twenty civilians. They are charged with the duty of rewarding or repaying helpers as the lines advance. The work is tremendous in scope. A total of 65,000 persons have aided our men . . . Clandestine organizations had to be activated from the day Italy capitulated until the Rome area was liberated. This small clandestine organization became the nucleus for the Allied Screening Commission. Experience has rendered them competent to effectively dispose of the varied claims.[3]

Fig. A3.2: Illustrated here is a blood chit that was sent home following the the war by Vernon H. Hanz. Also shown is the required paperwork, which refers to the chit as a "back flag", and the envelope in which it was sent. *International Blood Chit Museum*

Claims were screened for merit and then paid as determined by the commission. A report made in 1957 describes the functioning of a claims commission in Italy:

> The evidence, usually a blood chit [*sic*], is submitted to the Allied Military Government. The evidence is embodied and submitted in a claim to the Screening Commission. Services rendered are then determined and verified with the aid of . . . E&E Interrogation Reports. An Italian-speaking representative then holds open-house in that particular area for several days during which time people presenting claims are paid cash. Six or seven claims may be proffered on one evader. Each claim is honored.
>
> In some cases money is an insufficient reward because lives have been lost by those who aided the evader. A case record is made in meritorious cases and in due time both the British and U.S. Government still reward these people with the honors and various decorations commensurate with the services rendered.[4]

The term "Blood Chit" in the above quotation refers to a Promissory Note. The preparer of the above report was confused by the use of the word "chit" in World War II documents, which was often used in reference to Promissory Notes as well as in reference to actual Blood Chits.

In some areas when rewards were paid the serial numbers on Blood Chits were cancelled. This may also have been done when an airman rotating home wanted to keep his chit. This was done to prevent chits from being misused, or from paying a reward on it twice. Two chits which have been cancelled are shown in Fig. A3.1. A similarly cancelled chit is contained in the Special Collections Library of the U.S. Air Force Academy. The stamp was typically in purple ink, which seems to have been the color of choice in Asia, and it was applied over the area containing the serial number.

After the war ended, chits were declared "obsolete, expendable material" and could be brought home without first having the serial numbers cancelled. Fig. A3.2 shows a chit which was sent home by Vernon H. Hanz of the 67th FS, 18th FG, along with the required paperwork and envelope in which it was sent. The paperwork refers to the chit as a "back flag", one of many names by which chits were known in the China and India-Burma theaters.

The Allies' promises, printed on their Blood Chits and made by the men who carried them, were kept whenever possible, regardless of how much time may have passed following actual rescue. One of the few valid reasons for not paying a reward was to protect helpers from retribution by governments still hostile to the United States or to the other nations that participated in the conflict. Much of the information regarding payment of rewards remains classified today in order to protect those helpers who would face grave consequences should their identities be revealed. The many helpers who risked life and limb to rescue perfect strangers, whether named or unnamed, are the true heroes of the Blood Chit story.

Appendix IV:

HEADQUARTERS
51st Fighter Group, 10th U.S. Air Force
A.P.O. #629.

September 7, 1943

SUBJECT: Accountability of Chinese Personnel Identification Flags.

To : Commanding General, Hq., 10th U.S. Air Force,
A.P.O. #885. (THRU Commanding General,
Hq., A.A.C. No. 1.)

1. This headquarters has received two hundred and fifty (250) Chinese personnel identifications in August 1943. Accountability of these flags is given below.

No.	ISSUED TO:	No.	ISSUED TO:
04736	Gen. C.V. Haynes	04618	Lt. E.D. Atkins Jr.
04689	" " " "	04619	Lt. D.M. Colerich
04647	Lt. Shaper	04620	Lt. F.C. Huffman
04648	Lt. Jackson	04621	Lt. F.J. Altiere
04649	Lt. Slaughter	04622	Lt. S.J. Miller
04650	Sgt. Parish	04623	Lt. V.A. Thielhorn
04651	Sgt. Easley	04624	Lt. R.R. Santini
04652	Sgt. Moss	04625	Lt. J.R. Gring
04655	Maj. W.D. Miller	04626	Lt. V.O. Burge
04643	Maj. H.T. Wright	04627	Lt. J.J. Ferguson
04653	Capt. L.W. Goddard	04628	Lt. C.A. Whittington
04654	Capt. C.G. Goff	04629	Capt. T.B. Waldrip
04690	Col. J.F. Egan	04630	Lt. H.C. Adkins
04691	Maj. M.G. Marks*	04631	Lt. C.D. Evans
04657	" " " "	04632	Lt. J.C. Moore
04692	Maj. D.W. Wallace	04633	Lt. R.A. Coll
04602	Maj. N.W. Davis	04634	Lt. L.T. Boley
04603	Capt. E.M. Nollmeyer	04635	Lt. N. Bosich
04604	Lt. H.D. Halvey	04636	Lt. I.S. Wagner
04605	Lt. G.T. Grammes	04637	Lt. J.V. Wilke
04606	Lt. W.R. Rodgers Jr.	04638	Lt. A. Robinson Jr.
04607	Lt. L.C. Hicks Jr.	04639	Lt. J.W. Sharak
04608	Lt. W.A. Wendt.	04640	Lt. O.F. Smith Jr.
04609	Lt. E.C. Bishop Jr.	04641	Lt. J.K. Bavuso
04610	Lt. A.R. Duncan	04642	Lt. R.G. Williams
04611	Lt. L.O. Marshall	04643	Lt. J.L. Yantis Jr.
04612	Lt. E.L. Rathbun	04645	Lt. C.T. Streit
04613	Lt. J.P. Baglio	04646	Lt. A.L. Gregg
04614	Lt. D.H. Mc Donald	04658	Lt. J.J. Novak
04615	Lt. J.F. Cooman	04659	Lt. R.H. Bixby
04616	Lt. B.H. Ashmore	04800	Lt. J.D. Wooley
04617	Lt. V. Ellifrits		

No.	ISSUED TO:	No.	ISSUED TO:
04801	Lt. W.C. Hall	04827	Lt. R.D. Mc Reynolds
04802	Lt. T.M. Tracy	04828	Lt. D.F. Taylor
04803	Lt. J.R. Thorn	04829	Lt. M.G. Hopper
04804	Lt. R.N. Echols	04830	Lt. J.T. Irwin
04805	Lt. R.J. Bell	04831	Lt. A.D. Gaston
04806	Lt. P.S. Royer	04832	Lt. W.D. Livergood
04807	Capt. E.J. Harrington	04833	Lt. B.A. Stahl
04808	Maj. F.C. Dres	04835	Capt. I.M. Susky
04809	Lt. R.F. Cole	04836	Lt. D.E. Humbaugh
04810	Lt. H.T. Mc Coy	04837	Lt. Kenetasar
04811	Lt. K.C. Hoylman	04838	Lt. J.A. Larson
04812	Lt. J.G. Tew	04839	Lt. W.A. Southwell
04813	Lt. W.E. Burtram	04840	Lt. F.C. Forbes
04814	Lt. R.A. Mc Clung	04841	Lt. L.P. Mc Intosh
04815	Lt. E.J. Livesay	04842	Lt. T.S. Clark
04816	Lt. C. Kly	04843	Lt. R.E. Rhodes
04817	Lt. J.F. Keith	04844	Lt. D.R. Jones
04818	Lt. B.L.W. Mc Quillen	04845	Lt. C.C. Mitchell
04819	Lt. F.E. Moffitt	04846	Lt. R. Brateher
04820	Lt. R.A. Conden		
04821	Lt. S. Dare		
04822	Lt. L.S. Earnest		
04824	Lt. J.A. Hartswick		
04825	Lt. S.C. Kelley III		
04826	Lt. W.C. Orr		

2. Two Chinese flags, No's 04822 and 04834 are reported missing by the 25th Fighter Squadron.

3. The following flag serial numbers are being retained at this headquarters for future issue to flight personnel; no's are inclusive:

04659 to 04688	04767 to 04799
04693 to 04735	04847 to 04851

* Major Marks is now assigned to K.O.T.U.

JOHN F. EGAN
Colonel, Air Corps,
Commanding

424. 1st Ind. G/i-1
HEADQUARTERS AMERICAN AIR COMMAND NUMBER ONE, APO 629. 8 September 1943

TO: Commanding General, Tenth Air Force, U.S. Army, APO 885

W.A.W.

Appendix V:

RECAPITULATION OF CHINESE BLOOD CHITS

MIS-X SECTION, GHQ

PART A

TABLE VI — Date: 1 April 1945

ISSUED TO	NUMBERS
19th Bomb Sq., 22nd Bomb Gp.	34,126 - 34,174
Hqs. 22nd Bomb Gp.	34,175 - 34,375
MIS-X Section Display Board	50,001 -
Hqs. 5th Bomber Command	50,002 - 50,997
Lost in movement	50,998 - 50,999
Lt. Herlihy, MIS-X Section	51,000 -
Hqs. 5th Bomber Command	51,001 - 52,000
Hqs. 22nd Bomb Gp.	52,001 - 52,025
Hqs. 345th Bomb Gp.	52,026 - 53,325
Hqs. 38th Bomb Gp.	52,326 - 52,625
Hqs. 43rd Bomb Gp.	52,626 - 52,950
26th Photo Sq. Recon, 6th Photo Gp. Recon	52,591 - 52.975
25th Photo Sq. Recon, 6th Photo Gp. Recon	52,976 - 52,990
17th Photo Sq. Recon, 71st Tactical Recon Gp.	52,991 - 53,000
400th Bomb Sq., 90th Bomb Gp.	53,001 - 53,080
320th Bomb Sq., 90th Bomb Gp.	53,081 - 53,160
319th Bomb Sq., 90th Bomb Gp.	53,161 - 53,240
321st Bomb Sq., 90th Bomb Gp.	53,241 - 53,320
Hqs. 90th Bomb Gp.	53,321 - 53,325
Hqs. 380th Bomb Gp.	53,326 - 53,650
Hqs. 22nd Bomb Gp.	53,651 - 53,950
Flight Section 5th Bomber Command	53,951 - 54,000
VPB71	54,001 - 54,200
110 Tactical Recon Sq., 71st Tactical Recon Gp.	54,201 - 54,220
82nd Tactical Recon Sq., 71st Tactical Recon. Gp.	54,221 - 54,240
Hqs. 348th Fighter Gp.	54,241 - 54,246
341st Fighter Sq., 348th Fighter Gp.	54,247 - 54,276
340th Fighter Sq., 348th Fighter Gp.	54,277 - 54,306
460th Fighter Sq., 348th Fighter Gp.	54,307 - 54,336
3rd Emergency Rescue Gp.	54,337 - 54,356
342nd Fighter Sq., 348th Fighter Gp.	54,357 - 54,386
Hqs. 49th Fighter Gp.	54,387 - 54,390
7th Fighter Sq., 49th Fighter Gp.	54,391 - 54,420
8th Fighter Sq., 49th Fighter Gp.	54,421 - 54,450
9th Fighter Sq., 49th Fighter Gp.	54,451 - 54,480
Hqs. 308th Bomb Wing	54,481 - 54,486
MIS-X Section, On Hand	54,487 - 54,523
VPB20	54,524 - 54,748
VPB 111 & 117	54,749 - 55,000

- 1 -

ISSUED TO	NUMBERS
Hqs. 13th Air Force	55,001 - 57,000
Hqs. 90th Bomb Gp.	57,001 - 57,250
Hqs. 43rd Bomb Gp.	57,251 - 57,500
Hqs. 380th Bomb Gp.	57,501 - 57,625
Hqs. 22nd Bomb Gp.	57,626 - 57,750
Hqs. 5th Bomber Command	57,751 -
Hqs. 38th Bomb Gp.	57,752 - 57,851
Hqs. 345th Bomb Gp.	57,852 - 57,951
Hqs. 5th Bomber Command	57,952 - 58,000
VPB 71	58,001 - 58,075
Hqs. 35th Fighter Gp.	58,076 - 58,078
41st Fighter Sq., 35th Fighter Gp.	58,079 - 58,100
40th Fighter Sq., 35th Fighter Gp.	58,101 - 58,122
Hqs. 3rd Air Commando	58,123 - 58,125
3rd Commando Sq., 3rd Air Commando Gp.	58,126 - 58,147
4th Commando Sq., 3rd Air Commando Gp.	58,148 - 58,169
39th Fighter Sq., 35th Fighter Gp.	58,170 - 58,191
Hqs. 18th Fighter Gp.	58,192 - 58,194
70th Fighter Sq., 18th Fighter Gp.	58,195 - 58,216
12th Fighter Sq., 18th Fighter Gp.	58,217 - 58,238
44th Fighter Sq., 18th Fighter Gp.	58,239 - 58,260
Hqs. 86th Fighter Wing	58,261 - 58,267
Hqs. 308th Bomb Wing	58,268 - 58,282
Hqs. 71st Tachtical Recon Gp.	58,283 - 58,286
82nd Tactical Recon Sq., 71st Tactical Recon Gp.	58,287 - 58,308
110th Tactical Recon Sq., 71st Tactical Recon Gp.	58,308 - 58,330
Hqs. 38th Bomb Gp.	58,331 - 58,335
71st Bomb Sq., 38th Bomb Gp.	58,336 - 58,360
405th Bomb Sq., 38th Bomb Gp.	58,361 - 58,385
822nd Bomb Sq., 38th Bomb Gp.	58,386 - 58,410
823rd Bomb Sq., 38th Bomb Gp.	58,411 - 58,435
Hqs. 345th Bomb Gp.	58,436 - 58,643
76th Wing RAAF	58,644 - 58,925
Hqs. 85th Fighter Wing	58,926 - 58,937
VD-5	58,938 - 58,982
76th Wing RAAF	58,983 - 58,994
MIS-X Section, FEAF	58,995 - 58,997
Capt. E. W. Clark 393rd Air Service Sq.	58,998
Major Walcott Allied Geographical Service	58,999
Lt. Col., Hoover, G-2 USAFFE	59,000
Hqs. 91st Photo Recon Wing	59,001 - 59,150
Flight Section, 5th Fighter Command	59,151 - 59,165
Hqs. 8th Fighter Gp.	59,166 - 59,316
Hqs. 58th Fighter Gp.	59,317 - 59,466
Hqs. 5th Fighter Command	59,467 - 59,667
Hqs. 3rd Air Commando Gp.	59,668 - 59,718
Hqs. 91st Photo Recon Wing	59,719 - 59,865

Appendix VI:

AIR SEA RESCUE Bulletin

No. 11—MAY 1945

NAVCG 128

Information contained herein is assembled from various United States and foreign sources; it is disseminated in this Bulletin for information only to a limited list of addressees with interest in the field of air sea rescue. It will be apparent that the Bulletin may contain information which does not represent the policy of the Air Sea Rescue Agency or the Services represented on the Board for Air Sea Rescue.

Published monthly by the United States Coast Guard for

AIR SEA RESCUE AGENCY

WASHINGTON, D. C.

BLOOD CHITS

The "**Blood Chit**" is the aviators' IOU, issued in all theaters of operations by Allied governments. It is an agreement, or a pledge, from a government to a person or persons who are helpful to downed airmen.

The present "**Blood Chit**" may take the form of anything from the printed calling card type to the flag type used in China. Often they have been in the form of a penciled note on a piece of scrap paper. However, the principle is the same, and the purpose always is to aid in the survival and rescue of Allied personnel.

The "**Blood Chit**" sprang up in various areas from Europe to the South Pacific, but the name originated in China. Any China "hand" will explain that a "Chit" is a note, or an especially signed voucher for a small debt, such as a debt for food. It is the same as a "tab," or an IOU. It has been pointed out that the term "blood" could be interpreted as "life"—hence the term "**Blood Chit**," meaning a sort of IOU on one's life.

In the early days of the war, airmen who had been forced down found it possible to pay helpful natives with an IOU scribbled on a piece of paper, cloth, or anything else that happened to be handy or convenient. Subsequently, after wide semiunofficial use of the IOU or "**Blood Chit**," various governments recognized the practice officially and established payment as a general policy. At the same time more formal "Chits" began to spring up.

Not all of the "**Blood Chits**" read the same, but all of them put across the same general idea and speak for themselves. In effect they generally say:

I am an American airman.
My plane is destroyed.
I cannot speak your language.
I am an enemy of the Japanese.
Please nourish and protect me and take me to the nearest Allied military post.
My Government will reward you.

"**Blood Chits**" are worn according to the fancy of the airman. Some men wear them sewed on the inside of their flight jackets. Others sew them on the outside. Others wear them prominently attached to the uniform in various places.

Some men, however, regard the wearing of "**Blood Chits**" in view as dangerous, and carry them only in their pockets or in other places hidden from sight.

The "**Blood Chit**" today is a very definite part of the airman's equipment and is supplemented by the best available information concerning the territory he is about to fly over. This includes, in part, customs of local tribes or provinces, friendly and unfriendly territory, uses of money, forced landing procedure and much more. The Pointie-Talkie (AIR SEA RESCUE BULLETIN No. 7, January, 1945) is also used in conjunction with the **Blood Chit.**

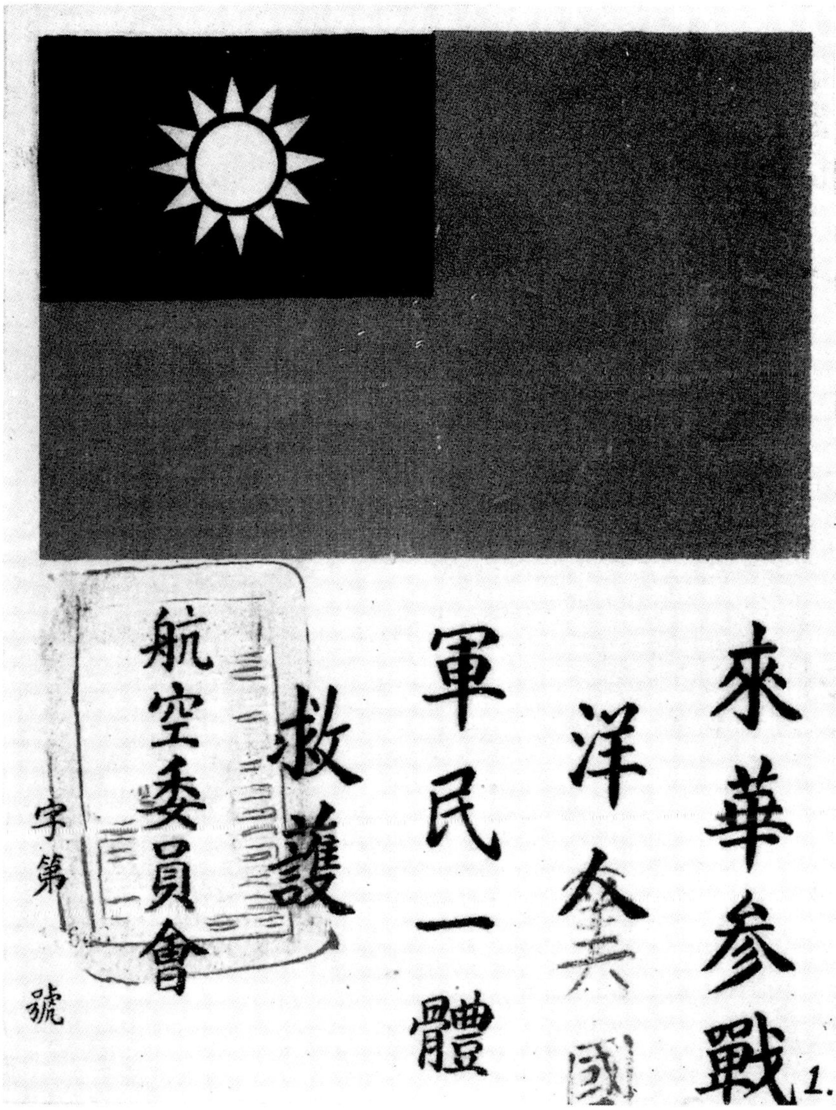

Chinese Flag Blood Chit above was printed in China in the early part of the war by the Chinese Intelligence for the use of General Chennault's famous "Flying Tigers." This organization was probably the first to make widespread use of the "Blood Chit"; however, there is no authentication of that fact. The "Flying Tigers" complement was made up of persons of many and varied nationalities and to allow for a more thorough identification, space was left on the Chit where the nationality of the bearer could be stamped. The two light characters at the bottom of the second vertical column from the right, inclosed in parentheses, means American. These were printed in purple whereas the rest of the characters were printed in black. This early Blood Chit instead of sporting the smooth silk finish of the present type was printed on a rough, coarse linen-type cloth. The official chop on the left side section is the chop of the Aeronautical Commission in Chungking. Many Blood Chits originated in the South Pacific area

BLOOD CHITS—*Continued*

American Flag Blood Chit printed in India. There is a similar Chit showing the British Flag instead of the American. The inset illustration shows the various ways the Blood Chit can be worn.

Continued on next page

BLOOD CHITS—*Continued*

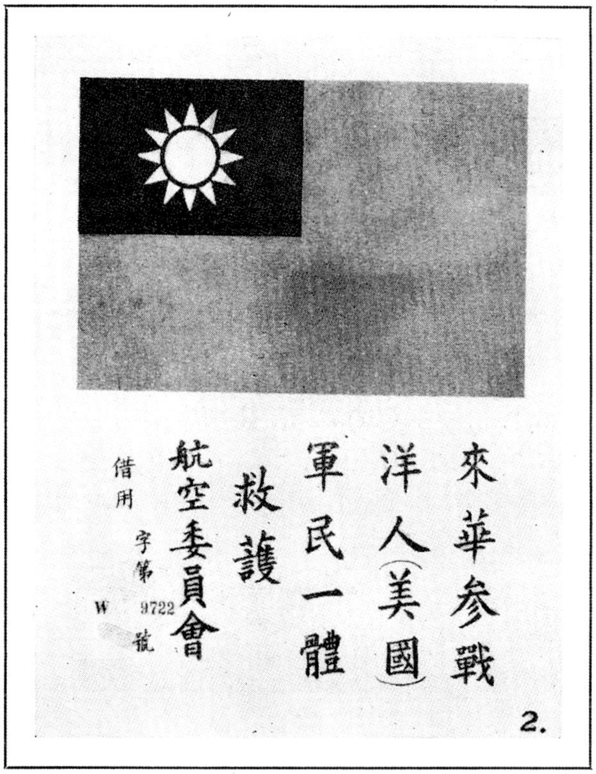

Chinese Flag Blood Chit 2 (upper left) is a copy of the original used by the "Flying Tigers" and was printed in America for U. S. Army Forces in China. The "W" preceding the serial number indicates that the chit was procured from Washington and also distinguishes it from a Chit produced in China which might have the same serial number. The chop used (not illustrated) is the Chop of the Chinese Embassy in Washington.

Chinese Flag Blood Chit 3 (upper right) has a message which is slightly different from that of number 2 but the chop is still that of the Chinese Embassy in Washington. Copies of the original Chit are in color, and these are printed on a waterproof silk-like fabric.

"Boy, the Chit!" is the cry that can be heard in all restaurants that are frequented by the "China Hand." *Continued*

BLOOD CHITS—*Continued*

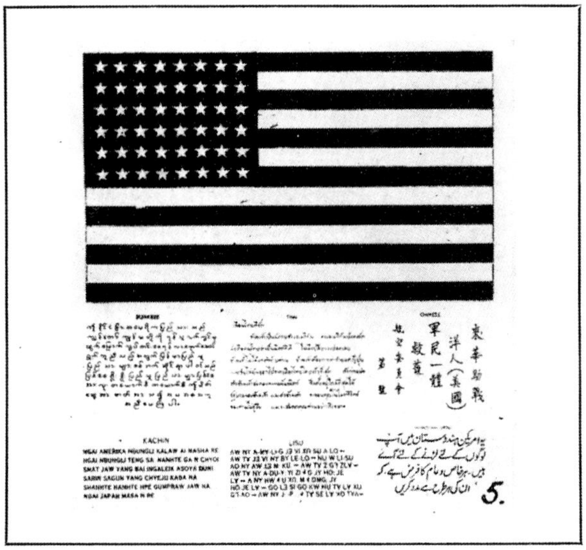

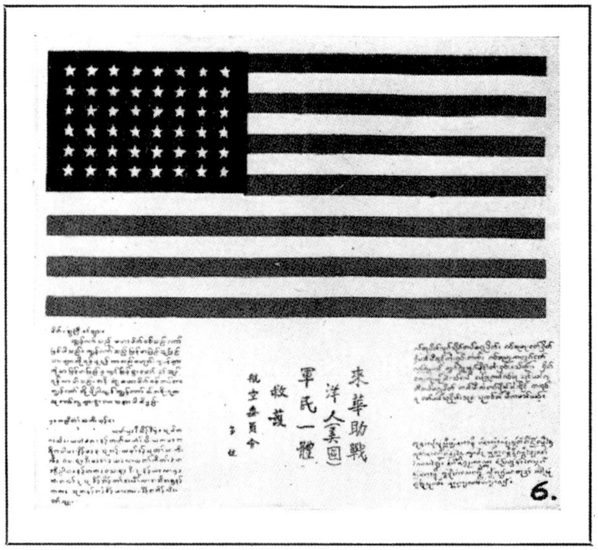

The MIS–X section, C. P. M. Branch, G2 Branch of G2 Section, War Department, has had much to do with the development of the American Flag **Blood Chit.** Accuracy in message and language has been the keynote of the thought behind the American Flag Chit and additions can be noted in the illustrations.

American Flag **Blood Chit** 5 (upper left) was printed in the United States about March 1944, and is the first **Blood Chit** put out by the MIS–X section, C. P. M. Branch, G-2 Branch for Airmen in China.

American Flag **Blood Chit** 6 (upper right) was thought to have been printed in India. The messages on it are in the following languages: upper left—Burmese; lower left—Sgaw Karen; center—Chinese; upper right—W. Shan; lower right—S. Shan.

American Flag **Blood Chit** 7 (lower left) is a copy of 6 but the text of the Chinese message has been changed slightly and the positions of the messages changed to the following: upper left—Burmese; lower left—S. Shan; center—Chinese; upper right—Sgaw Karen; lower right—W. Shan.

American Flag **Blood Chit** 8 (lower right) is a reprint of 7 and on this Chit the form of the message has been changed slightly for more economical use of space.

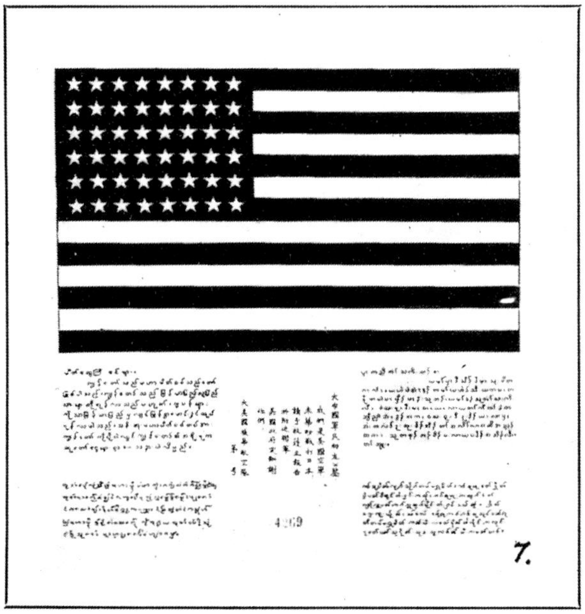

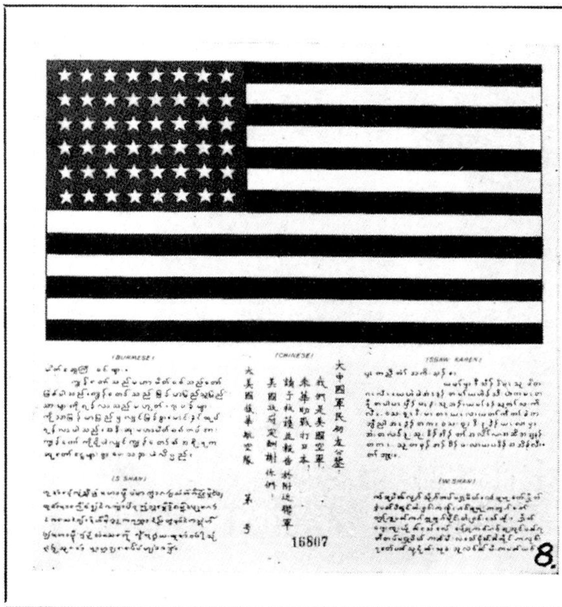

Continued on next page

BLOOD CHITS—*Continued*

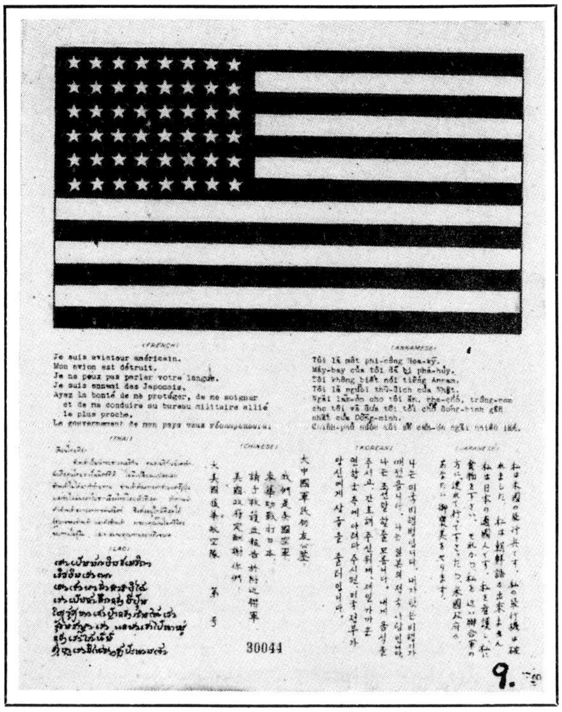

American Flag Blood Chit 9 is the latest one produced by MIS-X. Its message is in the languages found along the East Coast of Asia from French Indo-China to Manchuria and as far as 200 miles inland. This Chit is an example of the changes that must be made in order to keep up with the progress of war areas. It can also be noticed that the message in Japanese has been included (lower right). It was thought that there might be the possibility of an airman running into some Chinese who could read the language as a result of Japanese occupation.

Blood Chit 10 is the passport type used by airmen in the African Campaign. It contains a message from the President of the United States to the Arab peoples (illustrated below) to the effect that we (Americans) are friendly and if they (the Arabs) protect the bearer from harm and help him, they will be rewarded. It also includes a list of helpful words in Arabic, French, Spanish and Portuguese. This Blood Chit produced excellent results in the African Campaign and was responsible for the safe return of a large number of downed airmen.

Continued

18 METHODS PROCEDURES AND TECHNIQUES

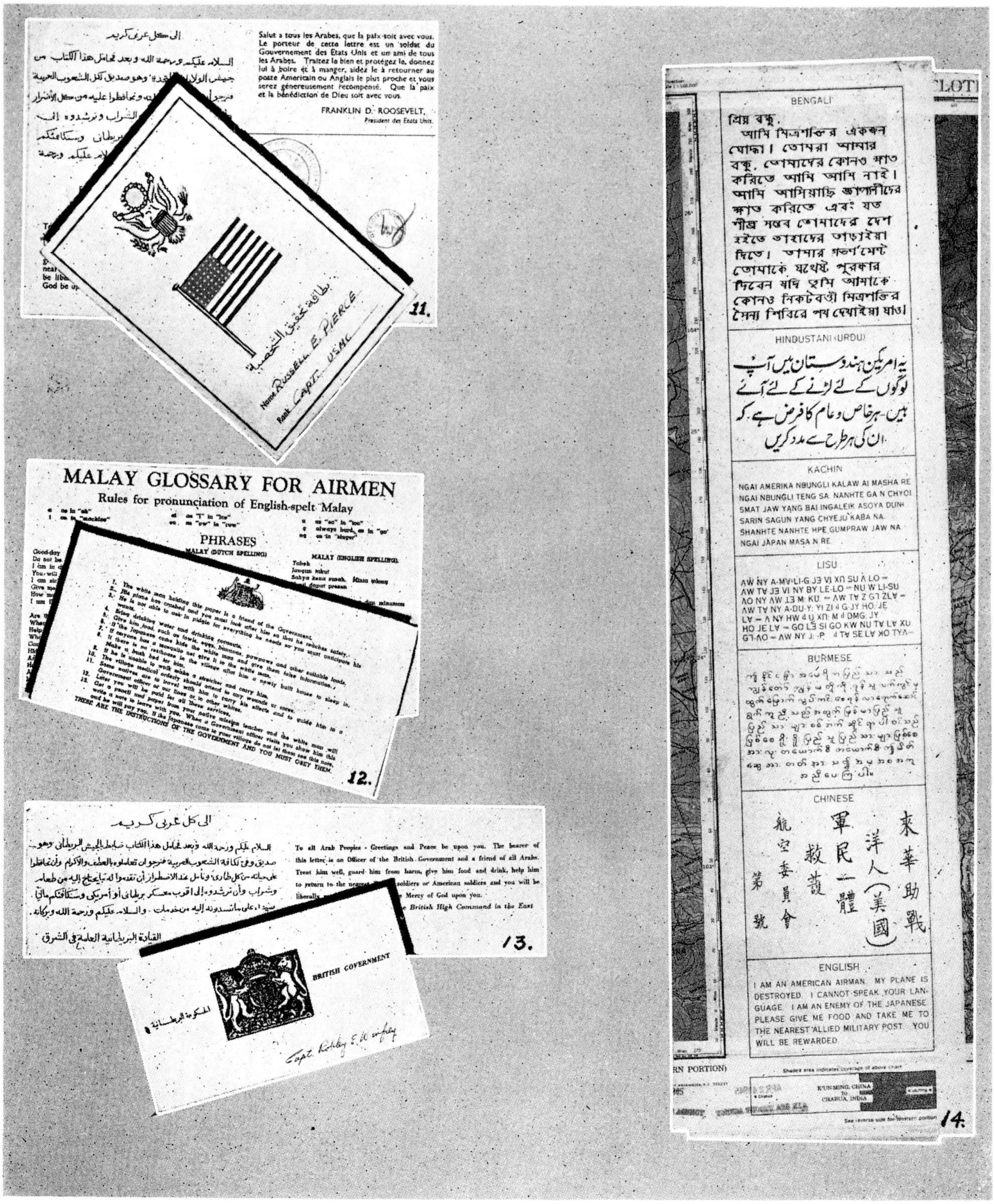

VARIOUS TYPES OF CARD AND MAP BLOOD CHITS: Number 11 is another passport type also used in the African campaign bearing a message from the President of the United States.

Number 12 is a Malay Glossary for airmen in "Pidgin" English phrases with Australian seal issued in the Southwest Pacific area.

Number 13 is a British card similar to the American bearing a message to the Arab peoples.

Blood Chit number 14 is different in that it is primarily a map (silk) K'unming China, to Chabua, India. It carries a strip on the right hand side bearing a message in seven languages including English. This is put out by the AAF Aeronautical Chart Service. *(Continued on next page.)*

BLOOD CHITS

Many Blood Chits Originated in The South Pacific Area

Blood Chits can be obtained for issue to combat units from:
MIS-X Section GHQ in SWPA—MIS-X Section JicPoa (advance and rear)—A. G. A. S., Hq., USAF, China.

HYDRAULIC HOISTS—*Continued*

on a davit for hoisting him aboard. Manpower or deck winch could be used for hoisting (illustration 6).

Life jackets are designed to float the survivor face up. The pick-up ring should be located where he can see and reach it without difficulty. The natural location for it is the chest. The ring should be either metal or cable for ease in hooking and should be of such lengths that it cannot possibly strike the wearer in the face. A fabric webbed type loop is not recommended because it would be difficult to hook into and would require a hook larger than necessary. This would in turn entail a larger float with attended danger of injury to personnel.

COMBINATION HARNESS

A combination belt and seat strap has been tried at Floyd Bennett Field with success. It was designed with the primary consideration in mind that the person being rescued would probably be in a weakened condition from exposure; hence the attachment must of necessity be simple, requiring only the minimum effort to hook on. A person may be picked up with the belt strap alone (illustration 7), which is sufficient for survivors who have not suffered bodily injury or with the seat pad snapped into rings attached to the belt (illustration 8). The latter harness is preferable in most cases as the survivor is supported in a natural position with no usual strain inflicted on the internal organs. The belt is sewed well down on the small of the back to prevent gripping the survivor under the arms and on the chest.

Many practice pick-ups have been made with the helicopter so far using a seat type harness incorporated in the regular aviator's jacket, without a single injury to personnel. In fact this harness is so comfortable that it has been suggested that the harness be used aboard ship in place of a boatswain's chair for working aloft or over the side. The wearer does not have to hang on as in a boatswain's chair so he has both hands free to work.

"Gibson Girl" Used Aloft Successfully

The story of how one B–17, its regular transmitter knocked out by enemy fire, used its "Gibson Girl" prior to ditching is told in a recent *MAYDAY in Action* story. The plane was badly mauled by flak, and the pilot knew that he would have to ditch. The problem of how to send out a distress signal was solved by the radio operator who attached the "Gibson Girl" to the plane's trailing antenna and cranked out an SOS. This same technique could also be used in a B–24.

RESCUE TIPS

A commander whose PBY was forced down in the Pacific among other more common suggestions, offered the following:

1. Everybody should check out on each individual piece of survival gear. All hands also should have a check out in rowing, sailing, knot-tying, and seamanship.
2. If you don't want sharks, don't trail bright objects over the side.

Appendix VII:

Table of Relative Scarcity

(Table of Language Aid Identification Numbers, Relative Scarcity, and other Information)

This table gives information regarding the aids shown in this book in a condensed form that may prove useful to many readers. These include a standardized numbering system, relative scarcity, time period of use, sponsoring nations, and areas of intended use. The scarcity of artifacts shown is based upon the relative frequency with which examples appear on the antique and collector markets in the United States; scarcity in the United Kingdom and Australia will differ from that shown. Note also that scarcity is rela*tive*, and that artifacts of even average scarcity may be difficult to obtain. Souvenir pieces, by definition, do not have a sponsoring nation or area of intended use. For the purposes of this table, World War II begins on 1 September, 1939, and ends on 1 September 1945.

Column Headings

ID NO.	Standardized Language Aid Identification Number
FIG.	Figure Number in this book
LOC	Location of item in Figure
R.S.	Relative Scarcity
ERA	Time period during which chit was used
SPONSOR	Sponsoring Nation (not necessarily Country of origin).
USED IN	Area in which aid was used
FABRIC	Materials of which aid is made
FLAG	Flag (or seal) appearing on aid
NO. LANG	Number of languages used in aid (including English)
S/N	Color of Serial Number
COMMENTS	Self explanatory

Figure Codes

N.S.	Not Shown

Location Codes

T	Top
B	Bottom
L	Left
C	Center
R	Right
Bkg	Background

Relative Scarcity

1	Very Common
2	Common
3	Scarce
4	Rare
5	Very Rare

Era Codes

Pre-WWI	Prior to World War I
WWI	World War I
B.Col	British Colonial
Pre-WWII	Between World War I and World War II
WWII	World War II

Country and Area Codes

Aus	Australia
Bur	Burma
Chi	China
Eth	Ethiopia
Eur	Europe
E.Eur	Eastern Europe
Fr	France
Ger	Germany
Grc	Greece
FIC	French Indo-China
Ind	India
Mal	Malaya
ME	Middle East
Med	Mediterranean
Meso	Mesopotamia
N.Afr	North Africa
N.Am	North America
NEI	Netherlands East Indies
N.Gui	New Guinea
POA	Pacific Ocean Area
Rus	Soviet Union (Russia)
SWPA	South West Pacific Area
UK	United Kingdom
USA	United States
W.Eur	Western Europe
?	Unsure or Unknown.

Fabric Codes

Cot	Cotton
Chc	Cheesecloth (gauze)
Crd	Cardstock
Burl	Burlap
Lin	Linen
Oil	Oilcloth
Pap	Paper
Pap/Lin	Paper glued to Linen
Pla	Plastic
Ray	Rayon
Ray/Cot	Rayon sewn to Cotton
Slk	Silk
Slk/Cot	Silk sewn to Cotton
T.Pap	Tissue Paper.

Flag and Seal Codes

Parentheses ( ) denote a Seal

Aus	Australian flag or seal
BG	British Government seal
BP	Burmese Peacock flag or seal
Eth	Emperor of Ethiopia's seal
Hol	Netherlands (Holland) Flag or Seal
NC	Nationalist China Flag
NWF	North West Frontier Province Seal
RAKC	Royal Coat of Arms with King's Crown
UK	Union Jack
US	Stars and Stripes
GSUS	Great Seal of the United States.

Color Codes

B	Blue
Bk	Black
BW	Black & White
P	Purple
R	Red
W	White

Explanation of Identification Code Logic

1xxx	British
2xxx	Chinese
3xxx	United States
4xxx	Other Countries
9xxx	Souvenirs (all time periods)
x0xx	Before World War I
x1xx	World War I up to World War II
x2xx	World War II
x3xx	Post World War II through the Korean War
x4xx	Post-Korean War through Vietnam War
x5xx	Post-Vietnam War through Dec. 31, 2000
x6xx-x8xx	For future use
x900	Miscellaneous language aids such as Phrase Cards and Pointie Talkies

Variations will be identified by numbers following a decimal point. For example, the numbers xxxx, xxxx.1, and xxxx.2 denote three variations of a single type of aid.

ID NO.	FIG.	LOC.	R.S.	ERA	SPONSOR	USED IN	FABRIC	FLAG	NO.	S/N	Comments
	2.1		3	Pre-WWI	UK	Eur/N.Am	Pap		6		UK FAI Certificate
	2.2		3	WWI	USA	Eur/N.Am	Pap		6		US FAI Certificate
	2.3	TL	2	WWI	USA	Fr	Pap		2		
	2.3	TR	2	WWI	USA	Fr	Pap		2		
	2.3	BL	2	WWI	USA	Fr	Pap		2		
	2.3	BR	2	WWI	USA	Fr	Pap		2		
	2.4	L	3	WWI		Eur	Slk	Fr	0		
	2.4	R	3	WWI		Eur	Slk	UK	0		
	2.6		2	WWI	USA	Fr	Pap		2		
REB 1110	3.1		5	B.Col	UK	Ind	Pap	(NWF)	2		
REB 1112	3.2		5	B.Col	UK	Ind	Pap	(NWF)	1		
REB 1113	3.3		5	B.Col	UK	Ind	Pap	(NWF)	1		
REB 1116	3.4		5	B.Col	UK	Ind	Pap	(NWF)	2		
REB 1118	3.5		5	B.Col	UK	Ind	Pap	(RAKC)	1		
REB 1120	3.6		5	B.Col	UK	Ind	Pap		2		
REB 1122	3.7		5	B.Col	UK	Meso	Pap		3		
REB 1124	N.S.		5	B.Col	UK	Meso	Pap		2		Iraq Arabic phrase card
REB 2110	4.4		5	Pre-WWII	Chi	Chi	Slk/Cot	NC	1	Bk	14th VBS
REB 1212	4.5		5	WWII	UK	Bur	Pap	(RA/BP)	2		AVG
REB 2212	4.6		5	WWII	Chi	Chi	Slk/Cot	NC	1	Bk	AVG
REB 2214	4.7		5	WWII	Chi	Chi	Slk/Cot	NC	1	Bk	AVG
REB 2216	4.8		5	WWII	Chi	Chi	Slk/Cot	NC	1	Bk	AVG
REB 2210	4.9		5	WWII	Chi	Chi	Slk/Cot	NC	1	Bk	AVG
REB 4210	5.1		5	WWII	Eth	Eth	Pap	(Eth)	1		
REB 1210	5.2		5	WWII	UK	Eth	Lin	(BG)	2		Heavy linen
REB 1210.1	N.S.		5	WWII	UK	Eth	Lin	(BG)	2		Thin coated linen
REB 4212	5.3		5	WWII	Eth	Eth	Pap	(Eth)	2		
REB 1214	5.5		4	WWII	UK	ME	Pap/Lin	(BG)	2		
REB 1216	5.6		4	WWII	UK	ME	Pap/Lin	(BG)	2		
REB 1218	5.7		4	WWII	UK	ME	Pap/Lin	(BG)	2		
REB 1220	5.8		4	WWII	UK	ME	Pap/Lin	(BG)	4		
REB 1222	5.9		3	WWII	UK	N.Afr	Crd	(RAKC)	2		
REB 4208	5.10		5	WWII	Ger.	N.Afr	Crd	(RAKC)	2		Counterfeit
REB 1226	5.11		4	WWII	UK	N.Afr	Pap/Lin	(RAKC)	2		
REB 1228	5.12	T	4	WWII	UK	ME	Pap	(RAKC)	6	Bk	Tear-out Chits
REB 1230	5.12	C	4	WWII	UK	ME	Pap	(RAKC)	7		Tear-out Chits
REB 1232	5.12	B	4	WWII	UK	ME	Pap	(RAKC)	7		Tear-out Chits
REB 1234	5.13		4	WWII	USA	N.Afr	Pap	(BG)	2		
REB 3210	5.14		4	WWII	USA	N.Afr	Pap	(BG)	2		
REB 3212	5.15		4	WWII	USA	N.Afr	Crd	US/(GSUS)	2		
REB 3214	5.16		4	WWII	USA	N.Afr	Pap	(GSUS)	5		
	5.19		3	WWII	USA	N.Afr/E.Eur	Pap			B	
REB 3216	5.20		4	WWII	USA	ME	Pap	(GSUS)	2		
	5.21	L	3	WWII	UK	ME					
	5.21	R	3	WWII	UK	ME					
REB 1910	6.1	T/BL	3	WWII	UK	Eur	Crd		4		
	6.1	BR	3	WWII	USA	Eur	Pap		1		
	6.1	BKG	4	WWII	UK	W.Eur	T.Pap				Tissue Map
REB 1911	6.2		4	WWII	UK	W.Eur	Crd		4		
REB 1912	6.3		3	WWII	UK	E.Eur	Crd		14		
REB 1910.1	6.4	T	2	WWII	UK	W.Eur	Pap		4		
	6.4	B	1	WWII	USA	W.Eur	Pap		1		
	6.5	T	2	WWII	UK	Eur	Burl/Pla	UK			
	6.5	B	1	WWII	USA?	Eur	Oil	US			

ID NO.	FIG.	LOC.	R.S.	ERA	SPONSOR	USED IN	FABRIC	FLAG	NO.	S/N	Comments
6.6	TL	3	WWII	USA	Eur	Chc	US	0			
	6.6	CL	3	WWII	USA	Eur	Cot	US	0		
	6.6	BL	3	WWII	USA	Eur	Oil	US	0		
	6.6	TR	3	WWII	USA	Eur	Cot	US	0		
	6.6	TCR	3	WWII	USA	Eur	Cot	US	0		
	6.6	BCR	1	WWII	USA	Eur	Cot	US	0		
	6.6	BR	3	WWII	USA	Eur	Cot	US	0		Appliqué
	6.8		4	WWII	UK	Fr					
	6.9		4	WWII	UK	Fr/It					
	6.10		3	WWII	UK	Eur					
	6.11		3	WWII	UK	Eur					
	6.12		2	WWII	UK	All					
	6.13		4	WWII	UK	E.Eur					
REB 1914	6.14	L	4	WWII	UK	E.Eur/Med?	Pap		5		
REB 1915	6.14	R	4	WWII	UK	E.Eur/Med?	Crd		6		
	6.15	TL	4	WWII	UK						RAF Tunic Button
	6.15	TLC	4	WWII	Can						RCAF Tunic Button
	6.15	TC	4	WWII	UK						RAF Cap Button
	6.15	TRC	4	WWII	US						USAAF Tunic Button
	6.15	TR	4	WWII	UK						RAF Collar Stud
	6.15	CL	3	WWII	UK						RAF Collar Stays
	6.15	CLC	3	WWII	UK						RAF Tunic Buckle
	6.15	CRC	3	WWII	UK						RAF Trouser Buttons
	6.15	CR	3	WWII	UK						RAF Blouse Button (Black)
	N.S.		3	WWII	UK						British Army Blouse Button (Brown)
	6.15	BL	4	WWII	UK						Pencil Clip
REB 1236	6.19		3	WWII	UK	Rus	Crd		2		
REB 1238	6.20	L	3	WWII	UK	Rus	Ray	UK	2		Blanket Stitch
REB 1238.1	6.20	R	3	WWII	UK	Rus	Ray	UK	2		Straight Stitch
REB 3218	6.21	L	3	WWII	UK	Rus	Ray	US	2		Blanket Stitch
REB 3218.1	6.21	R	3	WWII	UK	Rus	Ray	US	2		Straight Stitch
REB 1240	6.24	L	4	WWII	UK	Rus	Ray	UK	2		Paper-like Fabric
REB 3220	6.24	R	2	WWII	UK	Rus	Ray	US	2		Paper-like Fabric
REB 1242	6.26		5	WWII	UK	Rus	Crd	UK/US	1		
REB 3222	6.27		5	WWII	UK	Rus/E.Eur	Pap	US	8		
REB 3224	6.28		5	WWII	UK	E.Eur	Pap		10		
REB 1244	6.29		5	WWII	UK	Rus/E.Eur	Pap	UK	8		
REB 3922	6.31		5	WWII	USA	Rus	Pap	US	2		
	6.32	TL	3	WWII	USA	Fr/N.Afr/FIC	Pap		2		
	6.32	TR	1	WWII	USA	Eur	Pap		2		
	6.32	BL	1	WWII	USA	Eur	Pap		2		
	6.32	BR	1	WWII	USA	Eur	Pap		2		
REB 3910	6.35		3	WWII	USA	Eur	Pap		4		
	6.36		3	WWII	Grc	Grc	Pap		2		
REB 9010	6.37		4	WWII	UK	Eur	Cot		3		
	7.1		1	WWII	Chi	Chi	Pap				
	7.2		1	WWII	Chi	Chi	Pap				
REB 2220	7.3		5	WWII	Chi	Chi	Slk/Cot	NC	1	Bk	Centered Chop
REB 2222	7.4		4	WWII	Chi	Chi	Slk/Cot	NC	1	Bk	
REB 2224.1	7.5		3	WWII	Chi	Chi	Slk/Cot	NC	1	Bk	

ID NO.	FIG.	LOC.	R.S.	ERA	SPONSOR	USED IN	FABRIC	FLAG	NO.	S/N	Comments
REB 2224.2	7.6		3	WWII	Chi	Chi	Slk/Cot	NC	1	Bk	
REB 2224	7.7		3	WWII	Chi	Chi	Slk/Cot	NC	1	Bk	
REB 2226	7.8		3	WWII	Chi	Chi	Slk/Cot	NC	1	Bk	
	7.10		4	WWII	UK	CBI	Pap		3		
REB 1245	7.11		3	WWII	UK	Ind	Pap/Lin	(RAKC)	2		
REB 1246	7.12		3	WWII	UK	Bur	Pap/Lin	(RAKC)	5		
REB 1246.1	7.13		3	WWII	UK	Bur	Pap/Lin	(RAKC)	5		
REB 1247	7.14		4	WWII	UK	Thai	Pap/Lin	(RAKC)	2		
REB 3224	7.15		3	WWII	US	CBI	Pap/Canv		3		DEB-1
REB 3226	7.16		3	WWII	US	CBI	Pap		4		
REB 3228	7.17		4	WWII	US	CBI	Pap		4		
REB 3230	7.18		3	WWII	US	CBI	Ray		7		
REB 3232	7.19		3	WWII	US	CBI	Cot	US	6		First MIS-X Chit
REB 3234	7.20		3	WWII	US	CBI	Slk/Cot	US	5		DEB-10
REB 3234.1	7.21		4	WWII	US	CBI	Slk/Cot	US	5		DEB-10
REB 1252	7.22		4	WWII	UK	CBI,FIC	Slk	UK	17		Blue Text
REB 1253	7.23		4	WWII	UK	CBI,FIC	Slk	UK	17		Black Text
REB 3236	7.24		3	WWII	US	CBI,FIC	Slk	US	17		DEB-11, Blue Text
REB 3238	N.S.		3	WWII	US	CBI,FIC	Slk	US	17		DEB-11, Black Text
REB 3238.1	7.25		3	WWII	US	CBI,FIC	Slk	US	17		DEB-11, 2-sided
REB 3240	7.27		3	WWII	US	CBI,FIC	Ray	US	17		
REB 1256	7.29		4	WWII	UK	CBI,FIC	Slk	UK	17		
REB 1256.1	7.30		3	WWII	UK	CBI,FIC	Ray	UK	17		
REB 1258	7.31		4	WWII	UK	Bur	Slk/Cot	B.Bur	1		
REB 3242	7.32		3	WWII	US	CBI	Ray/Cot	US	5	Pur	
REB 3244	7.33		1	WWII	US	Chi	Ray	NC	1	Bk	Without "(1)"
REB 3246	7.34		1	WWII	US	Chi	Ray	NC	1	Bk	Without "(2)"
REB 3244.1	7.35		4	WWII	US	Chi	Ray	NC	1	Bk	With "(1)"
REB 3246.1	7.36		1	WWII	US	Chi	Ray	NC	1	Bk	With "(2)"
REB 3246.2	7.37		1	WWII	US	Chi	Ray	NC	1	Bk	With "(2)", S/N at Bottom
REB 3248	7.38		3	WWII	US	CBI	Ray	US	5	R	
REB 3250	7.39		1	WWII	US	CBI	Ray	US	7	R	Without "№"
REB 3250.1	7.40		1	WWII	US	CBI	Ray	US	7	R	With "№"
REB 3252	7.42		3	WWII	US	CBI	Slk/Cot	US	0		
REB 3254	7.43		4	WWII	US	CBI	Slk/Cot	US	1		
REB 3256	7.45		3	WWII	US	CBI	Slk?	US	0		
REB 3258	7.46		3	WWII	US	CBI	Ray	US	0		No borders on edge of Flag
REB 3260	7.47		3	WWII	US	CBI	Ray	US	0		Borders on edge of Flag
REB 3261	7.48		1	WWII	US		Ray?	US	0		
REB 3912	7.49	L	3	WWII	US	Bur	Pap		1		XEM-4
REB 3912.1	7.49	R	3	WWII	US	Bur	Pap		1		No "XEM" Notation
REB 1920	7.50	L	3	WWII	UK	Chi	Ray	3	2		
REB 1922	7.50	R	4	WWII	UK	Mal,NEI	Pap		2		
REB 3914	7.53	L	2	WWII	US	Chi	Pap	2	2		Jan. 1943 Printing
REB 3914.1	N.S.		4	WWII	US	Chi	Pap	2	2		Aug. 1943 Printing
REB 3915	7.53	R	3	WWII	US	Chi	Pap	2	2		GPO, 1944
REB 3916	7.56	L	3	WWII	US	Chi	Pap	2	9		OWI & 21st Recon Sqn
REB 3918	7.56	R	3	WWII	US	Chi	Pap	2	9		AGAS & OWI
REB 3920	7.59	L	3	WWII	US	Chi	Pap	2	9		Green Cloth Cover
REB 3920.1	N.S.		4	WWII	US	Chi	Pap	2	9		Cream Paper Cover
REB 3921	7.59	R	3	WWII	US	Chi	Pap	2	9		Cream Paper Cover
	7.61	TL	1	WWII	US	Chi	Pap		2		
	7.61	TR	1	WWII	US	Chi	Pap		2		

ID NO.	FIG.	LOC.	R.S.	ERA	SPONSOR	USED IN	FABRIC	FLAG	NO. S/N	Comments
	7.61	CL	1	WWII	US	Chi	Pap		2	
	7.61	CR	3	WWII	US	Chi	Pap		2	
	7.61	BL	1	WWII	US	Chi	Pap		2	
	7.61	BR	1	WWII	US	Chi	Pap		2	
REB 3924	7.63		4	WWII	US	Chi	Pap		2	
	7.64		3	WWII	US	Chi	Pap		1	CA-117
	7.65		3	WWII	US	Chi	Pap		1	CA-125
	7.66		3	WWII	US	Chi	Pap		1	CA-115B
	7.67		3	WWII	US	Chi	Pap		1	CA-132
	7.68		3	WWII	US	Chi	Pap		1	CA-19B
	7.69		3	WWII	US	Chi	Pap		1	CA-118
	7.70		3	WWII	US	Chi	Pap		1	CA-114
	7.71		3	WWII	US	Chi	Pap		1	CA-105
	7.72		3	WWII	US	Chi	Pap		1	CA-105B
	7.73		4	WWII	US	Chi	Pap		1	
	7.74		3	WWII	US	Bur	Pap		1	CBA-41
	7.75		3	WWII	US	FIC	Pap		1	AFA-110
	7.76		3	WWII	US	FIC	Pap		1	AFA-103
	7.77		3	WWII	US	FIC	Pap		1	AFA-5
REB 9011	7.79		4	WWII			Cot	NC	1	
REB 9012	7.80		4	WWII			Cot	NC	1	
REB 9013	7.81		4	WWII			Cot	NC	1	
REB 9014	7.82		3	WWII			Slk	NC	1	
REB 9015	7.83		3	WWII			Cot	NC	1	
REB 9016	7.84		3	WWII			Slk	NC	1	
REB 9017	7.85		3	WWII			Slk	US,NC	1	
REB 9018	7.86		3	WWII			Slk	US,NC	1	
REB 9019	7.87		3	WWII			Slk	NC	1	
REB 9020	7.88		3	WWII			Slk	US,NC	1	
REB 9021	7.89		3	WWII			Slk	NC	1	
REB 9022	7.90		3	WWII			Slk	NC	1	
REB 9023	7.91		3	WWII			Slk	NC	1	
REB 9024	7.92		3	WWII			Slk	NC	1	
REB 9025	7.93		3	WWII			Slk	NC	1	
REB 9026	7.94		3	WWII			Slk	NC	1	
REB 9027	7.95		2	WWII			Slk	NC	1	
REB 9028	7.96		3	WWII			Lth	NC	1	
REB 9029	7.97		3	WWII			Lth	US,NC	1	
REB 9030	7.98		3	WWII			Lth	US,NC	1	
REB 9031	7.100		3	WWII			Lth	US,NC	1	
REB 9032	7.101		4	WWII			Lth	NC,US	1	
REB 9033	7.102		3	WWII			Lth	US,NC	1	
REB 9034	7.103		3	WWII			Lth	US,NC	1	
REB 9035	7.104		4	WWII			Lth	US	1	
REB 9036	7.105		5	WWII			Lth	Bur	1	
REB 9037	7.106		3	WWII			Lth	US,NC	1	Embroidered Anchor
Chain										
REB 9038	7.107		3	WWII			Lth	US,NC	1	
REB 9039	7.108		3	WWII			Lth	US,NC	1	Vertical Anchor
REB 9040	7.109		3	WWII			Lth	US,NC	1	Tilted Anchor
REB 9041	7.110		3	WWII			Lth	NC	1	
REB 9042	7.111	L	5	WWII			Lth	NC	1	Small Patch
REB 9043	7.112		3	WWII			Lth	US	0	
REB 9044	7.113		3	WWII			Cbd	NC	1	Fiberboard?

ID NO.	FIG.	LOC.	R.S.	ERA	SPONSOR	USED IN	FABRIC	FLAG	NO.	S/N	Comments
REB 9044.1	7.114		3	WWII			Cbd	NC	1		Fiberboard?
REB 9045	7.115		3	WWII			Cbd	US,NC	1		Fiberboard?
	8.1		5	WWII	Aus	NEI					Complete Purse
REB 4236	8.2		4	WWII	Aus	NEI					
REB 4237	8.3		4	WWII	Aus	NEI					
REB 4214	8.4	T	3	WWII	Aus	NEI	Crd	(NL)	2		
REB 4216	8.4	B	4	WWII	Aus	P.Tim	Crd	US,Aus,NL	2		
REB 4218	8.5		3	WWII	Aus	NEI	Pap	(NL)	1		
REB 4220	8.6		3	WWII	Aus	NEI	Pap		2		
REB 4222	8.7		4	WWII	Aus	P.Tim	Pap		2		
REB 4224	8.8		4	WWII	Aus	SWPA	Pap				
	8.9		4	WWII	NL?	NEI					
REB 4238	8.10		4	WWII	Aus	NEI	Pap	Aus,US			
REB 4226	8.12		4	WWII	Aus	SWPA	Crd	(Aus)	2		
REB 4239	8.13	T	4	WWII	Aus	SWPA	Pap	Aus,US			
REB 4240	8.13	B	4	WWII	Aus	SWPA	Pap				
REB 4228	8.14		5	WWII	Aus	SWPA	Pap	(Aus)	2		Soft Cardboard
REB 4230	8.15		3	WWII	Aus	SWPA	Pap	(RA)	2		
REB 4232	8.16		4	WWII	Aus	SWPA	Pap	Aus,US	2		
REB 4234	8.17		4	WWII	Aus	SWPA	Pap		2		
	8.18		3	WWII	PI	PI					
REB 3246.3	8.20		5	WWII	US	Chi	Ray	NC	1	Bk	S/N Preceded by "B"
REB 3926	8.22		3	WWII	US	Chi	Pap		2		
REB 3928	8.23		3	WWII	US	Chi	Pap		1		
REB 3930	8.24		3	WWII	US	Chi	Pap		1		
	8.25		2	WWII	US		Pla				
REB 3932	8.26		5	WWII	US	Kor	Pap		2		
	8.27	TL	1	WWII	US	Mal,NEI	Pap				
	8.27	TR	4	WWII	Aus	Mal,NEI	Pap				
	8.27	CL	3	WWII	US	SWPA	Pap				
	8.27	CR	3	WWII	US	SWPA	Pap				
	8.27	BL	1	WWII	US	NEI	Pap				
	8.27	BR	1	WWII	US	N.Gui	Pap				
REB 3934	8.30		3	WWII	US	POA	Pap		2		
REB 3936	8.31		3	WWII	US	POA	Crd		2		
	A3.1		4	WWII							Stamped Canceled
	A3.2		4	WWII	Chi		Pap				w/ Papers
	ADD1		5	WWII	UK	N.Afr,It					Complete Belt
	ADD2	C	5	WWII	UK						Wee Compass
	ADD2	R	4	WWII	Ger						German Evasion Compass
	ADD3		5	WWII	US	N.Afr					Complete Kit
	ADD4		4	WWII	US	Eur					

The Next Chapter

The Blood Chit story does not end at the close of World War II. R. E. Baldwin is attempting to compile a second volume of *Last Hope* , covering the history of Blood Chits and related evasion aids from the immediate post-World War II era through the present day. This period includes the Korean War, the Cold War, the Cuban Missile Crisis, the Vietnam War, the Faulklands War, the Gulf War, and many smaller conflicts such as the Suez Crisis, Aden, and operations in Grenada, Panama, Somalia, and Bosnia. Much of the official government documentation from this period remains classified, so the assistance of veterans and others who have experience with evasion aids is more vital than ever if this history is to be preserved. Mr. Baldwin is also seeking information on the time period covered by this book so that corrections and additions to this volume can be made. Veterans and others who can provide assistance are encouraged to contact him at the location listed below.

Arrangements to see some of the artifacts illustrated in this book firsthand can be made through the International Blood Chit Museum. The museum is dedicated to preserving the artifacts and the documentary history of air crew evasion and escape in general, and of language-related evasion aids in particular. It is further dedicated to bringing that history to the public. The Museum can provide displays, slide shows, and lectures for aviation organizations, veterans reunions, and special events. Interested groups should contact the Museum at:

International Blood Chit Museum
R. E. Baldwin, Curator
PO Box 11131
Berkeley CA 94712-2131

Glossary

A

AAF: Army Air Force.
ACIC: Aeronautical Chart and Information Center.
ACIO: Air Combat Intelligence Officer.
AC of S: Assistant Chief of Staff.
ACS: Aeronautical Chart Service.
AEAF: Allied Expeditionary Air Forces.
A-Force: The section of MI9 charged with deception operations.
AGAS: Air Ground Aid Section. The section of MIS-X charged with Ameri can evasion and escape operations in China, headquartered in Kunming.
AIB: Allied Intelligence Bureau.
ATIS: Allied Translator and Interpreter Section.
ANGAU: Australian and New Guinea Administrative Unit.
ATC: Air Transport Command.
AVG: American Volunteer Group of the Chinese Air Force. A group of American airmen recruited from the U.S. armed forces to serve in the Chinese Air Force under the command of Claire L. Chennault from mid 1941 to 4 July 1942. This group is also known by its more popular nickname, the *Flying Tigers*.
A-2: Air forces Intelligence; air forces Intelligence Officer.

B

BAAG: British Army Aid Group. A British rescue organization headquartered in Kweilin, China.
Backflag: a Blood Chit.
Barter Kit: A kit containing barter items to be used by an evader to bribe or otherwise influence the natives of the country in which he is evading.
BEE: Bulletin on Escape and Evasion. A newsletter produced by the Twentieth Bomber Command which addressed evasion and escape issues.
Blood Chit: a device issued by a government to an air crew member bearing a message in one or more languages that identifies him as a member of a particular military establishment, and promises a reward for his return to friendly control.

C

CATF: China Air Task Force. The official US air arm in China from 4 July 1942 to 10 March 1943.
CBI: China-Burma-India Theater of Operations, later split into the China Theater of Operations and the India-Burma Theater of Operations.
CCP: Chinese Communist Party.
Chop: a seal, usually carved from stone, and typically stamped onto docu ments in red ink.
Cloth Chart: Cloth maps used by aircrew personnel as navigational aids while evading.
CMF: Central Mediterranean Forces.
CN: Chinese National Yuan, the monetary unit of the Nationalist Chinese government.
CPM: Captured Personnel and Material branch of G-2.
CRB: The monetary unit of the Japanese puppet government in China.
CTO: China Theater of Operations. See also CBI.

D

DMAAC: Defense Mapping Agency Aerospace Center.

E

E&E: Evasion and escape.
E-Group: Evasion & escape section of MI9 in India.
Emergency Sustenance Kit: Kits containing such items as first aid kits, emergency rations and survival equipment designed to sustain the life of aircrew personnel in such emergencies as forced landings in friendly or hostile isolated regions or ditching in open waters.
Escape: The act of gaining freedom from enemy custody by force or stealth.
Escapee: see Escaper.
Escaper: Any person in the military service who gains freedom from enemy custody by means of force or stealth.
Escape Flag: a Blood Chit.
ETO: European Theater of Operations.
Evader: Any person in the military service who has been cut off or isolated within hostile territory who is avoiding capture.
Evasion: The avoidance of capture by military personnel who have been cut off or isolated within hostile territory.
Evasion Aids: Equipment specifically designed for evasion and furnished to combat crew personnel in addition to their normal equipment. Examples of evasion aids are Blood Chits, Cloth Charts, Pointie Talkies, Evasion currency, and Barter Kits.
Evasion and Escape Intelligence: Evaluated information obtained through research, evasion and escape experiences, or through normal intelligence collection channels, that is necessary in improving evasion and escape plans or activities.
Evasion currency: Monetary medium of exchange which may be coins, gold bars, or Government notes of any country to be used by an evader to bribe or otherwise influence the natives of the country in which he is evading.
Evasion Nets: Established facilities within enemy territory organized for the purpose of assisting evaders' return to friendly military control.
Evasion Routes: Recommended routes of travel through enemy controlled territory prepared from research and use of evasion and escape intelligence.

F

FAI: Federation Aeronautique International.
FELO: Far East Liaison Office.

G

GBT: Gordon-Bernard-Tan Intelligence Network in French Indo-China.
Goolie: a testicle (slang).
Goolie Chit: a Blood Chit (British).
GPO: Government Printing Office.
G-2: Intelligence section, War Department.
G-3: Operations section, War Department.

H

HQ: Headquarters.

I

IBTO: India-Burma Theater of Operations. See also CBI.
ICD-ATC: India-China Division of the Air Transport Command.
Indigenous Aid: Assistance received from natives of the country through which the evader is passing.
IO: Intelligence Officer.
IS9: Intelligence School 9. The school used to train British Intelligence Officers. Later, organization supervising evasion and escape operations in the European Theater.

K

KMT: Kuomintang (Nationalist) Party, China.

L

LRDG: Long Range Desert Group.

M

ME: Middle East.
MEAF: Middle East Air Force.
MEF: Middle East Forces.
MIS: Military Intelligence Service.
MIS-X: Military Intelligence Service, Evasion and Escape Section. The secret section of the Captured Personnel and Materiel Branch of G-2 created to oversee American evasion and escape operations, headquartered at Fort Hunt, Virginia.
MIS-Y: Military Intelligence Service, Prisoner of War Section. The section of the Captured Personnel and Materiel Branch of G-2 created to oversee the interrogation of enemy Prisoners of War on American soil, headquartered at Fort Hunt, Virginia.
MTO: Mediterranean Theater of Operations.
MI9: Military Intelligence 9. The branch of British Military Intelligence created to oversee British and Commonwealth evasion and escape operations, headquartered at Wilton Park, England.

N

NACI-HO: Naval Air Combat Intelligence, Hydrographic Office
NEI: Netherlands East Indies.
N-Section: Evasion and escape section of A-Force

O

OCI: Office of the Coordinator of Information
Operational Rescue: Return of military personnel from hostile territory by military forces employing land, sea or air media.
Operational Return: Return of military personnel from hostile territory by evasion nets or other organized mechanisms, unassisted by conventional military forces.
OSS: Office of Strategic Services.
OWI: Office of War Intelligence.

P

POA: Pacific Ocean Area.
Pointie Talkies: Sheets or booklets containing basic phrases printed in both English and a foreign language or languages, to be used by aircrew personnel in communicating with people of foreign tongues by pointing to the phrases.
P/W&X: A staff branch of Headquarters, European Theater of Operations - U.S. Army, created to oversee American evasion and escape interests in the European Theater.

R

RAAF: Roayal Australian Air Force
RAF: Royal Air Force.
Ransom Note: an early name for a Blood Chit (British).
RAMP: Recovered Allied Military Personnel. This acronym is sometimes used as a verb. See also Recovered Personnel.
RCAF: Royal Canadian Air Force.
Recovered Personnel: All individuals returning or returned to friendly control from enemy-controlled or dominated territory by any means.
Repatriate: A recovered person who is returned voluntarily by the enemy during hostilities.
Rescue: Recovery of personnel in distress on land or at sea by use of search and rescue facilities.
Rescue Patch: the name by which Blood Chits were called in China.
Returnee: An evader who, having received no organized assistance, has returned to his nation's control through his own efforts.
RFC: Royal Flying Corps.

S

SAAF: South African Air Force.
SACO: Sino American Cooperative Organization. A joint Chinese and American force created by an alliance of the United States Navy and the Nationalist Chinese military.
SAS: Special Air Service.
SERE: Survival, Evasion, Resistance, and Escape.
SOE: Special Operations Executive.
Survival Kit: see Emergency Sustenance Kit.
SWPA: South West Pacific Area.
S-2: Ground forces Intelligence; ground forces Intelligence Officer.

U

U.K.: United Kingdom of Great Britain and Northern Ireland.
U.S.: United States of America.

V

14th VBS.: The Fourteenth Volunteer Bombardment Squadron. An international group of aviators who served under the command of Claire L. Chennault in China during the winter and spring of 1937-38.

W

WEA: Western European Area.

X

XX BC: Twentieth Bomber Command.

Notes

Chapter 1

1. Masculine nouns and pronouns such as he, him, airman, and crewman are used in this work. The increasingly active role of women in modern military forces is recognized and appreciated, but the masculine gender is used in this work for purposes of brevity and simplicity. There is no depreciatory connotation.

2. Letter from David B. Nash, Imperial War Museum, to R. E. Baldwin, 24 November 1988.

3. Joyce M. Hawkins and Robert Allen, Editors, *The Oxford Encyclopedic English Dictionary*, (Oxford, Clarendon Press, Oxford University Press, 1991), p. 258. Used by permission of Oxford University Press.

4. *Training Intelligence Report No. 3451, General Series No. 473.* Training Intelligence Service, Army Air Forces Training Command, 1945, RESTRICTED.

5. Woodford Agee Heflin, Editor, *The United States Air Force Dictionary*, (Air University Press, 1956), p. 83.

6. R. E. Baldwin's definition of a classic Blood Chit for use in this book. This is not an official military definition.

7. "Pointie Talkies." *Air Sea Rescue Bulletin*, No. 7. Air Sea Rescue Agency, Washington, D.C., January 1945, RESTRICTED.

8. *Bulletin on Escape and Evasion (BEE).* HQ XX BC - Intelligence Section, Bulletin No. 10, 8 September 1944, SECRET.

Chapter 2

1. Bill Gunston, Editor, *Chronicles of Aviation* (London, Chronicle Communications Ltd., 1992), p.17.

2. Text of Blanchard's "passport" as given on an exhibit card in display at the National Museum of American History, Washington, D.C., March 1995.

3. Statement attributed to Blanchard as given on an exhibit card in display at the National Museum of American History, Washington, D.C., March 1995.

4. William H. Northracker, "Afghan Death March" in *Soldier of Fortune*, Soldier of Fortune Magazine, Inc., February 1989, pp. 64-85.

5. Lee Siwecki, in a conversation with R. E. Baldwin.

6. Letter from Bill Robie to Thomas Wm McGarry, 5 January 1993.

7. Lee Kennett, *The First Air War: 1914-1918*. (New York, The Free Press, 1991), p. 18, copyright © 1991 by Lee Kennett.

8. *The First Air War: 1914-1918* by Lee Kennett, pp. 18-19. Copyright © 1991 by Lee Kennett. Reprinted with permission of The Free Press, a division of Simon & Schuster.

9. C. G. Sweeting, *Combat Flying Equipment, U.S. Army Aviators' Personal Equipment 1917-1945* (Washington, D.C., Smithsonian Institution, 1989), p. 178.

10. U.S. Air Force Museum, case no. N527, March 1995.

11. Arch Whitehouse, *The Legion of the Lafayette*, (New York, Doubleday, 1962). Reprinted by permission of Doubleday, a division of Bantam Doubleday Dell Publishing Group, Inc.

12. Robie letter to McGarry.

Chapter 3

1. *Training Intelligence Report No. 3451*, General Series No. 473, Training Intelligence Service, Army Air Forces Training Command, 1945, RESTRICTED.

2. Norman L.R. Franks, *First In The Indian Skies*, (Lincoln, U.K., Life Publications Ltd, 1981), p. 14. Reprinted by permission of Life Publications Ltd./31 Squadron RAF Association

3. Letter from Richard Simpson to R.E. Baldwin, 28 November 1991.

4. Joyce M. Hawkins and Robert Allen, editors, *The Oxford Encyclopedic Dictionary* (Oxford, Clarendon Press, Oxford University Press, 1991), p. 609. Used by permission of Oxford University Press.

5. Franks, *First In The Indian Skies*, p. 14. Reprinted by permission of Life Publications Ltd.

6. Air Vice Marshall A.G. Dudgeon, *The Luck of the Devil* (Shrewsbury, U.K., Airlife Publishing Ltd, 1985) pp. 61-62. Reprinted by permission of Airlife Publishing Ltd.

7. Eric Nicholls, "North West Frontier", *FlyPast Special*, April 1993, p. 14. Reprinted by permission of FlyPast.

8. Letter from Tony Dudgeon to Thomas Wm. McGarry, 8 June 1992.

9. Dudgeon letter to McGarry.

10. Franks, *First In The Indian Skies*, p. 20.

11. Franks, *First In The Indian Skies*, p. 20. Reprinted by permission of Life Publications Ltd./31 Squadron RAF Association.

12. Illustration prepared from a photocopy from the Imperial War Museum, courtesy of Jeffrey D. Guidry. The museum was unable to find the artifact at the time photos were requested, and did not object to use of the photocopy subject to this note.

13. Information courtesy of Jeffrey D. Guidry.

14. Translation by Amir Kaifi, courtesy of Homayun Karim.

15. Information courtesy of Jeffrey D. Guidry.

16. Translation by Amir Kaifi, courtesy of Homayun Karim.

17. From a photocopy of the note provided to R. E. Baldwin by the Imperial War Museum.

18. Translation by Amir Kaifi, courtesy of Homayun Karim.

19. Translation by Amir Kaifi, courtesy of Homayun Karim.

20. Illustration from a photocopy from the Imperial War Museum, courtesy of Jeffrey D. Guidry. The museum was unable to find the artifact at the time photos were requested, and did not object to use of the photocopy subject to this note.

21. Phillip Jarrett, "By Day and By Night: Wildbeest and Vincent" in *Aeroplane Monthly*, February 1995, p. 20.

22. S. J. Carr, *You Are Not Sparrows*, (Ian Allen Ltd., Shepperton, Surrey, 1975), p. 36. Reprinted by permission of Ian Allen Ltd.

23. Chaz Bowyer, *RAF Operations 1918-1938*, (London, Wm. Kimber, 1988), p. 24.

24. T. E. Lawrence, *Revolt in the Desert*, (New York, George H. Doran Company, 1927), p. 121, and T. E. Lawrence, *Seven Pillars of Wisdom*, (Garden City, New York, Doubleday, Doran & Company, Inc., 1937), p. 321.

25. Letter from Don R. Neate to Baldwin and McGarry, 2 February 1993.

26. Carr, *You Are Not Sparrows*, p. 51. Reprinted by permission of Ian Allen Ltd.

27. Letter from K.S. Batchelor to R. E. Baldwin, 12 August 1993.

28. K.S. Batchelor, quoted in letter from M.R.D. Foot to R.E. Baldwin, 2 March 1993. Quoted by permission of K.S. Batchelor.

29. K.S. Batchelor, quoted in Foot letter to Baldwin. Quoted by permission of K.S. Batchelor.

30. Illustration is from a photocopy courtesy of Donald R. Neate.

31. Translation by Salem Attiga, courtesy of Haidook Hovanesian.

32. Information courtesy of David Vincent.

Chapter 4

1. Ray Wagner, Prelude to Pearl Harbor The Air War in China *1937-1941* (San Diego, CA, San Diego Aerospace Museum, 1991), p.4.

2. The Pinyin system for the anglicization of Chinese, preferred by the current regime in China, is not used in this work. For consistency, the Wade system is used in this work because it was used during the historical period covered by this work, it is used in the written material from which this work was compiled, and it is used by the veterans who served in China that contributed to this work.

3. Claire Lee Chennault, *Way of a Fighter*, (New York, G.P. Putnam's Sons, 1949), p. 41

4. Ken Jernstedt in a speech at the AVG Reunion, Portland, Oregon, August 1995

5. Jack Sampson, *Chennault*, (New York, Doubleday, 1987), p. 27

6. Witold Rodzinski, *The Walled Kingdom, A History of China from Antiquity to the Present* (New York, The Free Press, 1984), p. 352

7. Art Chin in a conversation with R. E. and Shirlaine Baldwin, Portland, Oregon, August 1995

8. William Labussiére and Gerry Beauchamp, "The Truth About the 14th Squadron Foreign Volunteers In China" in AAHS Journal, Number 4, American Aviation Historical Society, Winter 1987, p.285.

9. Agnes Chan in a conversation with R. E. Baldwin.

10. Edward L. Leiser, "Memoirs of Pilot Elwyn H. Gibbon, The Mad Irishman", *AAHS Journal*, Volume 23, Number 1, American Aviation Historical Society, 1st Quarter 1978, p.10

11. Malcolm Rosholt, *Flight in the China Air Space*, (Rosholt, WI, Rosholt House, 1984), p. 106.

12. Leiser, "Memoirs of Pilot Elwyn H. Gibbon", pp.12-16

13. Labussiére/Beauchamp, "The Truth About the 14th Squadron", p.288.

14. Translation by Agnes Chan

15. Translation by Agnes Chan

16. Translation by Agnes Chan

17. Translation by Agnes Chan

18. Letter from Gary Hammill to Thomas Wm. McGarry, 16 August 1994.

19. Charles Bond and Terry Anderson, *A Flying Tiger's Diary*, (College Station, TX, Texas A&M University Press, 1984, p. 34. Reprinted by permission of Texas A&M University Press.

20. Letter from J.R. "Dick" Rossi to R. E. Baldwin, 30 March 1992.

21. Illustration is from a photocopy courtesy of J.R. "Dick" Rossi.

22. Rossi letter to Baldwin

23. Erik Shilling, in conversations with R.E. Baldwin, Oshkosh, Wisconsin, July 1991, and Portland, Oregon, August 1995.

24. Ken Jernstedt in a conversation with R.E. Baldwin, July 1992.

25. Jernstedt conversation with Baldwin.

26. Bond and Anderson, *A Flying Tiger's Diary*, p.67.

27. Jernstedt conversation with Baldwin.

28. R. T. Smith, *Tale of a Tiger*, (Van Nuys, CA, Tiger Originals, 1986), p. 184. Reprinted by permission of R.T. Smith.

29. Rossi letter Baldwin.

30. Translation by Agnes Chan.

31. Translation by Agnes Chan.

32. Translation by Agnes Chan.

33. Rossi letter to Baldwin.

34. Translation by Agnes Chan.

35. Agnes Chan in a conversation with R. E. Baldwin.

36. Agnes Chan in a conversation with R. E. Baldwin.

37. Don Moser et al, *China-Burma-India*, (Alexandria, VA, Time-Life Books, 1978), p. 70.

38. Rossi letter to Baldwin.

39. Larry M. Pistole, *The Pictorial History of the Flying Tigers*, (Orange, Virginia, Moss Publications, 1981), color plates.

40. Jernstedt conversation with Baldwin.

41. D.L. "Tex" Hill in a conversation with R.E. Baldwin, Oshkosh, WI, July 1991.

42. Jernstedt conversation with Baldwin.

43. Stuart Leuthene and Oliver Jensen, *High Honor, Recollections by Men and Women of World War II Aviation* (Washington, D.C, Smithsonian Institution Press, 1989) pp. 30-31.

44. Bond and Anderson, *A Flying Tiger's Diary*, p. 167. Reprinted by permission of Texas A&M University Press.

45. Letter from Charles Bond to R. E. Baldwin, 9 January 1995.

46. Hans Onderwater, *Second To None* (London: UK, Airlife Publishing Ltd., 1992), p. 31.

47. Jernstedt conversation with Baldwin.

48. Jernstedt conversation with Baldwin.

Chapter 5

1. *Information Concerning MI9 and P/W Matters*, undated WWII British Document, stamped CONFIDENTIAL at bottom, SECRET handwritten across top.

2. *Information Concerning MI9 and PW Matters.*

3. M.R.D. Foot and J.M. Langley, *MI9, Escape and Evasion 1939-1945*, (Boston-Toronto, Little, Brown & Company, 1977), p.77-86.

4. *MIS-X Manual on Evasion, Escape and Survival*, (Washington, D.C., U.S. War Department, February 1944 with revisions through 1945), SECRET.

5. Geoffrey Perret, *Winged Victory The Army Air Forces in World War II*, (New York, Random House, 1993), p. 176-186.

6. *Prisoners of War*, 3-3926 AF, Washington, D.C., U.S. Army Air Forces, undated, RESTRICTED.

7. *MIS-X Manual*

8. *MIS-X Manual*

9. *Capture Intelligence, Escape Bulletin No. 5*, Northwest African Air Forces, A-2 Section, 4 May 1943, SECRET.

10. *Prisoners of War.*

11. *Capture Intelligence, Escape Bulletin No.5.*

12. *Capture Intelligence, Escape Bulletin No. 11*, Northwest African Air Forces, A-2 Section, 17 August 1943, SECRET.

13. *Prisoners of War.*

14. *Capture Intelligence, Escape Bulletin No.5.*

15. Charles Cruickshank, *Deception in World War II*, (New York, Oxford University Press, 1980) p. 20. Reprinted by permission of Mrs. Cruickshank.

16. Robert J. Jakeman, The Divided Skies (Tuscaloosa, AL, University of Alabama Press, 1992) p. 23.

17. Translation by Tsion Arega, courtesy of Shirlaine Baldwin.

18. Sqn Ldr R. A. Collins DFC and Sqn Ldr F. Bavin Smith DFC, "Operation Wingate" in *Aeroplane Monthly*, September 1993, page 28.

19. Illustration is from a photocopy courtesy of A. C. Rawlinson.

20. Letter from A. C. Rawlinson to Thomas Wm. McGarry, 18 May 1993.

21. Letters from Ian Glen to Baldwin and McGarry, 10 July 1993 and 16 August 1993.

22. Letter from Jack Butterfield to Baldwin and McGarry, 15 May 1992.

23. Illustration is from a black-and-white photo courtesy of Don Townend.

24. Letters from Don Townend to Baldwin and McGarry, 19 July 1992 and 2 May 1993.

25. Letter from Gabriel Taschereau to R.E. Baldwin, 11 August 1993.

26. Illustration is from a photocopy appearing in *Falling Leaf*, quarterly magazine of the Psywar Society, Vol. XXI, No. 1, Whole No. 84 (St. Albans, Herts., U.K., The Psywar Society, 1979). Reprinted by permission of R. G. Auckland, Hon. Editor.

27. Peter H. Robbs, "Blood Chits" in *Falling Leaf*, quarterly magazine of the Psywar Society, Vol. XXI, No. 1, Whole No. 84 (St. Albans, Herts., U.K., The Psywar Society, 1979) p. 5. Reprinted by permission of R. G. Auckland, Hon. Editor.

28. Letters from Donald Townend to Baldwin and McGarry, 19 July 1992 and 2 May 1993.

29. Photocopy of photo album and letter from William Fulcher to R.E. Baldwin, 12 December 1994.

30. Ralph Barker, *Down In The Drink*, (London, Chatto & Windus, 1955), P. 208.

31. Barker, *Down In The Drink*, p. 210. Reprinted by permission of Ralph Barker.

32. Letter from Bob Gilroy to Baldwin and McGarry, 15 April 1992.

33. Letter from David Underwood to Baldwin and McGarry, 5 March 1993.

34. Letter from D. G. Upward to Baldwin and McGarry, 28 April 1992.

35. Foot and Langley, *MI9, Escape and Evasion 1939-1945*, p.26-27.

36. Lloyd R. Shoemaker, *The Escape Factory - The Story of MIS-X*, (New York, St. Martin's Press, 1990), pp. 9-11.

37. Shoemaker, *The Escape Factory*, pp. 11-17.

38. Directive of 6 October 1942 quoted in *Memorandum to The Chief of Staff*, 11 December 1942 over the signature of Geo. V. Strong, Major General, A.C. of S., G-2, SECRET.

39. Shoemaker, *The Escape Factory*, p.12.

40. *MIS-X Manual.*

41. Letter from Joseph Donato to R.E. Baldwin, 18 December 1991.

42. *The National Geographic Magazine*, National Geographic Society, Washington, D. C., June 1943, pp. 771 and 774.

43. Letters from William F. Beck to Baldwin and McGarry, 12 February 1992 and 4 March 1992.

44. This letter has been reprinted with permission from the Spring 1991 issue of Foundation magazine. Letter to the editor, *Foundation*, Museum of Naval Aviation, Pensacola, FL., Volume 12, Number 1, Spring 1991.
45. S/Ldr P. G. Hering, *Customs and Traditions of the Royal Air Force* (Aldershot, Hants, U.K., Gale & Polden Ltd.)

Chapter 6

1. Lloyd R. Shoemaker, *The Escape Factory - The Story of MIS-X*, (New York, St. Martin's Press, 1990), pp. 9-11.
2. M.R.D. Foot and J.M. Langley, *MI9, Escape and Evasion 1939-1945*, (Boston and Toronto, Little Brown and Company, 1979), p. 159.
3. Mitchell G. Bard, Forgotten Victims, (Boulder CO, Westview Press, Inc., 1994) p.44.
4. *Information Concerning MI-9 and PW Matters*, undated WWII British document, SECRET.
5. *MIS-X Manual on Evasion, Escape, and Survival*, (Washington D.C., U.S. War Department, February 1944 with revisions through 1945) SECRET.
6. *MIS-X Manual*.
7. Brig. Gen. Partridge, *Combat Crew Handbook*, USAAF, 8th AF, 3rd Bomb Division, undated, SECRET.
8. *MIS-X Manual*.
9. *E&E Bulletin #46: EVASION*, HQ, 12th AF, Office of A. C. of S., A-2, 21 July 1944, SECRET.
10. Tom Moon, *This Grim and Savage Game: O.S.S. and The Beginning of U.S. Covert Operations in WW II*,(Garden Grove, CA, Burning Gate Press, 1991), p. 278.
11. *MIS-X Manual*.
12. "Luftwaffe Bomber Pilot, Summer 1940" in *Militaria Magazine*, No. 8, September 1994, p. 20.
13. *Anecdotes on Evasion and Escape for the use of Intelligence Officers*, Headquarters, ETOUSA, Office of A. C. of S., G-2, MIS Detachment, 1 August 1943, SECRET.
14. *Memorandum to Lt. Col Holt*, P/W&X Detachment, MIS-X, ETOUSA, HQ, A-2, 11 July 1943, SECRET.
15. *E&E Bulletin #28*, HQ, 12th AF, A. C. of S., A-2 Section, 29 December 1943, SECRET.
16. John L. Lowden, *Silent Wings at War: Combat Gliders in World War II*, (Washington, D.C, Smithsonian Institution Press, 1992) photo caption.
17. *Information concerning MI 9 and PW Matters*.
18. *Memorandum to Lt. Col Holt*.
19. Major James M. Inks, *Eight Bailed Out* (New York, W. W. Norton & Company, Inc., 1954), p. 12.
20. Edi Selhaus, *Evasion and Repatriation, Slovene Partisans and Rescued American Airmen in World War II*, (Manhattan, KS, Sunflower University Press, 1993) p. 118-119.
21. *MIS-X Manual*.
22. *History and Operational Use of Blood Chits, World War II and Korea*, Research Studies Institute, Air University, Maxwell AFB, undated, circa 1957, SECRET.
23. "Chapter 2: E Group Escape Aids", *E Group Bulletin*, Undated.
24. "Chapter 2: E Group Escape Aids".
25. *Escape Bulletin # 26*, HQ, 12th AF, A. C. of S., A-2 Section, 29 December 1943, SECRET.
26. *E&E Bulletin #46: EVASION*.
27. *Blood Chits Distributed to 1st Bomb Wing*, Memo to A-2, 1st Bomb Wing from HQ, ETOUSA, Office of A.C. of S., G-2, P/W and X Detachment, MIS, 14 September 1943.
28. Foot and Langley, *MI9*, p. 184
29. Jean de Pange, "Normandie-Niemen" in *Icare*, No. 62, Spring/Summer 1972, p. 156.
30. *MIS-X Manual*.
31. "Escape and Evasion, General Information" in *The History of "Frantic" - American Shuttle-Bombing to and from Russian Bases*, SECRET.
32. "Intelligence Annex No. 3, Operation Frantic" by Major Vernon E. Fairbanks, A.C. of S., A-2, in *The History of "Frantic" - American Shuttle-Bombing to and from Russian Bases*, SECRET.
33. Letter from Jack Ilfrey to R.E. Baldwin, 9 December 1991.
34. Letter from J. Wesley Coe to Baldwin and McGarry, 16 May 1992.
35. Letter to the Editor from W. W. Wallace, *FlyPast*, July 1991.
36. John A. Scott, in a conversation with R. E. Baldwin, May 1995.
37. Letter from L. D. Volkmer to Baldwin & McGarry, 3 May 1994.
38. Illustration is from color photocopies, courtesy of J. F. Notman.
39. Letter from J. F. Notman to authors, 1993.
40. Translation by R. E. Baldwin.
41. Flt/Sgt William Young, "Missing: Believed Killed" in *Royal Air Force Flying Review*, June 1954, Vol. 9. No.9.
42. Letter from Ben L. Donahue to R.E. Baldwin, 3 January 1992.
43. Letter from Marvin C. Bigelow to R.E. Baldwin, 14 November 1991.
44. Letter from Malcom E. Corum to Baldwin and McGarry, 20 February 1992.
45. *Equipment Requirements*, to I.S.9 (z) (Attn: S/Ldr Whitehead) from Maj. Irvin H. Luiten, HQ, ETO, Det. C, P/W&X Detachment, MIS, 12 March 1945.
46. *War Diary For August and September 1945*, MIS-X Technical Section, Prepared by Leo H. Crosson, Maj., Inf., Chief Technical Section, SECRET.

Chapter 7

1. *Specimen Lecture on Evasion and Escape in the Far East; All Services, All Arms; For the Guidance of Intelligence Officers*, Main HQ, E-Group, South East Asia and India Commands, New Delhi, February 1945.
2. M.R.D. Foot and J.M. Langley, *MI9, Escape and Evasion 1939-1945*, (Boston-Toronto, Little, Brown and Company, 1979), p. 283.
3. Foot and Langley, *MI9*, p. 264 265.
4. Lloyd R. Shoemaker, *The Escape Factory - The Story of MIS-X*, (New York, St. Martin's Press, 1990), pp. 220-221.
5. *MIS-X Manual on Evasion, Escape and Survival*, (Washington, D.C., U.S. War Department, February 1944 with revisions through 1945), SECRET.
6. *MIS-X Manual*.
7. *MIS-X Manual*.
8. Foot and Langley, *MI9*, p.270.
9. Maj. Richard Vernon Hill, ret., *My War With Imperial Japan - Escape and Evasion*, (New York, Los Angeles, Chicago, Vantage Press, 1989), p. 303.
10. Vice Admiral Milton E. Miles and Hawthorne Daniel, *A Different Kind of War*, (New York, Doubleday & Co., 1967), Chapter 6.
11. Foot and Langley, *MI9*, p. 271.
12. Miles and Daniel, *A Different Kind of War*, p. 119.
13. Geoffrey T. Barker, *A Concise History of U.S. Army Special Operations Forces with Lineage and Insignia*, (Tampa, Anglo-American Publishing Company,1988), pp. 1-2.
14. Foot and Langley, *MI9*, p. 272.
15. Claire Lee Chennault, *Way of a Fighter*, (New York, G.P. Putnam's Sons, 1949), p. 214.
16. *What You Should Know About China*, (Calcutta, Indian Press Ltd., 1945), p. 26.
17. *MIS-X Manual*.
18. *MIS-X Manual*.
19. *MIS-X Manual*.
20. *MIS-X Manual*.
21. *MIS-X Manual*.
22. *MIS-X Manual*.
23. *MIS-X Manual*.
24. *MIS-X Manual*.
25. Sources will be quoted verbatim in this work, including those from the World War II period containing offensive terms such as "Jap". Likewise, illustrations that may stereotype some nationalities will be used without alteration. The difficult decision not to censor the material was made in the belief that the unsavory parts of our past should not be covered up, that readers deserve access to references, and that readers are intelligent enough to separate the prejudices of the past from relevant historical material. No offense is intended by the inclusion of this material.
26. *MIS-X Manual*.
27. *Report, AGAS-China*, from HQ AGAS-China to Col. Russell H. Sweet, Chief CPM Branch, Attn: Col. J. E. Johnston, G.S.C., 22 August 1944, TOP SECRET.
28. *History and Operational Use of Blood Chits, World War II and Korea*, prepared by Research Studies Institute, Air University, Maxwell Air Force Base, Alabama, at the request of the Strategic Air Command dated 27 June 1957, SECRET.
29. Miles and Hawthorne, *A Different Kind of War*, p. 59.
30. *MIS-X Manual*.
31. *Air Notes From China*, ComNavGp China, April 1945.
32. *MIS-X Manual*.

33. "Bailout - Robert V. Reynolds" in *China Airlift - The Hump*, (Poplar Bluff, Missouri, China-Burma-India Hump Pilots Association, Inc., 1980) p. 267.
34. Hill, *My War With Imperial Japan*, p. 312 & 372.
35. *What You Should Know About China*, p. 41.
36. *Material for "E" Vest*, AGAS-China, HQ AGAS-China, 20 August 1944, SECRET.
37. "Bailout - Joseph R. Whelan" in *China Airlift - The Hump*, p. 272.
38. *Walkout Report on 1st Lt. Freeling H. Clower*, HQ 89th Fighter Squadron, Office of the Intelligence Officer, 17 January 1944, SECRET.
39. Gen. Curtis E. LeMay and Bill Yenne, *Superfortress The B-29 and American Air Power* (New York, McGraw-Hill Book Company, 1988), p. 86-87.
40. *Walkout Report on 1st Lt. Harold W. Robbins*, to Commanding Officer, 89th Fighter Squadron, from HQ 89th Fighter Squadron, Office of the Intelligence Officer, 1 May 1944, SECRET.
41. *Walkout Report of Capt. Ransom W. Rideout*, USAAF, 31 July, 1945, SECRET.
42. "Chapter 2: E-Group Escape Aids" from an undated *E-Group Bulletin*.
43. "Chapter 2", undated *E-Group Bulletin*.
44. *Specimen Lecture on Evasion and Escape in the Far East*.
45. *MIS-X Manual*.
46. Translation by Agnes Chan.
47. Translation by Agnes Chan.
48. Letter from C. H. Anderson to Baldwin and McGarry, 13 August 1993.
49. Letter from Jimmy Bell to Baldwin and McGarry, 20 October 1993.
50. Letter from D.W. Walker to Baldwin and McGarry, 5 May 1992.
51. *Code for "E" Material*, undated, GS 1e, MOST SECRET.
52. *Code for Describing Escape Aids*, HQ India Task Force, 26 August 1943, SECRET.
53. Memorandum for Major A. R. Wichtrich, MIS-X, China, 24 February 1944.
54. Letter from Russell A. Bales to R. E. Baldwin, undated, ca June 1992.
55. "Blood Chits" in *Air Sea Rescue Bulletin*, No. 11, May 1945, RESTRICTED.
56. Letter from Everett E. Swanson to R. E. Baldwin, 7 May 1992.
57. Letter from Horace O. Bedsole to R. E. Baldwin, 17 January 1992.
58. *War Diary for February 1945*, HQ IBT, MIS-X India, SECRET.
59. Letter from William A. Rooney to R. E. Baldwin, 30 October 1991.
60. *War Diary for February 1945*.
61. Letter from D. W. Walker to Baldwin and McGarry, 5 May 1992.
62. *War Diary* for 1945, HQ IBT, MIS-X India, SECRET.
63. Letter from Bob Gilroy to Baldwin and McGarry, 15 April 1992.
64. Letter from Merv. M. McGrane to Thomas Wm. McGarry, undated.
65. Translation by Vince Lau.
66. *War Diary for February 1945*.
67. *Memo to A. C. of S., G-2, HQ Forward Echelon , USAF, CBI Theater*, Attn: Maj. A. R. Wichtrich, MIS-X Section, from Russell H. Sweet, Col., GSC, Chief CPM Branch, 9 September 1944, SECRET.
68. Letter from Russell A. Bales to R. E. Baldwin, undated, ca June 1992.
69. *Memo to Col. Russell H. Sweet, G.S.C. Chief CPM Branch*, from HQ AGAS-China, 20 August 1944, SECRET.
70. *Report, AGAS-China*, from HQ AGAS-China to Col. Russell H. Sweet, Chief CPM Branch, Attn: Col. J. E. Johnston, G.S.C., 22 August 1944, TOP SECRET.
71. *Memo to A. C. of S., G-2, HQ Forward Echelon , USAF, CBI Theater*, Attn: Maj. A. R. Wichtrich, MIS-X Section, from Russell H. Sweet, Col., GSC, Chief CPM Branch, 9 September 1944, SECRET.
72. Translation by Agnes Chan.
73. Translator Agnes Chan, in a conversation with R. E. Baldwin.
74. *Memo from HQ AGAS-China to Chief, CPM Branch, MID*, Attn: Col. J. E. Johnston, G.S.C., 9 October 1944, SECRET.
75. *Memo to A. C. of S., G-2, HQ Forward Echelon , USAF, CBI Theater*, Attn: Maj. A. R. Wichtrich, MIS-X Section, from Russell H. Sweet, Col., GSC, Chief CPM Branch, 9 September 1944, SECRET.
76. *Letter to All Concerned, Translation of Message on Blood Chits*, General Headquarters, United States Army Forces, Pacific, MIS-X SECTION, 19 JULY 1945, RESTRICTED.
77. Translation by Agnes Chan.
78. Translator Agnes Chan, in a conversation with R. E. Baldwin.
79. Letter from William A. Rooney to R. E. Baldwin, 30 October 1991.
80. Letter from Roger Perreault to R. E. Baldwin, 10 January 1990.
81. *Memo to A. C. of S., G-2, HQ USAF*, from Russell H. Sweet, Colonel, GSC, Chief CPM Branch, 31 July 1945, CONFIDENTIAL.
82. Letter from William A. Rooney to R. E. Baldwin, 30 October 1991.
83. *Letter to All Concerned, Translation of Message on Blood Chits.*.
84. Translation by R. E. Baldwin.
85. *Memo to Commanding Officer*, 80th Fighter Group and Commanding Officer, 311th Fighter Group, from Headquarters, 5320th Air Defense Wing (Provisional), 25 January 1944, RESTRICTED.
86. *Memo to Col. Russell H. Sweet*, G.S.C., Chief, CPM Branch, from HQ AGAS-China, 20 August 1944, SECRET.
87. Letter from Horace O. Bedsole to R.E. Baldwin, 17 January 1992.
88. *Bulletin on Escape and Evasions (BEE)*, HQ XX BC, Office of A.C. of S., A-2, Bulletin No. 9, 2 August 1944, SECRET.
89. Letter from John L. Cella to R. E. Baldwin, 25 November 1991.
90. *Memo to Col. Russell H. Sweet*, G.S.C. Chief CPM Branch, from HQ AGAS-China, 20 August 1944, SECRET.
91. Memo, Case # 45, HQ China Theater Search Detachment, 15 January 1946.
92. *Letter to A. C. of S., G-2, HQ, USAF*, Attn: Lt. Col. A. R. Wichtrich, HQ, AGAS from Russell H. Sweet, Col.. GSC, Chief, CPM Branch, 15 February 1945.
93. *War Diary for August and September 1945*, MIS-X Technical Section, Prepared by Maj. Leo H. Crosson, Chief of Technical Section, SECRET.
94. *War Diary for August and September 1945*.
95. Letter from William Dreeland to R. E. Baldwin, 13 January 1990.
96. *Bulletin on Escape and Evasion (BEE)*, HQ XX BC, Office of A.C. of S., A-2, Bulletin No 8, 26 June 1944, SECRET.
97. "Chapter 2", *E-Group Bulletin*.
98. *Walkout Report on 1st Lt. Harold W. Robbins*.
99. Letter from Bob Gilroy to Baldwin and McGarry, 15 April 1992.
100. *AGAS Journal* for 1944, SECRET.
101. *AGAS Journal* for 1944.
102. *Report, AGAS-China*, from HQ AGAS-China to Col. Russell H. Sweet, Chief CPM Branch, Attn: Col. J. E. Johnston, G.S.C., 22 August 1944, TOP SECRET.
103. *Memo to A. C. of S., G-2, HQ Forward Echelon , USAF, CBI Theater*, Attn: Maj. A. R. Wichtrich, MIS-X Section, from Russell H. Sweet, Col., GSC, Chief CPM Branch, 9 September 1944, SECRET
104. *Intelligence Summary No. 269*, 26 April 1945, SECRET.
105. *War Diary for August and September 1945*, MIS-X Technical Section, Prepared by Maj. Leo H. Crosson, Chief, Technical Section, SECRET.
106. *War Diary for August and September 1945*.
107. *Walkout Report of Capt. Ransom W. Rideout*.
108. Translation by Vince Lao.
109. Translation by Vince Lau.
110. Translation by Agnes Chan.
111. Translation by Vince Lau.
112. Translation by Vince Lau.
113. Translation by Vince Lau.
114. Translation by Agnes Chan.
115. Translation by Agnes Chan.
116. Translation by Vince Lau.
117. Translation by Agnes Chan.
118. Translation by Paul Tran, courtesy of Matt Baldwin.
119. Translation by R. E. Baldwin.
120. Agnes Chan, in a conversation with R. E. Baldwin.
121. Agnes Chan, in a conversation with R. E. Baldwin.
122. Translation by Agnes Chan.
123. Letter from Russell A. Bales to R. E. Baldwin, undated, ca June 1992.
124. Translation by Agnes Chan.
125. Translation by Agnes Chan.
126. Agnes Chan in a conversation with R. E. Baldwin.
127. Translation by Agnes Chan.
128. Translation by Agnes Chan.
129. Translation by Agnes Chan.
130. Translation by Agnes Chan.
131. Agnes Chan in a conversation with R. E. Baldwin..
132. Translation by Agnes Chan.
133. *History and Operational Use of Blood Chits*.
134. "Bailout - Allen B. Robbins" from *China Airlift: The Hump, Vol.*

II (Poplar Bluff, MO, China-Burma-India Hump Pilots Association, Inc.) p. 128.

135. *China Walkout, Escape and Evasion Bulletin # 17*, HQ Tenth Air Force, Office of the A.C. of S., A-2, Air Jungle Rescue Section, 1 April 1945, SECRET.

136. *Walkout Report of Captain Kenneth E. George*, 76th Fighter Squadron, 23rd fighter Group, Office of the Intelligence Officer, 25 February 1944, SECRET.

137. *Bulletin on Escape and Evasion (BEE)*, No. 10, HQ XX BC-Intelligence Section, 8 September 1944, SECRET.

138. "B-24 Bailout of Lt. Ralph W. Gould" in *China Airlift-The Hump*, p. 236.

139. "Bailout - W.F.'Bill' Aulenbach", *China Airlift-The Hump*, p.236.

140. Letter from John J. Munro to Thomas Wm. McGarry, 2 March 1992.

Chapter 8

1. John W. Dower, *War Without Mercy: Race & Power in the Pacific War* (New York, Pantheon Books, 1986), p. 48.

2. *United States Pacific Fleet and Pacific Ocean Areas-Report of Intelligence Activities in the Pacific Ocean Areas*, CINCPAC-CINCPOA, 15 October 1945, SECRET.

3. *Report of MIS-X Section, SWPA*, February 1944, SECRET.

4. *Report of MIS-X Section, SWPA.*

5. *Report of MIS-X Section, SWPA.*

6. *Memorandum for Cmdr. John L. Riheldaffer, USN, Subject: Naval Officers Trained by MIS-X*, from Lt. Basil D. Henning, Navy Liaison Officer, MIS-X, 5 July 1944, SECRET.

7. *Monthly Report, MIS-X Activities, SWPA*, GHQ, SWPA, 18 May 1945, SECRET.

8. *Brief History of the G-2 Section, GHQ, SWPA*, undated post-war report.

9. *MIS-X Manual on Evasion, Escape and Survival*, U.S. War Department, Washington, D.C., February 1944, with revisions through 1945, SECRET.

10. *United States Pacific Fleet and Pacific Ocean Areas Evasion and Escape Manual*, CINCPAC-CINCPOA, 30 January 1945, SECRET.

11. *MIS-X Manual.*

12. *MIS-X Manual.*

13. *MIS-X Manual.*

14. *Castaway's Baedeker To the South Seas*, prepared by the Objective Data Section, Intelligence Center, Pacific Ocean Areas (1942), p. 29.

15. *Report of MIS-X Section, SWPA.*

16. *Report of MIS-X Section, SWPA.*

17. *Report of MIS-X Section, SWPA.*

18. *MIS-X Manual.*

19. *MIS-X Manual.*

20. *MIS-X Manual.*

21. *Survival on Land and Sea (NAVAER 13-1-501)*, Publications Branch, Office of Naval Intelligence, U.S. Navy, 1944.

22. *MIS-X Manual.*

23. *Equipment Distributed During Month of April, 1945*, Report, MIS-X, SWPA, SECRET.

24. *Funds for MIS-X Section*, Memo from HQ Allied Air Forces to G-2 GHQ, 22 February 1944, SECRET.

25. *MIS-X Manual.*

26. *MIS-X Manual.*

27. *MIS-X Procurement Authority and Fund Requirements*, Memo from HQ Allied Air Forces to GHQ G-2, 26 February 1944, SECRET.

28. Illustration from a color photocopy courtesy of Jan Madarasz.

29. Letter from Jan Madarasz to R. E. Baldwin, 14 January 1994.

30. *MIS-X Manual.*

31. Letter from Maj. Donald J. Gibb to Baldwin and McGarry, undated.

32. Letter from John F. Russell to Thomas Wm. McGarry, 17 May 1994.

33. *Report of MIS-X Section, SWPA.*

34. *Memorandum for Lt. Cmdr. Ward Cheney, USNR, OP-16-V* from Lt. Basil D. Henning, Navy Liaison Officer, MIS-X, 4 July 1944, SECRET.

35. *United States Pacific Fleet and Pacific Ocean Areas - Report of Intelligence Activities in the Pacific Ocean Areas*

36. *Survival on Land and Sea.*

37. *United States Pacific Fleet and Pacific Ocean Areas Evasion and Escape Manual.*

38. M.R.D. Foot and J.M. Langley, *MI 9, Escape and Evasion 1939-1945*, (Boston-Toronto, Little, Brown and Company, 1979), p. 289.

39. *Final Report*, MIS-X Section, POA, September 1945, SECRET.

40. Letter from Mrs. Walter D. Innis to Baldwin and McGarry, 21 September 1993.

41. Untitled Navy manual, stamped CONFIDENTIAL.

42. *United States Pacific Fleet and Pacific Ocean Areas Evasion and Escape Manual.*

43. Steve Hasegawa in a conversation with R. E. Baldwin, October 1995.

44. *United States Pacific Fleet and Pacific Ocean Areas Evasion and Escape Manual.*

45. Letter from H. Paul Brehm to Baldwin and McGarry, 12 April 1992.

46. Letter from Roger Perreault to R. E. Baldwin, 10 January 1990.

47. Letter from Robert S. McCarter to R. E. Baldwin, 2 June 1992.

48. Letter from Robert S. McCarter to R. E. Baldwin, 27 January 1993.

49. Translation by Vince Lau.

50. Translation by Vince Lau.

51. Letter from Eldon Ray Ford to R. E. Baldwin. 29 April 1992.

52. *Training and Indoctrination of 201st Mexican Fighter Squadron*, memo from HQ V Fighter Command to Commanding Officer, 58th Fighter Group, 4 May 1945, SECRET.

53. Letter from Santiago A. Flores to Thomas Wm. McGarry, 17 August 1995.

54. Letter from Mark Hardisty to Baldwin and McGarry, January 1995.

55. Letter from Roger Perreault to R. E. Baldwin, 10 January 1990.

56. Letter from Eldon Ray Ford to R. E. Baldwin. 29 April 1992.

57. Letter from Maj. G.L. Ferguson to Baldwin and McGarry, 2 February 1994.

58. Untitled Navy manual, stamped CONFIDENTIAL.

59. Letter from Lloyd R. Shoemaker to R. E. Baldwin, 28 September 1992.

60. Perreault letter to Baldwin.

61. Lloyd R. Shoemaker, *The Escape Factory - The Story of MIS-X*, (New York, St. Martin's Press, 1990), pp. 1-2.

Appendix II

1. Barbara A. Bond, "Silk Maps: The Story of MI9's excursion into the world of cartography 1939-45" in *The Cartographic Journal*, Vol. 21 No. 2, December 1984, pp. 141-143. Reprinted by permission of the British Cartographic Society.

2. Clayton Hutton, *Official Secret*, (New York, Crown Publishers, Inc., 1961), p.24. Reprinted by permission of Crown Publishers, Inc.

3. John G. Doll, "Cloth Maps of World War II" in *Information Bulletin*, Western Association of Map Libraries, Vol. 20, No. 1, November 1988, pp. 23-35.

4. Doll, "Cloth Maps"

5. Alfred H. Burton, *Conquerors of the Airways: A Brief History of the USAF-ACIC and Aeronautical Charts*, (St. Louis, MO, USAF/ACIC, 1953), pp. 155-159.

6. Burton, *Conquerors of the Airways.*

7. Burton, *Conquerors of the Airways.*

8. Burton, *Conquerors of the Airways.*

9. Burton, *Conquerors of the Airways.*

10. Doll, "Cloth Maps".

11. *Letter to Chief, C P and M, MIS-X Branch*, 2b 677 Pentagon, Attn: Technical Section, from HQ AGAS-China, 23 December 1944, CONFIDENTIAL.

12. *Memo to A. C. of S., G-2, USAF*, Attn Lt. Col A. R. Wichtrich, HQ AGAS, from Russell H. Sweet, Colonel GSC, Chief, CPM Branch, 18 July 1945.

13. *Memo to Chief, CPM Branch, MIS*, from HQ AGAS-China, 8 August 1945.

Appendix III

1. *Memorandum for Record*, Office of the Secretary, Department of the Air Force, 14 August 1957.

2. *Memorandum for Record.*

3. *History and Operational Use of Blood Chits, World War II and Korea*, Research Studies Institute, Air University, Maxwell AFB, Alabama, 1957.

4. *History and Operational Use of Blood Chits, World War II and Korea.*

Bibliography and Recommended Reading

Readers are encouraged to pursue the books and articles listed below. Some of them contain only general background information, while others provide highly detailed specifics on a variety of topics touched upon in this work. The endnotes will assist readers in determining which references may be of greatest value for their specific interests.

For readers interested in Allied evasion and escape operations during World War II, four books are especially recommended. *MI 9, Escape and Evasion 1939-1945* by M. R. D. Foot and J. M. Langley reveals the breadth and depth of British operations, which were the best organized and the longest running of the war. *Official Secret* by Clayton Hutton, is the story of the development Britain's E&E aids, told by the man who invented them. *The Escape Factory - The Story of MIS-X* by Lloyd R. Shoemaker provides a detailed look at MIS-X and its efforts to aid American flyers in the European Theater. *My War With Imperial Japan, Escape and Evasion* by Richard V. Hill is the story of a member of the Air-Ground Aid Section of MIS-X in China.

All government documents used in the preparation of this book are unclassified. Their original classifications (restricted, classified, secret, etc.) are given, when known.

Books

Bard, Mitchell G. *Forgotten Victims*. Boulder, CO.: Westview Press Inc., 1994.

Barker, Ralph. *Down In The Drink, True Stories of the Goldfish Club*. London: Chatto & Windus, 1955.

Barrett, David. D. *Dixie Mission: The United States Army Observer Group in Yenan, 1944*. Berkeley, CA: Center for Chinese Studies, University of California, 1970.

Bates, Charles C. and Fuller, John E. *America's Weather Warriors 1814-1985*. College Station, TX: Texas A&M University Press, 1986.

Blank, Stephen J., Grinter, Lawrence E., Magyar, Karl P., Ware, Lewis B., and Weathers, Bynum E. *Conflict, Culture, & History*. Maxwell AFB, AL: Air University Press, 1993.

Bond, Charles and Anderson, Terry. *A Flying Tiger's Diary*. College Station, TX: Texas A&M Press, 1984.

Bowyer, Chaz. *RAF Operations 1918-1938*. London: Wm. Kimber, 1988.

Burton, Alfred H. *Conquerors of the Airways: A Brief History of the USAF-ACIC and Aeronautical Charts*. St. Louis, MO: U.S. Air Force Aeronautical Chart and Information Center, 1953.

Castaway's Baedeker to the South Seas. Objective Data Section, Intelligence Center, Pacific Ocean Areas, 1942.

Chennault, Claire Lee. *Way of a Fighter*, New York: G.P. Putnam's Sons, 1949.

China Airlift - The Hump. Poplar Bluff, MO: China-Burma-India Hump Pilots Association, Inc., 1980.

China Airlift: The Hump, Vol. II. Poplar Bluff, MO: China-Burma-India Hump Pilots Association, Inc.

Cruickshank, Charles. *Deception in World War II*. New York: Oxford University Press, 1980.

Carr, S. J. *You Are Not Sparrows*. Shepperton, Surrey: Ian Allen Ltd., 1975.

Clayton, Anthony. *The British Empire as Superpower 1919-1939*. Athens, GA: University of Georgia Press, 1986.

Craven, Wesley F. and Cates, James L., editors. *Army Air Forces in World War II, Vol. 3, Services Around the World*. Chicago: University of Chicago Press, 1958.

Daws, Gavan. *Prisoners of the Japanese*. New York: William Morrow and Company, Inc., 1994.

Dower, John W. *War Without Mercy: Race & Power in the Pacific War*. New York, Pantheon Books, 1986.

Dudgeon, A. G. *The Luck of the Devil*. Shrewsbury, UK: Airlife Publishing Ltd, 1985.

Franks, Norman L.R. *First In The Indian Skies*. Lincoln, UK: Life Publications Ltd, 1981.

Foot, M. R. D. and Langley, J. M. *MI 9, Escape and Evasion 1939-1945*. Boston-Toronto: Little, Brown and Company, 1979.

Ford, Daniel. *Flying Tigers, Claire Chennault and the American Volunteer Group*. Washington,D.C.: Smithsonian Institution Press, 1991.

Goldman, Ben. *German Treatment of American Prisoners of War In World War II*. Published by the U.S. Air Force.

Gunston, Bill, Editor. *Chronicles of Aviation*, London: Chronicle Communications Ltd., 1992.

Harries, Merion and Susie *Soldiers of the Sun*. New York: Random House, 1991.

Hawkins, Joyce M. and Allen, Robert, Editors. *The Oxford Encyclopedic English Dictionary*, Oxford: Clarendon Press, Oxford University Press, 1991.

Heflin, Woodford Agee, Editor. *The United States Air Force Dictionary*. Maxwell AFB, AL: Air University Press, 1956.

Hering, P. G., *Customs and Traditions of the Royal Air Force*, Aldershot, Hants, U.K.: Gale & Polden Ltd.

Hill, Richard V. *My War With Imperial Japan, Escape and Evasion*. New York: Vantage Press, 1989.

Howard, James H. *Roar of The Tiger*. New York: Orion Books, 1991.

Hutton, Clayton. *Official Secret*. New York: Crown Publishers, Inc., 1961.

Inks, James M. Eight Bailed Out. New York: W. W. Norton & Company, 1954.

Jackson, Robert. *The Secret Squadrons: Special Duty Units of the RAF and USAAF in the Second World War*. London: Robson Books, Ltd., 1983.

Jakeman, Robert J. *The Divided Skies*. Tuscaloosa, AL: University of Alabama Press, 1992.

Kennedy, Paul *The Rise and Fall of the Great Powers*. New York, Random House, 1987.

Kennett, Lee. *The First Air War*. New York: The Free Press, 1991.

Kogon, Eugene. *The Theory and Practice of Hell*. New York: Berkley Books, 1980.

Lawrence, T. E. *Revolt in the Desert*. New York: George H. Doran Company, 1927.

Lawrence, T. E. *Seven Pillars of Wisdom*. Garden City, New York: Doubleday, Doran & Company, Inc., 1937.

LeMay, Curtis E. and Yenne, Bill. *Superfortress The B-29 and American Air Power*. New York: McGraw-Hill Book Company, 1988.

Leuthene, Stuart and Jensen, Oliver. *High Honor, Recollections by Men and Women of World War II Aviation*. Washington, D.C.: Smithsonian Institution Press, 1989.

Lindstrom, Lamont and White, Geoffrey M. *Island Encounters, Black and White Memories of the Pacific War.* Washington & London: Smithsonian Institution Press, 1990.

Lowden, John L. *Silent Wings at War: Combat Gliders in World War II.* Washington, D.C.: Smithsonian Institution Press, 1992.

McMillan, Peter. *The Greatest Flight.* Atlanta, GA: Turner Publishing, Inc., 1995.

Miles. Milton E. and Daniel, Hawthorne. *A Different Kind of War.* Garden City, NY: Doubleday & Co., 1967.

Mishler, Clayton. *Sampan Sailor.* Washington, D.C.: Brassey's, 1994.

Molesworth, Carl. *Sharks Over China, The 23rd Fighter Group in World War II.* London and Washington, D.C.: Brassey's, 1994.

Moon, Tom. *This Grim and Savage Game: O.S.S. and The Beginning of U.S. Covert Operations in WW II.* Garden Grove, CA: Burning Gate Press, 1991.

Moser, Don, et al. *China-Burma-India.* Alexandria, VA: Time-Life Books, 1978.

Onderwater, Hans. *Second To None.* London: Airlife Publishing Ltd., 1992.

Perret, Geoffrey, *Winged Victory The Army Air Forces In World War II.* New York: Random House, 1993.

Pistole, Larry M. *The Pictorial History of the Flying Tigers.* Orange, Virginia: Moss Publications, 1981.

Ride, Edwin. *BAAG, Hong Kong Resistance 1942-1945.* Hong Kong: Oxford University Press, 1981.

Rodzinski, Witold. *The Walled Kingdom, A History of China from Antiquity to the Present.* New York: The Free Press, 1984.

Rosholt, Malcolm. *Flight in the China Air Space.* Rosholt, WI: Rosholt House, 1984.

Sampson, Jack. *Chennault.* New York: Doubleday, 1987.

Selhaus, Edi. *Evasion and Repatriation, Slovene Partisans and Rescued American Airmen in World War II.* Manhattan, KS: Sunflower University Press, 1993.

Shoemaker, Lloyd R. *The Escape Factory - The Story of MIS-X.* New York: St. Martin's Press, 1990.

Smith, R. T. *Tale of a Tiger.* Van Nuys, CA: Tiger Originals, 1986.

Stettinius, Edward R., Jr. *Lend Lease, Weapon for Victory.* New York, The Macmillan Company, 1944.

Survival on Land and Sea (NAVAER 13-1-501). Publications Branch, Office of Naval Intelligence, U.S. Navy, 1944.

Wagner, Ray. *Prelude to Pearl Harbor The Air War in China 1937-1941.* San Diego, CA: San Diego Aerospace Museum, 1991.

What You Should Know About China. Calcutta: Indian Press Ltd., 1945.

Whitehouse, Arch. *The Legion of the Lafayette.* New York: Doubleday, 1962.

Articles

Bond, Barbara A. "Silk Maps: The Story of MI9's excursion into the world of cartography 1939-45." *The Cartographic Journal.* Vol. 21 No. 2, December 1984.

"Blood Chits." *Air Sea Rescue Bulletin.* No. 11, May 1945. RESTRICTED.

Doll, John G. "Cloth Maps of World War II." *Information Bulletin.* Western Association of Map Libraries, Vol. 20, No. 1, November 1988.

Fischer, Joseph R. "Cut From a Different Cloth: The Origins of the U.S. Army Special Forces." *Special Warfare.* Vol. 8, No. 2, April 1995.

Holcomb, Mal. "Flying Behind The Twelve Pointed Star." *Wings.* Vol. 13, No. 5. Granada Hills, CA: Sentry Books, October. 1983.

Nicholls, Eric. "North West Frontier." *FlyPast Special,* April 1993

Jarrett, Philip. "By Day and By Night: Wildbeest and Vincent." *Aeroplane Monthly.* February, 1995.

Labussière, William and Beauchamp, Gerry. "The Truth About the 14th Squadron Foreign Volunteers In China." *AAHS Journal.* Number 4, American Aviation Historical Society, Winter 1987.

Leiser, Edward L. "Memoirs of Pilot Elwyn H. Gibbon, The Mad Irishman." *AAHS Journal.* Volume 23, Number 1, American Aviation Historical Society, 1st Quarter 1978.

Northracker, William H. "Afghan Death March." *Soldier of Fortune,* Soldier of Fortune Magazine, Inc., February 1989.

Olmsted, Merle. "The Yoxford Boys and Operation 'Frantic', Part II." *Friends Journal.* Vol. 15, No. 3. Dayton, OH: Air Force Museum Foundation.

Parsons, David W. "British Air Control, A Model for the Applications of Air Power in Low Intensity Conflict?" *Air Power Journal,* Vol VIII, No. 2, Summer 1994

"Pointie Talkie." *Air Sea Rescue Bulletin,* No. 7, Air Sea Rescue Agency, Washington, D.C., January 1945, RESTRICTED.

Robbs, Peter H. "Blood Chits." *Falling Leaf Magazine.* Vol. XXI, No. 1, Whole No. 84. St. Albans, Herts., U.K.: The Psywar Society, 1979.

Stell, Max Welton. Letter to the editor. *Foundation.* Museum of Naval Aviation, Pensacola, Florida, May 1993.

Young, William. "Missing: Believed Killed." *Royal Air Force Flying Review,* Vol. 9. No.9. June, 1954.

Wallace, W.W. Letter to the Editor. *FlyPast.* July, 1991.

Government Documents

AGAS Journal for 1944. SECRET.

Air Notes From China. ComNavGp China, April 1945.

Anecdotes on Evasion and Escape for the use of Intelligence Officers. Headquarters, ETOUSA, Office of A. C. of S., G-2, MIS Detachment, 1 August 1943. SECRET

Blood Chits Distributed to 1st Bomb Wing. Memo to A-2, 1st Bomb Wing from HQ, ETOUSA, Office of A.C. of S., G-2, P/W and X Detachment, MIS, 14 September 1943.

Brief History of the G-2 Section, GHQ, SWPA. Undated post-war report.

Bulletin on Escape and Evasion (BEE). HQ XX BC, Office of A.C. of S., A-2, Bulletin No. 8, 26 June 1944. SECRET.

Bulletin on Escape and Evasion (BEE). HQ XX BC - Intelligence Section, Bulletin No. 10, 8 September 1944, SECRET.

Bulletin on Escape and Evasion (BEE). HQ XX BC, Intelligence Section, Bulletin No. 12, 11 December 1944. SECRET.

Capture Intelligence, Escape Bulletin No. 5. Northwest African Air Forces, A-2 Section, 4 May 1943. SECRET.

Capture Intelligence, Escape Bulletin No. 11. Northwest African Air Forces, A-2 Section, 17 August 1943. SECRET.

Case # 45. Memo, HQ China Theater Search Detachment, 15 January 1946.

"Chapter 2: E Group Escape Aids". *E Group Bulletin,* Undated British WWII document.

"China Walkout." *Escape and Evasion Bulletin # 17.* HQ Tenth Air Force, Office of the A.C. of S., A-2 Secton, Air Jungle Rescue Section, 1 April 1945. SECRET.

Code for Describing Escape Aids. HQ India Task Force, 26 August 1943. SECRET.

"Escape and Evasion, General Information." *The History of "Frantic" - American Shuttle-Bombing to and from Russian Bases.* SECRET

Escape and Evasion Bulletin No. 8. HQ Tenth Air Force, Office of the Assistant C. of S., A-2, MIS-X Section, 10 December 1944. SECRET.

Escape Bulletin # 26. HQ, 12th AF, A. C. of S., A-2 Section, 29 December 1943. SECRET

E&E Bulletin #28. HQ, 12th AF, A. C. of S., A-2 Section, 4 March 1944. SECRET

E&E Bulletin #46: Evasion. HQ, 12th AF, A. C. of S., A-2 Section, 21 July 1944. SECRET

Equipment Distributed During Month of April, 1945. Report, MIS-X, SWPA. SECRET.

Equipment Requirements. Memo to I.S.9 (z) (Attn: S/Ldr Whitehead) from Maj. Irvin H. Luiten, HQ, ETO, Det. C, P/W&X Detachment, MIS, 12 March 1945.

Final Report. MIS-X Section, POA, September 1945. SECRET.

Functions of E&E Section. HQ 12th AF, Office of A.C. of S., A-2, E&E Subsection, 20 May 1945. SECRET.

Funds for MIS-X Section. Memo from HQ Allied Air Forces to G-2 GHQ, 22 February 1944. SECRET.

History and Operational Use of Blood Chits, World War II and Korea. Research Studies Institute, Air University, Maxwell AFB, undated, circa 1957. SECRET.

Information Concerning M.I.9 and P/W Matters. Undated WWII British document, stamped CONFIDENTIAL at bottom, SECRET handwritten across top.

"Intelligence Annex No. 3, Operation Frantic" by Major Vernon E. Fairbanks, A.C. of S., A-2, *The History of "Frantic" - American Shuttle-Bombing to and from Russian Bases.* SECRET

Intelligence Summary No. 269. 26 April 1945. SECRET.

Intelligence Summary No. 279. July 1945. SECRET.

Letter to A. C. of S. G-2, HQ, USAF, Attn: Lt. Col. A. R. Wichtrich, HQ, AGAS from Russell H. Sweet, Col.. GSC, Chief, CPM Branch, 15 February 1945.

Letter to All Concerned, Translation of Message on Blood Chits. General Headquarters, United States Army Forces, Pacific, MIS-X Section, 19 JULY 1945. RESTRICTED.

Letter to Chief, C P and M, MIS-X Branch, 2b 677 Pentagon, Attn: Technical Section, from HQ AGAS-China, 23 December 1944. CONFIDENTIAL.

Memo to Chief, CPM Branch, MIS, from HQ AGAS-China, 8 August 1945

Memo, from HQ AGAS-China to Chief, CPM Branch, MID, Attn: Col. J. E. Johnston, G.S.C., 9 October 1944. SECRET.

Memo to A. C. of S., G-2, HQ USAF, from Russell H. Sweet, Colonel, GSC, Chief CPM Branch, 31 July 1945. CONFIDENTIAL.

Memo to A. C. of S., G-2, HQ Forward Echelon , USAF, CBI Theater, Attn: Maj. A. R. Wichtrich, MIS-X Section, from Russell H. Sweet, Col., GSC, Chief CPM Branch, 9 September 1944. SECRET.

Memo to A. C. of S., G-2, USAF, Attn Lt. Col A. R. Wichtrich, HQ AGAS, from Russel H. Sweet, Colonel GSC, Chief, CPM Branch, 18 July 1945.

Memo to Col. Russell H. Sweet, G.S.C. Chief CPM Branch, from HQ AGAS-China, 20 August 1944. SECRET.

Memorandum for Cmdr. John L. Riheldaffer, USN, Subject: Naval Officers Trained by MIS-X, from Lt. Basil D. Henning, Navy Liaison Officer, MIS-X, 5 July 1944. SECRET.

Memorandum for Lt. Cmdr. Ward Cheney, USNR, OP-16-V, from Lt. Basil D. Henning, Navy Liaison Officer, MIS-X, 4 July 1944. SECRET.

Memorandum for Major A. R. Wichtrich, MIS-X, China. 24 February 1944.

Memorandum for Record. Office of the Secretary, Department of the Air Force, 14 August 1957.

Memorandum to Lt. Col Holt. P/W&X Detachment, MIS-X, ETOUSA, HQ, A-2, 11 July 1943. SECRET.

Memorandum to The Chief of Staff. Geo. V. Strong, Major General, A.C. of S., G-2, 11 December 1942. SECRET.

MIS-X Manual on Evasion, Escape and Survival. Washington, D.C.: U.S. War Department, February 1944, with additions through 1945. SECRET.

MIS-X Procurement Authority and Fund Requirements, Memo from HQ Allied Air Forces to GHQ G-2, 26 February 1944. SECRET.

Monthly Report, MIS-X Activities, SWPA. GHQ, SWPA, 18 May 1945. SECRET.

Brig. Gen. Partridge, *Combat Crew Handbook*. USAAF, 8th AF, 3rd Bomb Division, undated. SECRET.

Prisoners of War, 3-3926 AF. Washington, D.C., U.S. Army Air Forces, undated WWII document. RESTRICTED.

Report, AGAS-China, from HQ AGAS-China to Col. Russell H. Sweet, Chief CPM Branch, Attn: Col. J. E. Johnston, G.S.C., 22 August 1944. TOP SECRET.

Report from HQ 51st Fighter Group, 10th Air Force, to Commanding General, HQ 10th Air Force, 7 September 1943.

Report of MIS-X Section, SWPA. February 1944. SECRET.

Report of MIS-X SWPA Section. 12 May 1945. SECRET.

Specimen Lecture on Evasion and Escape in the Far East; All Services, All Arms; For the Guidance of Intelligence Officers. Main HQ, E Group, South East Asia and India Commands, New Delhi, February 1945.

Specimen Lecture on Evasion and Escape in the Far East; All Services, All Arms; For the Guidance of Intelligence Officers. Main HQ, E Group, South East Asia and India Commands, New Delhi, February 1945.

Training and Indoctrination of 201st Mexican Fighter Squadron, memo from HQ V Fighter Command to Commanding Officer, 58th Fighter Group, 4 May 1945. SECRET.

Training Intelligence Report No. 3451, General Series No. 473. Training Intelligence Service, Army Air Forces Training Command, 1945. RESTRICTED.

United States Pacific Fleet and Pacific Ocean Areas Evasion and Escape Manual. CINCPAC-CINCPOA, 30 January 1945. SECRET.

United States Pacific Fleet and Pacific Ocean Areas-Report of Intelligence Activities in the Pacific Ocean Areas. CINCPAC-CINCPOA, 15 October 1945. SECRET.

Untitled manual, stamped CONFIDENTIAL.

Walkout Report on 1st Lt. Freeling H. Clower. HQ 89th Fighter Squadron, Office of the Intelligence Officer, 17 January 1944. SECRET.

Walkout Report of Captain Kenneth E. George. 76th Fighter Squadron, 23rd fighter Group, Office of the Intelligence Officer, 25 February 1944. SECRET.

Walkout Report of Captain Ransom W. Rideout. Office of the Intelligence Officer, 31 July, 1945. SECRET.

Walkout Report on 1st Lt. Harold W. Robbins. 89th Fighter Squadron, from HQ 89th Fighter Squadron, Office of the Intelligence Officer, 1 May 1944. SECRET.

War Diary for February 1945. HQ India-Burma Theater, MIS-X India. SECRET.

War Diary For August and September 1945, MIS-X Technical Section. Prepared by Leo H. Crosson, Maj., Inf., Chief Technical Section. SECRET.

Index

also from the publisher

MiG PILOT SURVIVAL:
Russian Aircrew Survival Equipment and Instruction

Alan R. Wise

Siberia – infamous for its brutal winters, and larger than the entire United States – is not the only wilderness within the former Soviet Union. Harsh southern deserts, arctic islands, disputed border regions, and minority populations belligerent towards the their present government are spread all across the gigantic nation. Flying over Russia presents risks few other aircrews in time of peace must face. And while Russian combat aircraft are world renowned for their reliable performance, what happens when something does go wrong? Given their exceptional egress systems – odds are the pilot will eject safely, but how does he survive and advance his rescue under such potentially dire circumstances? *MiG Pilot Survival: Russian Aircrew Survival Equipment and Instruction* explores the components and details of Russian survival science with color photographs, in depth descriptions, and a full translation of the exact manual – with original illustrations intact – as used by Russian aircrews in time of crisis. Alan R. Wise is a consultant, writer, and photographer who has done work for the Department of Defense, Department of State, U.S. Army, U.S. Air Force, U.S. Air Force Museum, and the Kansas Cosmosphere and Space Center. He has served as a military advisor and has written numerous articles for magazines such as *National Defense*, *Journal of Defense and Diplomacy*, *Armor*, and *Behind the Lines: The Journal of Special Operations*. He is also the coauthor with Michael S. Breuninger of *Jet Age Flight Helmets* (see page 5 of this catalog).

Size: 8 1/2" x 11" over 180 color photographs and line drawings
96 pages, hard cover
ISBN: 0-7643-0130-6 $19.95
Available October

GEAR UP! Flight Clothing & Equipment of USAAF Airmen in WWII

Jon A. Maguire.

This new work from Jon A. Maguire provides an in-depth examination of personal flight clothing and equipment used by United States Army Air Force flyers in World War II. Included are: flight helmets, oxygen masks, goggles, clothing, boots, gloves, flak armor, paper items and emergency equipment such as parachutes and "Mae Wests."

Size: 8 1/2" x 11" over 600 b/w and color photos
184 pages, hard cover
ISBN: 0-88740-744-7 $45.00